# EARTH SCIENCE WORK-TEXT

## Revised Edition

## Constantine Constant

*Bridgeport Science
Technical Assistant Consultant,
Scientist in Residence,
Bridgeport Public School,
Bridgeport, Connecticut*

AMSCO

AMSCO SCHOOL PUBLICATIONS, INC.,
a division of Perfection Learning®

The author wishes to acknowledge the helpful contributions of the following consultants in the preparation of this book:

Jonathan Kolleeny, Geologist
New York State Department of Environmental Conservation

Phil Kuczma
Teacher, Earth Science
Bronxville High School, Bronxville, N.Y.

Sy Rifkin
Teacher, General Science
Simon Baruch Junior High School 104, New York City

Paul Speranza
Teacher, Earth Science
Lafayette High School, New York City


*Science, Technology and Society* features written by Carl Proujan.

Cover Photo: Sloth Canyon, Lake Powell, AZ. Gavriel Jecan / The Stock Market

Text Photo Credits: Figs. 3-2, 3-3, 3-4, 3-14, 3-18, and 3-19, negs. # 297258, 297316, 297287, 297296, 297279, and 297284, by Dwight Bentel, courtesy of Dept. of Library Services, American Museum of Natural History; Figs. 3-5, 5-11, and 20-12, The Bettmann Archive; Fig. 4-3, F.C. Calkins, courtesy of U.S. Geological Survey (USGS); Fig. 4-4, E.S. Bastin, USGS; Fig. 5-12, N.H. Darton, USGS; Fig. 8-6, #67, USGS; Fig. 3-15, courtesy of The British Tourist Authority; Fig. 5-13, courtesy of Arizona Office of Tourism; Fig. 7-7, courtesy of Swiss National Tourist Office; Fig. 17-12, negs. # 124940, 124970, 124988, 124996, 335747, and 335749, courtesy of Dept. of Library Services, American Museum of Natural History.

*Science, Technology and Society* feature photos: p. 15, courtesy of Jim Castelein, RIT Communications; pp. 59, 154, 226, and 276, UPI/Bettmann; pp. 86 and 257, Uniphoto, Inc.; p. 120, R.E. Wallace, U.S. Geological Survey; p. 196, Tony Craddock, Science Source/Photo Researchers, Inc.; p. 318, Walter Dawn, Science Source/Photo Researchers, Inc.; pp. 240 and 295, The Bettmann Archive.

Please visit our Web sites at:
*www.amscopub.com* and *www.perfectionlearning.com*


When ordering this book, please specify:
13589 *or* EARTH SCIENCE WORK-TEXT

ISBN 978-0-87720-190-8

Printed in the United States of America

8 9 10      15 14 13

# Contents

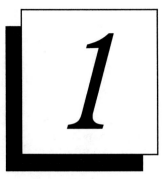

# *Science and Scientists*

## LABORATORY INVESTIGATION

### *MEASURING THE DENSITY OF ROCKS*

In earth science, measuring and calculating are basic investigative activities. Mass is measured by using a *triple beam balance scale.* Before measuring the mass of an object, you must make sure that the arrow attached to the beams is positioned exactly in the middle of the *post scale.* If the arrow is not exactly in the middle, you can adjust it by turning the *balance nut* (see Fig. 1-1). Then, place the mass on the pan to the left of the *pivot.* The *riders* to the right of the pivot have known masses hanging from them and are moved until the beam is balanced. The sum of the numbers indicated by the riders equals the mass of the object on the pan.

Density is a measurement of how closely packed particles of matter are. The *density* of an object is calculated by dividing its mass by its volume (density = mass/volume). You can measure volume by using a *graduated cylinder.* When the mass is broken apart, do the pieces of rock, or rock fragments, have different densities from the whole rock? In this laboratory investigation, you will calculate the density of a small rock, break it apart, calculate the densities of the

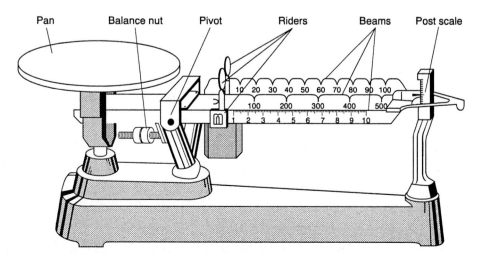

**Fig. 1-1. Triple beam balance.**

fragments, and then compare the density of the whole rock to the density of its fragments. (*Note:* Every pure substance has a characteristic density; for example, a block of foam rubber is less dense than an equal-sized block of marble.)

**A.** Your teacher will give you a rock for this investigation. Measure the mass of the rock (to the nearest tenth of a gram [g]) with a triple beam balance scale. Record the mass in Table 1-1 below.

**B.** Add 50 milliliters (ml) of water to the 100-ml graduated cylinder. In Table 1-1, record the volume of water in cubic centimeters (cc): 1 ml = 1 cc. Tie a 25-centimeter (cm) length of sewing thread around the rock. Carefully lower the rock into the graduated cylinder. In Table 1-1, record the new water level in the graduated cylinder (see Fig. 1-2).

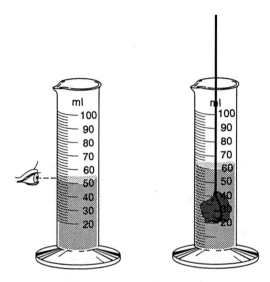

**Fig. 1-2. Measuring volume with a graduated cylinder.**

**C.** Find the volume of the rock by subtracting the original volume of the water from the volume of the water containing the rock. In Table 1-1, record the volume of the rock.

**D.** Determine the density of the rock by dividing the mass of the rock by the volume of the rock. (Round figure to the nearest 0.1 g/cc.) In Table 1-1, record the density.

**E.** Remove the rock from the graduated cylinder. Dry the rock as thoroughly as possible, and then wrap it in a heavy cloth or bath towel. Use the geologic hammer to break the rock into several fragments.

**(CAUTION: Wear safety goggles, and make sure that the rock is wrapped securely before breaking it apart with the geologic hammer.)**

**F.** Select the three largest rock fragments. Calculate the mass of one (sample A) of the fragments. In Table 1-1, record its mass.

**G.** Add enough water to the graduated cylinder to bring the level back up to 50 ml. In Table 1-1, record the volume of water. Tie a 25-cm length of sewing thread around sample A, and gently lower it into the graduated cylinder. Calculate the density of sample A by dividing its mass by its volume, and enter your result in Table 1-1.

**H.** Repeat steps F and G for the other two rock fragments (sample B and sample C), and record your results in Table 1-1.

**TABLE 1-1. MEASURING MASS, VOLUME, AND DENSITY**

| Rock Samples | Mass (g) | Volume of Water (cc) | Volume of Rock and Water (cc) | Volume of Rock (cc) | Density of Rock (g/cc) |
|---|---|---|---|---|---|
| Whole rock | | | | | |
| Sample A | | | | | |
| Sample B | | | | | |
| Sample C | | | | | |

**1.** Is there any difference between the mass of the whole rock and that of the rock fragments? Explain. _____

_____

_____

**2.** Is there any difference between the volume of the whole rock and that of the rock fragments? Explain. _____

_____

_____

**3.** How did you find the volume of the rock fragments?

_____

_____

_____

**4.** Is there a difference between the density you found for the whole rock and the density you found for each of the rock fragments? Explain. _____

_____

# Science: A Search for Understanding

For thousands of years, humans have tried to understand the world around them. Long ago, they invented myths and legends to explain such natural phenomena as volcanic eruptions, thunderstorms, and the change of seasons. In time, people began to make logical connections between the possible causes and effects of everyday events.

## SCIENTIFIC KNOWLEDGE

The observations of early scientists did not always lead to correct explanations. For example, until the beginning of the sixteenth century, most people believed that the sun and all of the planets revolved around Earth. As a result of the observations and measurements made by scientists such as Galileo and Copernicus, the truth (that all planets revolve around the sun) was discovered (see Chapter 20). In time, the careful observations and logical reasoning of scientists led to the acquisition of *scientific knowledge* about the world.

## THE NATURE OF SCIENCE

*Science* is a body of knowledge that can be grouped into four main branches and numerous subbranches. The four major branches of science are:

1. *Earth Science*—the study of Earth and its place in the universe.
2. *Physics*—the study of the interactions of matter and energy.
3. *Chemistry*—the study of the structure and properties of matter.
4. *Biology*—the study of living things.

The main branches of science often overlap or are broken down into more specialized areas of study, called *subbranches*. These include meteorology, biochemistry, and paleontology. Table 1-2 lists some of the subbranches of science and their areas of study.

## HOW A SCIENTIST WORKS

A basic characteristic of all scientists is curiosity. After scientists observe natural events that they do not understand, they ask probing questions. By asking many relevant questions and doing research, scientists have been able to explain heat, light, the movement of the sun, and many other natural events.

**Theories and Laws.** After studying facts, making observations, and performing experiments, scientists may develop a theory. A **theory** is a scientific explanation of natural events. It often takes years of research by many people throughout the world before a theory is generally accepted.

For example, research by earth scientists throughout the world during the past 30 years has led to wide acceptance of the *theory of plate tectonics*. This theory has helped people better understand how continents move (see Chapter 8). In earth science, however, some theories that are hard to test may never be widely accepted, such as those concerning the origin of the universe or the kinds of elements in Earth's core. After a theory has been proven true by being successfully tested many times, scientists may call it a **law**.

## TABLE 1-2. THE MAIN BRANCHES AND SUBBRANCHES OF SCIENCE

| Branch | Subbranch | Areas of Study |
|---|---|---|
| Biology | Genetics<br>Botany<br>Zoology<br>Paleontology<br>Microbiology | Heredity<br>Plants<br>Animals<br>Fossils<br>Microscopic life forms |
| Chemistry | Organic chemistry<br>Polymer chemistry<br>Nuclear chemistry<br>Biochemistry | Carbon compounds<br>Very large molecules<br>The atom's nucleus<br>Chemistry of living things |
| Physics | Astrophysics<br>Nuclear physics<br>Mechanics<br>Optics<br>Acoustics | Stars and objects in space<br>Atomic structure and power<br>Forces<br>Light<br>Sound |
| Earth Science | Seismology<br>Meteorology<br>Oceanography<br>Geology<br>Astronomy | Earthquakes<br>Weather<br>Oceans<br>Rocks<br>Stars and galaxies |

# THE SCIENTIFIC METHOD

Science is also an activity in which questions are investigated and scientific knowledge is accumulated in an orderly and systematic manner. The process scientists use to perform problem-solving research is called the **scientific method** and includes the following main steps (see Fig. 1-3):

1. Stating the problem
2. Collecting information
3. Forming a hypothesis
4. Testing the hypothesis
5. Recording observations
6. Checking results
7. Drawing conclusions
8. Communicating results

**Stating the Problem.** Before beginning a scientific investigation, a scientist should be able to state the exact problem to be researched in the form of a question. By stating the problem as a question, the scientist can focus on a particular cause-and-effect relationship he or she wishes to investigate. For example, a researcher may ask: Does water containing a lot of soil, or muddy water, warm up faster than water containing little or no soil?

**Collecting Information.** Once a scientific problem has been clearly stated, a scientist collects all available information that relates to the problem by doing research. Such research includes making additional observations, collecting physical evidence, and referring to books and scientific journals to learn how other people have investigated the problem. By doing research, scientists become better prepared to solve a specific problem.

**Forming a Hypothesis.** A possible solution or explanation of a scientific problem is called a **hypothesis.** A hypothesis is often referred to as an educated guess. However, a hypothesis must be supported by either experiments or physical evidence. For example, a scientist might form a hypothesis stating: The greater the quantity of soil dissolved in water, the more rapidly the water temperature will rise.

**Testing the Hypothesis.** A hypothesis must be able to be tested. One way a scientist tests a hypothesis is by designing and performing a **controlled experiment.** In a controlled experiment, there are usually two groups that are tested, the **experimental group** and the **control group.** The experimental group differs from the control group in that one specific fac-

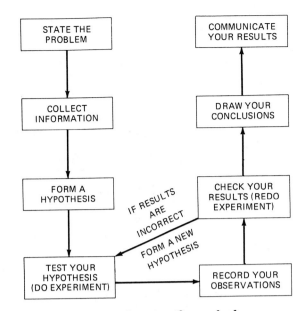

Fig. 1-3. **The scientific method.**

tor in the experimental group has been changed. The experimental and control groups are exposed to identical conditions except for the particular factor being changed, called the **variable.** In this case, the hypothesis would propose that the amount of soil is the variable that causes the water to heat up more rapidly.

How might a scientist test the hypothesis that muddy water heats up faster than water that contains little or no soil? First, the scientist would obtain four beakers of the same size and made of the same material. All of the beakers would be filled to the same level with water from the same source. Then, three of the beakers—the *experimental* group—would receive varying amounts of soil.

Identical thermometers would then be placed in each of the beakers, and four lamps, each equipped with a 200-watt bulb, would be turned on directly over each of the beakers. In this experiment, the amount of soil added to each of the three beakers is the *variable* being tested. The beaker that does not receive any soil is the *control.*

**Recording Observations.** Scientists must keep a careful record of their scientific observations, or **data,** to better understand their experimental results. Such data may include charts and diagrams, as well as information about the experimental procedure. In addition, there should be a record of the materials and equipment used. This record enables other scientists to repeat the experiment so that they can compare results if necessary.

**Checking Results.** Scientists repeat controlled experiments several times before drawing any conclusions. By performing an experiment several times, a researcher can be fairly certain that the results obtained are valid, or correct. In this way, a hypothesis is either supported or disproved.

Another way to test a hypothesis is by collecting evidence. For example, a paleontologist would find support for a hypothesis concerning the age of a particular dinosaur by collecting its bones in rock layers dated to that age. If the evidence agrees with the scientist's hypothesis, it has been supported. However, it usually takes a great deal of evidence collected by many people before a hypothesis becomes widely accepted.

**Drawing Conclusions.** After recording, graphing, and analyzing the results of a controlled experiment, a scientist draws conclusions. The conclusions drawn from specific experiments and investigations often provide useful data for future scientific research. For example, after repeating the above experiment several times, a scientist might conclude that the amount of soil in a beaker of water does affect the rate at which water can be heated. This information may help other scientists who are studying lakes and rivers that have much sediment in them.

**Communicating Results.** A scientist communicates the results of his or her experiments to other scientists. In this way, other scientists can repeat the experiment to either confirm or disprove the hypothesis. A scientist may communicate the results of an experiment or investigation by publishing a report in a scholarly journal or by giving a lecture at a scientific conference. Communication of scientific knowledge often inspires other scientists to think of new problems to investigate.

## MEASUREMENT

Scientists often have to measure the length, mass, volume, or temperature of various substances. All measurements, however, are somewhat *uncertain*, or not precise. That is because the instruments used are not perfectly made, and the ability of a scientist to observe a measurement is not perfect. Measurements, however, are made as precisely as possible during experiments or while collecting evidence.

**The English System.** The system of measurement still widely used in the United States is called the **English system.** The English system uses the *foot* as the unit of length, the *pound* as the unit of mass, and the *quart* as the unit of volume. A problem with the English system is that large units cannot be converted easily into smaller units. For example, in units of length:

$$1 \text{ foot} = 12 \text{ inches}$$
$$1 \text{ yard} = 3 \text{ feet}$$
$$1 \text{ mile} = 1760 \text{ yards}$$

### TABLE 1-3. COMMON METRIC UNITS OF LENGTH, MASS, AND VOLUME

*Common Metric Units of Length*

| | | |
|---|---|---|
| kilometer (km) | = | 1000 meters |
| decameter (dkm) | = | 10 m |
| decimeter (dm) | = | .1 m |
| centimeter (cm) | = | .01 m |
| millimeter (mm) | = | .001 m |

*Common Metric Units of Mass*

| | | |
|---|---|---|
| kilogram or kilo (kg) | = | 1000 grams |
| decagram (dgm) | = | 10 g |
| decigram (dg) | = | .1 g |
| centigram (cg) | = | .01 g |
| milligram (mg) | = | .001 g |

*Common Metric Units of Volume*

| | | |
|---|---|---|
| kiloliter (kl) | = | 1000 liters |
| decaliter (dkl) | = | 10 l |
| deciliter (dl) | = | .1 l |
| centiliter (cl) | = | .01 l |
| milliliter (ml) | = | .001 l |
| cubic centimeter (cc) | = | .001 l |

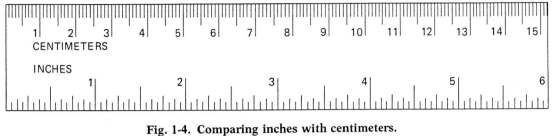

**Fig. 1-4. Comparing inches with centimeters.**

**The Metric System.** The metric system was first used in France in 1837. Today, the **metric system** is used in most countries and by most scientists throughout the world. In the metric system, the *meter* is the basic unit of length, the *kilogram* is the unit of mass, and the *liter* is the unit of volume.

Because the metric system is based on multiples of 10, it is easier to convert a particular measure into either larger or smaller units. For example, when measuring length, a distance of 3 meters (m) may also be expressed as 300 cm, because 1 m is equal to 100 cm. The following examples show how easily large units can be converted into smaller units in the metric system:

$$1 \text{ kilometer} = 1000 \text{ meters}$$
$$1 \text{ meter} = 100 \text{ centimeters}$$
$$1 \text{ centimeter} = 10 \text{ millimeters}$$

Likewise, a length of 10,000 m may be more conveniently expressed as 10 kilometers (km). Table 1-3 shows common prefixes and units of the metric system for length, mass, and volume. (*Note: kilo* = thousand; *deca* = ten; *deci* = tenth; *centi* = hundredth; and *milli* = thousandth.)

Although scientists make calculations using the metric system, they must occasionally convert units in the English system to metric units (see Fig. 1-4 to compare inches and centimeters). Some common equivalents in the two systems are shown in Table 1-4. In both the metric system and the English system, the unit of time is the second.

Most scientists use the *Celsius (C) scale* to measure temperature. On the Celsius scale, the freezing point of water is 0°C, and the boiling point of water is 100°C. In the United States, the *Fahrenheit (F) temperature scale* is widely used. On the Fahrenheit scale, the freezing point of water is 32°F, and the boiling point of water is 212°F (see Fig. 1-5).

**TABLE 1-4. EQUIVALENTS: ENGLISH AND METRIC SYSTEMS**

| English Unit | Metric Equivalent |
| --- | --- |
| 1 inch | 2.54 centimeters |
| 39.4 inches | 1 meter |
| 1 pound | 454 grams |
| 2.2 pounds | 1 kilogram |
| 1.06 quarts | 1 liter |

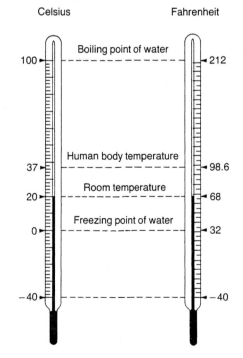

**Fig. 1-5. Fahrenheit and Celsius temperature scales.**

## THE INSTRUMENTS OF SCIENCE

Scientists use many tools, or *instruments*, during experiments and investigations. Some instruments, such as rulers and balance scales, allow scientists to measure quantities of length, mass, and volume. Others enable scientists to extend their senses. For example, the use of a telescope or a microscope permits a scientist to see and describe objects not possible to view with unaided vision.

Besides instruments for measurement and observation, scientists use other tools in the laboratory and outdoors. These include audiovisual equipment for filming and recording objects and natural events; various containers for holding and transferring substances, such as beakers, flasks, test tubes, and vials; tools for obtaining rock specimens, such as a geologic hammer; and apparatus for heating substances, such as a Bunsen burner. In your laboratory investigation on the density of rocks, you used a graduated cylinder to measure volume and a balance scale to measure mass. Fig. 1-6 shows some common tools used by earth scientists.

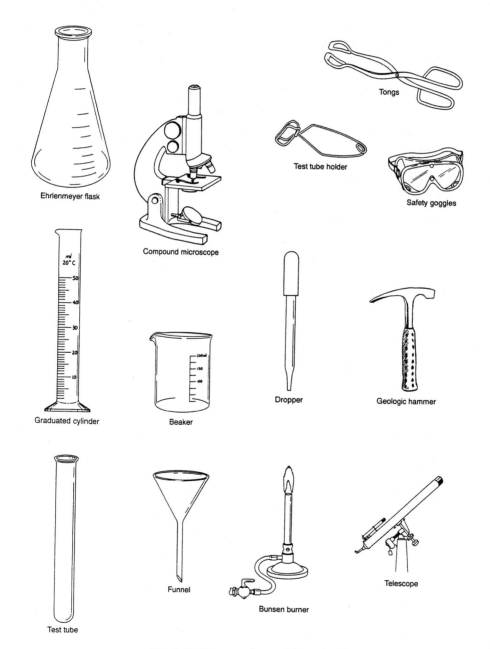

Ehrlenmeyer flask

Compound microscope

Tongs

Test tube holder

Safety goggles

Graduated cylinder

Beaker

Dropper

Geologic hammer

Test tube

Funnel

Bunsen burner

Telescope

**Fig. 1-6. Some tools used by scientists.**

# Laboratory Safety and Skills

In the science laboratory, you must handle scientific equipment and instruments correctly and follow certain safety procedures. For example, when heating a substance in a test tube, you should always use a test tube holder and point the open end of the test tube away from yourself and others in the laboratory.

In case of an accident, you must be prepared to take quick action. If you spill an acid, you should clean it up immediately and then dispose of the wastes properly. It is also important to clean all instruments and equipment before and after using them.

You will be prepared to perform laboratory work safely by paying careful attention to the following list of laboratory safety rules:

1. Tie back long hair, secure loose-fitting clothing, and remove dangling jewelry.
2. Wear safety goggles, gloves, and aprons when working with chemicals and when heating substances.
3. Never touch or taste chemicals; never mix chemicals without your teacher's permission; never use chemicals from unlabeled containers; never directly inhale chemical vapors.
4. When heating substances in a test tube, point the open end of the test tube away from yourself and others in the laboratory.
5. Never eat or drink anything in the science laboratory.
6. If you burn your skin or spill a chemical on your body, immediately rinse the affected area with plenty of cold water, and report the accident to your teacher.
7. Do not directly handle heated or broken glassware (wear laboratory gloves). (Use only Pyrex or similar glassware for heating substances.)
8. Observe proper cleanup procedures by storing all chemicals, instruments, and equipment in their proper places.

# CHAPTER REVIEW

## *Science Terms*

*The following list contains all of the boldfaced words found in this chapter and the page on which each appears.*

control group (p. 5)
controlled experiment (p. 5)
data (p. 5)
English system (p. 6)
experimental group (p. 5)
hypothesis (p. 5)

law (p. 4)
metric system (p. 7)
scientific method (p. 5)
theory (p. 4)
variable (p. 5)

# Matching Questions

*On the blank line, write the letter of the item in column B that is most closely related to the item in column A.*

| Column A | Column B |
|---|---|
| ____ 1. scientific explanation of natural events | *a.* hypothesis |
| ____ 2. organized process of problem solving | *b.* experimental group |
| ____ 3. record of scientific observations | *c.* control group |
| ____ 4. possible solution to a scientific problem | *d.* Fahrenheit |
| ____ 5. all factors remain the same in this group | *e.* scientific method |
| ____ 6. one factor is changed in this group | *f.* data |
| ____ 7. the experimental factor that is tested | *g.* Celsius |
| ____ 8. measurement system based on units of 10 | *h.* theory |
| ____ 9. scale on which water boils at 100° | *i.* metric system |
| ____ 10. scale on which water boils at 212° | *j.* scientific knowledge |
| | *k.* variable |

# Multiple-Choice Questions

*On the blank line, write the letter preceding the word or expression that best completes the statement or answers the question.*

**1.** The density of an object is calculated by dividing its mass by its
   *a.* weight   *b.* length   *c.* width   *d.* volume                          1 ____

**2.** Scientific knowledge is based on
   *a.* myths and legends                 *c.* educated guesses only
   *b.* careful observations and reasoning   *d.* imaginative explanations      2 ____

**3.** The subbranch of earth science that studies the weather is called
   *a.* seismology   *b.* astronomy   *c.* oceanography   *d.* meteorology      3 ____

**4.** An orderly and systematic approach to solving scientific problems is called
   *a.* a theory   *b.* a law   *c.* a hypothesis   *d.* the scientific method   4 ____

**5.** The last step of the scientific method is to
   *a.* draw conclusions            *c.* state the problem
   *b.* test the hypothesis          *d.* communicate results                   5 ____

**6.** A possible solution or explanation of a scientific problem is known as a
   *a.* law   *b.* theory   *c.* hypothesis   *d.* control                      6 ____

**7.** A scientist performs an experiment or collects evidence to
   *a.* test a hypothesis           *c.* collect information
   *b.* form a hypothesis           *d.* state the problem                      7 ____

**8.** In an experiment, the group that is exposed to the variable being tested is called the
   *a.* control group   *b.* theory group   *c.* experimental group   *d.* hypothesis group   8 ____

9. The system of measurement that uses the pound as the unit of weight and the quart as the unit of volume is the
   *a.* scientific system  *b.* metric system  *c.* English system  *d.* decimal system  9 ____

10. The system used by most scientists throughout the world to measure length, volume, and mass is the
    *a.* Celsius scale  *b.* Fahrenheit scale  *c.* English system  *d.* metric system  10 ____

11. In the metric system, a milligram equals
    *a.* 1000 grams  *b.* 100 grams  *c.* 1/100 gram  *d.* 1/1000 gram  11 ____

12. On the Celsius scale, the boiling point of water is
    *a.* 212°  *b.* 32°  *c.* 100°  *d.* 50°  12 ____

13. One pound equals approximately
    *a.* 10 grams  *b.* 100 grams  *c.* 450 grams  *d.* 1000 grams  13 ____

14. A liter is slightly greater in volume than a
    *a.* quart  *b.* gallon  *c.* pint  *d.* half gallon  14 ____

15. The metric system is based on multiples of
    *a.* 10  *b.* 12  *c.* 25  *d.* 36  15 ____

16. A length of 20,000 meters can be expressed as
    *a.*  2 kilometers          *c.*  20 kilometers
    *b.*  10 kilometers          *d.*  200 kilometers  16 ____

# *Modified True-False Questions*

*In some of the following statements, the italicized term makes the statement incorrect. For each incorrect statement, write the term that must be substituted for the italicized term to make the statement correct. For each correct statement, write the word "true."*

1. The mass of an object can be measured with a *graduated cylinder.*  1 ____

2. You can adjust the arrow on a triple beam balance scale with the *balance nut.*  2 ____

3. A scientific explanation that becomes widely accepted by scientists after many years of research is called a *hypothesis.*  3 ____

4. The subbranch of earth science that studies earthquakes is called *seismology.*  4 ____

5. The process scientists use to perform research is referred to as the *experimental group.*  5 ____

6. The particular factor that is changed in a controlled experiment is called the *control.*  6 ____

7. The basic unit of length in the metric system is the *liter.*  7 ____

8. An inch equals about *2.5* centimeters.  8 ____

9. Most scientists use the *Celsius scale* to measure temperature.  9 ____

10. On the Fahrenheit scale, the boiling point of water is *100°.*  10 ____

# *Testing Your Knowledge*

**1.** Describe two ways that a scientist would test a hypothesis.

_____

_____

**2.** Explain the difference between the two terms in each of the following pairs:
   *a.* experimental group and control group

_____

_____

   *b.* hypothesis and theory

_____

_____

   *c.* metric system and English system

_____

_____

   *d.* Celsius and Fahrenheit

_____

_____

**3.** Why is it useful for a scientist to state the problem before beginning an investigation?

_____

_____

_____

_____

**4.** Why does a scientist repeat a controlled experiment several times before drawing any conclusions?

_____

_____

_____

**5.** Why do scientists prefer using the metric system rather than the English system to make measurements?

_____

_____

_____

_____

**6.** Why is measurement always considered to be uncertain?

_____

_____

_____

**7.** Why do you think it might be dangerous to use chemicals from unlabeled containers?

_____

_____

_____

**8.** Convert the following measurements in the English system to measurements in the metric system:

    *a.* 44,000 grams    = _____ kilograms

    *b.* 390 centimeters  = _____ meters

    *c.* 1000 centigrams = _____ grams

    *d.* 100 meters      = _____ centimeters

    *e.* 21 decaliters    = _____ liters

# Energy from the Sun: Should We Make It Work for Us?

You might think that the nuclear power plant upriver from you or the power plant down the road that runs on coal or oil are your best sources of electricity. If that's what you're thinking, think again, say many concerned scientists and citizens.

For one thing, nuclear power plants can be dangerous. As you may know, accidents have occurred in both the former Soviet Union and the United States that threatened or took human lives. As for coal- or oil-burning power plants, both pollute the atmosphere when the coal or oil is burned. Besides, coal and oil won't last forever. Earth holds a limited supply of these resources.

Electricity, however, is a form of energy we must have. We light our homes, office buildings, and streets with it. We use it to make most of our products. Electricity runs all our appliances, from refrigerators to air conditioners. Without it, we wouldn't be able to tune in to our favorite radio stations, TV channels, or watch a movie in a theater.

So if we must turn our backs on electricity generated from nuclear energy, coal, and oil, where can we turn instead? We can turn to our nearest star, the sun! But, you might say, *that* power plant is some 150 million kilometers away. How can we get electricity from so far a distance?

The sun may be far away but its energy is right here. You see it in the form of light. And you feel it in the form of heat. If we could harness all the solar energy striking Earth, it would provide 15,000 times more energy than all of the people in the world use!

But how can we harness a beam of light and turn it into electricity? There are a number of answers to this question. Let's look at two.

If you've ever used a magnifying glass to burn a hole through a sheet of paper, you have an idea of how you can harness the sun's energy. You haven't made electricity, of course, but you've demonstrated a principle that can be used to make electricity, namely, that the sun's energy can be concentrated by focusing it.

That's what's being done at Harper Lake in southern California. Built in the late 1980s, rows of troughlike structures called solar collectors, each 100 meters long, stretch across the desert landscape. Shaped in the form of a parabola, each solar collector is lined with mirrors that focus the heat of the sun onto pipes filled with oil.

The oil gets hot enough to turn water into steam, which is exactly what happens when the heated oil is piped to a nearby power plant. There the steam hisses into turbines which, as they turn, power an electric generator. The drawback, however, is that this electricity costs a lot more than that produced by regular power plants. But the price is being brought down as efficiency is increased, say engineers.

Part of the cost problem is caused by the various steps involved in converting sunlight into electricity by this process. There is, however, another process that does this directly. It is called *photovoltaic electricity*, the kind that runs certain solar-powered calculators.

In this case, sunlight strikes panels that contain substances that respond by giving off electrons. These electrons can be made to flow in an organized way. Since a flow of electrons is electric current, the panels have turned sunlight directly into usable electricity.

Many houses in the United States are now equipped with solar panels. Only about 40 square meters are needed to supply all of the electricity for a house. So how about building really huge collections of solar panels to provide electricity for the entire country? It could be done, but it would mean covering 34,000 square kilometers of the U.S. with panels. Although this seems like a lot, it represents only about 0.37% of the country's land area—too much to turn over to solar power plants, according to some people. Besides, the electricity produced by photovoltaic panels costs much more than the electricity your family currently buys. On the other hand, these panels produce no pollution or danger to people.

Should we keep trying to find ways to turn sunlight into electricity? What do you think?

*1. The average distance from Earth to the sun is about*

    a. 1.5 million kilometers.
    b. 15 million kilometers.
    c. 150 million kilometers.
    d. 1500 million kilometers.

*2. An electric current consists of a flow of*

    a. sunlight.    c. nuclear energy.
    b. oil.    d. electrons.

*3. In North America, a slanting roof equipped with solar panels should face in what direction?*

    a. north.    c. east.
    b. south.    d. west.

*4. What are the steps used in producing electricity at Harper Lake?*

_____

_____

_____

*5. On a separate sheet of paper, discuss the pros and cons of switching from current fossil fuel and nuclear power plants to ones powered by the sun. End by presenting your opinion of what should be done, and why.*

# 2

# Earth and Its Minerals

## LABORATORY INVESTIGATION

### THE HARDNESS OF MINERALS

**A.** You will need the following materials at your desk before beginning this investigation: a glass square, a steel nail, a copper penny, and labeled samples of the minerals talc, gypsum, calcite, fluorite, apatite, feldspar, quartz, topaz, and corundum.

When one mineral sample is scraped against another, a scratch often appears on the surface of the softer mineral. You can feel this scratch with the edge of your fingernail.

**B.** Scrape the copper penny against a smooth, clean surface of the calcite sample as shown in Fig. 2-1 below.

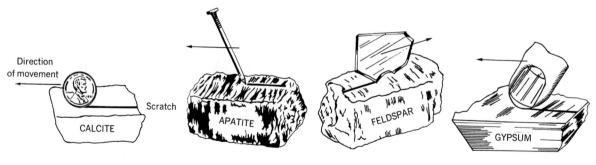

**Fig. 2-1.**

**1.** What do you observe?

_____

_____

**2.** Explain your observation.

_____

_____

**C.** First, scrape the quartz sample against the sample of calcite. Then, scrape the calcite sample against the quartz sample. Use the edge of your fingernail to determine which of the mineral samples was scratched.

**3.** What do you observe? _____

_____

**4.** Which is the harder mineral? _____

_____

**D.** You can determine the order of hardness of all the mineral samples by scraping them against each other. The copper penny, steel nail, glass square, and your fingernail will help you determine the relative hardness of the mineral samples.

Use the following information to help you in this part of the investigation:

*a.* The softest minerals can be scratched by your fingernail.

*b.* The copper penny can scratch a few of the minerals that cannot be scratched by your fingernail.

*c.* The steel nail can scratch a few of the minerals that cannot be scratched by the copper penny.

*d.* The edge of the glass square can scratch some of the mineral samples that cannot be scratched by the steel nail. Other mineral samples can scratch the glass square.

**E.** Use the procedure described in parts B and C and the information provided in part D to determine the hardness of the mineral samples as follows:

*a.* Use the glass square, steel nail, copper penny, and your fingernail to determine the relative hardness of each mineral. Then, group the mineral samples according to whether they can be scratched by your fingernail, the copper penny, the steel nail, or the glass square.

*b.* Determine which mineral sample is the hardest and which one is the softest among the group of minerals that you can scratch with your fingernail. Then, arrange the mineral samples by relative hardness. On a separate sheet of paper, list the softest mineral as 1, the next softest as 2, and so forth.

*c.* Repeat this procedure with the minerals that you have classified into the other groups. When you have completed this investigation, all mineral samples should be arranged in order of increasing hardness. If you have classified the minerals correctly, you will discover that sample 2 will scratch sample 1, sample 3 will scratch sample 2, sample 4 will scratch sample 3, and so forth.

**5.** List the mineral samples below according to relative hardness (softest = 1; hardest = 9).

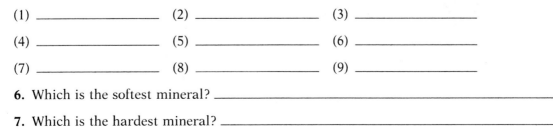

(1) _____  (2) _____  (3) _____

(4) _____  (5) _____  (6) _____

(7) _____  (8) _____  (9) _____

**6.** Which is the softest mineral? _____

**7.** Which is the hardest mineral? _____

**8.** Which minerals are harder than the glass square? _____

_____

**9.** Which minerals are softer than the copper penny? _____

## THE SURFACE OF EARTH

From outer space, Earth looks like a smooth, blue-green sphere. In fact, Earth resembles the globe found in many classrooms. Even the world's highest mountain (Mount Everest) appears flat to an astronaut. At a height of about 8.8 kilometers (km), Mount Everest is slightly more than 1/1400 of Earth's diameter (about 12,800 km).

An astronaut orbiting Earth also can see how little of our planet's surface (about 1/4) consists of land. The remainder of Earth's surface is covered by water, as shown in Fig. 2-2 below. Surrounding Earth is an envelope of air—often filled with great masses of floating clouds.

As you probably know, this envelope of air is called the **atmosphere.** All bodies of water on Earth—oceans, lakes, and rivers—make up the **hydrosphere.** And the solid part of Earth is called the **lithosphere.** In each of these words, the prefixes that precede *sphere* (also derived from a Greek word, meaning "ball") are derived from the following Greek words: *atmos-* (meaning "vapor"), *hydro-* (meaning "water"), and *litho-* (meaning "rock").

The lithosphere contains most of the substances that support human beings and other living things. For example, fruits, vegetables, and grains are all grown in the topsoil, found on the surface of the lithosphere. And miners dig deep into the lithosphere, extracting minerals needed to maintain industrialized societies throughout the world. Factories and refineries convert these minerals into metal alloys and other materials used to make bridges, trucks, airplanes, tools, bottles, cans, bricks, television sets, and thousands of other items that people use every day.

**Changes on Earth's Surface.** Earth's land surface is continually reshaped by two opposing sets of forces. External forces, such as running water and moving ice, tend to wear down Earth's surface. Internal forces, which can cause earthquakes and volcanic eruptions, tend to raise the land. The interplay of these two types of forces produces the landscape features, or *landforms*, of Earth's surface. Major landforms include mountains, valleys, plains, and plateaus.

## THE STRUCTURE OF EARTH

People have not been able to drill very deep into Earth's interior. To date, the deepest drill hole, located in Russia, reaches down about

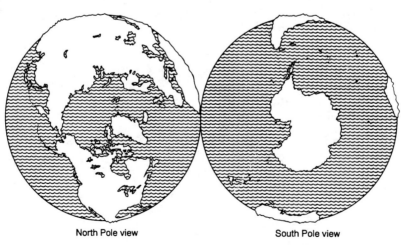

North Pole view          South Pole view

**Fig. 2-2. About three fourths of Earth's surface is covered with water.**

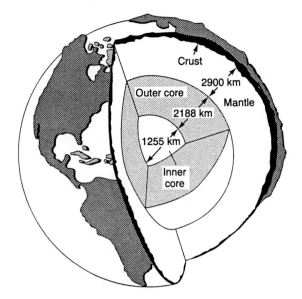

**Fig. 2-3. The major zones of Earth. (Thickness of the crust is exaggerated.)**

13 km. Yet, **geologists,** or earth scientists, have learned much about Earth's structure. Today, geologists even have ideas about the structure and composition of Earth's center.

Geologists have learned about the structure of Earth's interior by observing and recording earthquake vibrations (see Chapter 8). Scientists have invented instruments that record these vibrations as wavy lines on a sheet of graph paper. As a result of studying and interpreting these recordings, geologists have concluded that Earth consists of four major layers, or *zones* (Fig. 2-3). The major zones of Earth are the *crust, mantle, outer core,* and *inner core.*

**Crust.** The **crust,** Earth's outermost layer, is a relatively thin layer of solid rock that ranges from 5 to 90 km in thickness (Fig. 2-4). In fact, the crust is the thinnest of Earth's major zones. The crust is thickest under the continents and thinnest under the oceans.

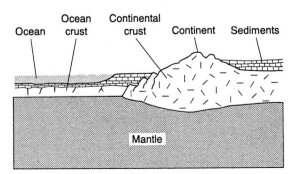

**Fig. 2-4. Cross section of Earth's crust.**

The continuous, unbroken rock of the crust is called the *bedrock.* In most places, the bedrock is covered by boulders, pebbles, sand, and soil. These loose materials were originally part of the bedrock. In some places, the bedrock is exposed at Earth's surface. Exposed portions of bedrock are called *outcrops.* Fig. 2-5 shows the surface of the crust.

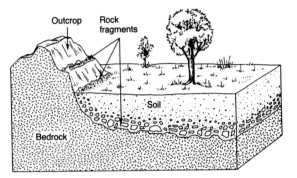

**Fig. 2-5. The surface of the crust.**

**Mantle.** The **mantle,** located directly beneath the crust, is the thickest of Earth's major zones. Geologists have determined that the mantle is about 2900 km thick, making up over 80 percent of Earth's volume. Rocks of the mantle are more dense and contain more iron than rocks of the crust.

**Outer and Inner Cores.** The deepest part of Earth's interior consists of two zones: an **outer core** and an **inner core.** The combined thicknesses of both zones is about 3440 km. Both the inner and outer core are probably composed of iron and nickel.

Based on the interpretation of earthquake vibrations, geologists think that the outer core is in a molten (hot liquid) state. The inner core is probably solid because of the great pressure exerted upon it by the surrounding zones. The temperature at Earth's center is estimated to be about 5000°C.

## COMPOSITION OF EARTH'S CRUST

Geologists think that during the early period of Earth's formation, the newly formed planet was completely made of molten material. As the molten material cooled and became solid, it formed the rocks and minerals that make up Earth's crust.

**Rocks and Minerals.** Some rocks are made up of only one mineral. Most rocks, however, consist of two or more minerals. Minerals consist of chemical elements or chemical compounds. A **chemical element** is the simplest kind of substance. A chemical element cannot be broken down into simpler substances by ordinary means. Gold, silver, oxygen, and carbon are examples of chemical elements.

A **chemical compound** is a substance that consists of two or more elements combined in fixed proportions by weight. For example, water, table salt, and natural gas (methane) are chemical compounds. Water consists of one part hydrogen and eight parts oxygen, by weight. All minerals can be broken down into various chemical elements and compounds.

**Elements in Earth's Crust.** All of the rocks and minerals in Earth's crust are combinations of 88 different elements. Yet, more than 98 percent of these rocks and minerals consist of only 8 of the 88 elements (refer to Table 2-1 below). The chemical symbols shown to the right of each element are a shorthand method of writing the names of these elements.

**TABLE 2-1. EIGHT MAJOR ELEMENTS IN EARTH'S CRUST**

| Element | Symbol | Approximate Percentage (by weight) |
|---------|--------|-----------------------------------|
| Oxygen | O | 47 |
| Silicon | Si | 28 |
| Aluminum | Al | 8 |
| Iron | Fe | 5 |
| Calcium | Ca | 4 |
| Sodium | Na | 3 |
| Potassium | K | 3 |
| Magnesium | Mg | 2 |

## THE COMPOSITION OF MINERALS

A **mineral** is a substance that has a definite chemical composition. Some minerals are composed of only one element. For example, graphite and diamond are minerals that are composed solely of carbon (C). Other minerals consist of two or more elements chemically combined and are, therefore, chemical compounds.

Halite and quartz are two common minerals that are compounds of only two elements chemically combined. Halite—commonly called *rock salt*—is a mineral composed of the elements sodium (Na) and chlorine (Cl). You know this mineral as table salt, which has the chemical name *sodium chloride (NaCl)*. Quartz is a compound of the elements silicon (Si) and oxygen (O). Quartz is also known as *silica*, as well as by its chemical name, *silicon dioxide* ($SiO_2$).

Most minerals, however, are chemical compounds that consist of more than two elements. Calcite and feldspar are two common minerals that are composed of more than two elements. Both of these minerals are major components in many kinds of rock. Calcite is a compound of calcium (Ca), carbon (C), and oxygen (O). The chemical name of calcite is *calcium carbonate ($CaCO_3$)*. A common type of feldspar is a compound of potassium (K), aluminum (Al), silicon (Si), and oxygen (O). The chemical formula for this feldspar is $KAlSi_3O_8$.

## MINERAL IDENTIFICATION

You can identify most minerals by studying their characteristics, or properties. These properties can be either chemical or physical. Each kind of mineral has its own distinctive chemical and physical properties. Chemical properties of minerals are described on page 24. A *physical property* of a mineral is a property that involves its appearance, as well as its behavior when heated, broken, or scratched. Hardness is an example of a physical property. Other physical properties of minerals include color, streak, cleavage, fracture, crystal shape, luster, and specific gravity.

**Hardness.** The **hardness** of a mineral is determined by observing how easily it scratches other minerals or how easily other minerals scratch it. As you discovered in your laboratory investigation, talc is a very soft mineral; it can be scratched by your fingernail. Glass, a relatively hard common material, also will easily scratch talc. However, the minerals quartz, topaz, and corundum can scratch glass, and thus are harder than glass.

A German mineralogist, *Friedrich Mohs* (1773–1839), invented a method for comparing the hardness of different minerals. His method is called the *Mohs scale of hardness*. In the Mohs scale, minerals are classified according to their hardness. By scratching minerals against each other, Mohs selected several minerals as standards and assigned hardness values to them. Thus, the softest mineral (talc) is assigned the number 1. The hardest mineral (diamond) is assigned the number 10. Table 2-2 below lists the standard minerals used by Mohs and their assigned hardness values.

You should not confuse hardness with brittleness. A brittle mineral cracks or breaks easily. For example, although diamond is the hardest mineral, diamond is quite brittle. In fact, a diamond will crack rather easily when struck sharply by a hammer. In contrast, an iron nail is softer than a diamond, but an iron nail is not brittle. You can strike an iron nail again and again with a hammer without cracking or breaking the nail.

### TABLE 2-2. THE MOHS SCALE OF HARDNESS

| | |
|---|---|
| 1. Talc | 6. Feldspar |
| 2. Gypsum | 7. Quartz |
| 3. Calcite | 8. Topaz |
| 4. Fluorite | 9. Corundum |
| 5. Apatite | 10. Diamond |

**Color.** A mineral sample often contains impurities, or small amounts of foreign substances. These foreign substances can change the natural color of the mineral sample. As a result, a specific kind of mineral may have several different colors, depending on the kind of impurities it contains.

Because the colors of many minerals can vary, color is not by itself a positive way of identifying a mineral. However, the colors of some minerals always remain the same. For example, the mineral malachite is always green, the mineral cinnabar is always red, and the element sulfur is always yellow.

The surface of a mineral tends to change color when exposed to air and water. Thus, you should break off a small fragment of the mineral before trying to identify it. The freshly broken surface of the mineral will reveal its true color.

**Streak.** The **streak** is the color of a mineral when it is ground into a powder. As shown in Fig. 2-6 above, rubbing a mineral against a piece of dull (unglazed) ceramic tile will powder enough of the mineral to enable you to identify its streak. The ceramic tile is called a *streak plate*.

A mineral may have several different colors, depending on the impurities it contains. However, the color of its streak rarely varies. In general, metallic minerals leave dark gray or black streaks. Nonmetallic minerals commonly leave streaks that are colorless or much

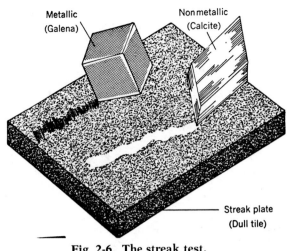

**Fig. 2-6. The streak test.**

lighter than the color of the particular mineral.

Some minerals leave streaks that are much different in color from the color of the mineral itself. The mineral hematite, which contains iron, may be reddish brown, silvery, or black. The streak of hematite, however, is always red. The mineral pyrite (fool's gold) has a brassy yellow color, but the streak of pyrite is always greenish black.

**Cleavage.** Some minerals split apart neatly and cleanly, or *cleave*, into two or more pieces. The split pieces form *plane* surfaces, or surfaces that are smooth and flat. Different minerals split apart neatly and cleanly along two or more planes that form fixed angles in relation to each other. The physical property that causes these minerals to split apart along planes that meet at characteristic angles is called **cleavage.**

A mineral splits apart in a certain way because zones of weakness, or *cleavage zones,* exist inside the mineral. The arrangement of atoms and molecules within a mineral is responsible for its cleavage zones. The atoms and molecules in a mineral form a regular pattern, much as the bricks in a brick wall form an orderly pattern. A mineral will split more easily along a cleavage zone than in any other direction. Because the atoms and molecules within each mineral form their own distinctive patterns, each mineral has its own distinctive cleavage zones.

The knowledge of cleavage zones has economic importance. For example, a diamond cutter must know the cleavage zones within a diamond. When a diamond cutter wishes to split a large diamond, a shallow groove is first cut into the surface of the diamond. This groove must be cut parallel to one of the cleavage zones inside the diamond. The diamond cutter then places a steel wedge in the groove and taps the wedge sharply with a hammer. If the cleavage zone has been determined correctly, the diamond will split neatly and cleanly into two pieces. However, if the direction of the cleavage zone has been misjudged, the diamond may shatter into many small, less valuable pieces.

Some minerals, such as graphite, talc, and mica, have only one cleavage direction. As a result, these minerals split easily into thin sheets (see Fig. 2-7 below). In fact, you can peel off thin flakes of talc, graphite, or mica with a knife. Other minerals, such as feldspar, galena, and halite, have cleavage in two or three directions.

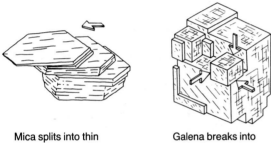

Mica splits into thin sheets because it has one direction of cleavage

Galena breaks into cube-shaped pieces because it has three directions of cleavage at right angles

Fig. 2-7. Examples of cleavage.

**Fracture.** Many minerals do not have regular zones of weakness. When a mineral does not split along a zone of weakness, the break is called a **fracture.** Thus, these minerals do not exhibit cleavage, but instead *fracture* when broken. For example, when you break a rock in two, the rough, irregular surface produced on each piece is a fracture.

Some minerals and rocks have a special kind of fracture. For example, when you strike a piece of quartz sharply with a hammer, the broken pieces may have smooth, spiraled surfaces that resemble the inside of a seashell. Because the surfaces of the break resemble a conch (a marine snail that has a spiral shell), this kind of break is called a *conchoidal* fracture. You probably have seen conchoidal fractures in the chipped surfaces of thick glass windows. A glasslike rock called *obsidian* usually exhibits conchoidal fracture (see Fig. 2-8 above).

Fig. 2-8. Conchoidal fracture in obsidian.

**Crystal Shape.** A mineral **crystal** is a solid that has flat surfaces arranged in a definite geometric shape. Hundreds of different minerals form crystals. Some minerals form several different kinds of crystals. Because of their many beautiful and unusual shapes, crystals have been called the "flowers of the mineral world." Several different crystal shapes are shown in Fig. 2-9.

Mineral crystals form in two ways. A crystal usually forms when a mineral cools slowly from molten rock deep inside Earth's crust. However, crystals also form when the water in which a mineral is dissolved slowly evaporates. In either case, the shape of the resulting crystals is determined by the arrangement of the particles (atoms or molecules) that make up the mineral.

If table salt (sodium chloride) is dissolved in water and the water is allowed to evaporate slowly, crystals of salt will form. The slow evaporation of water allows enough time for the sodium atoms and chlorine atoms to form cube-shaped crystals of sodium chloride, as shown in Fig. 2-10 below. If the water evaporates very slowly, the individual salt crystals will grow large enough for you to observe with your naked eye. However, if the water evaporates more rapidly, you will need a magnifying glass to see the salt crystals.

You can easily grow salt crystals at home. First, make a saturated solution by adding table salt to a dish of water until no more salt can be dissolved in the water. Then, place the dish containing the saturated salt solution in a place where it will not be disturbed. After all of the water has evaporated, use a magnifying glass to examine the salt crystals left on the bottom of the dish.

You also can grow crystals of borax, Epsom salts, and sodium bicarbonate (baking soda) at home. First, make saturated solutions of Epsom salts, borax, and sodium bicarbonate in the same way that you made the saturated

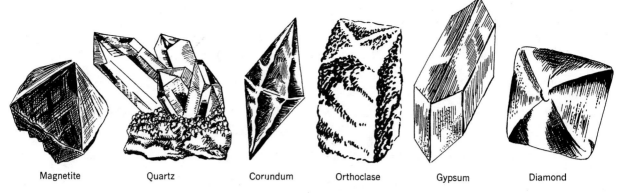

Magnetite  Quartz  Corundum  Orthoclase  Gypsum  Diamond

**Fig. 2-9. Mineral crystals are the "flowers of the mineral world."**

saltwater solution. Empty the saturated solution of each material into a separate bottle. Then, suspend a crystal of each material in the appropriate bottle of saturated solution. Over several days, as the water evaporates from each of the solutions, the crystals will slowly increase in size (see Fig. 2-11 below).

**Luster.** **Luster** describes the way a mineral shines, or reflects light. Minerals are described as having two main types of luster: metallic and nonmetallic.

*Metallic luster* is the shine of a freshly polished metal surface. A shiny aluminum pot, a gold or silver bracelet, and a chrome bumper on a car are common examples of metallic luster. Galena and pyrite are minerals that have a metallic luster.

*Nonmetallic luster* is the shine of all minerals that do not have a metallic luster. The two most common types of nonmetallic luster are called *glassy* and *dull*. A windowpane, a crystal decanter, and a wineglass have a glassy luster. Quartz, feldspar, and calcite are minerals that have a glassy luster. A sheet of writing paper is an example of a familiar object that has a dull luster. Minerals that have a dull luster include bauxite and kaolinite.

**Specific Gravity.** If you hold a sample of the mineral galena in one hand and a sample of quartz of equal size in the other hand, the galena will feel much heavier. If you were to determine the mass of each sample, you would find that the sample of galena is about three times heavier than the sample of quartz. Sci-

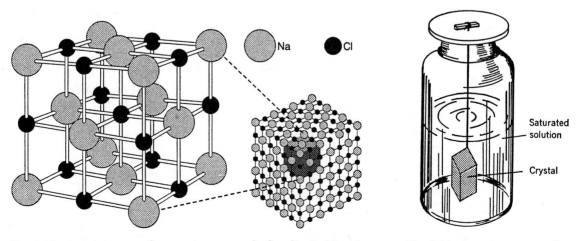

Na  Cl

Saturated solution

Crystal

**Fig. 2-10. Arrangement of atoms in a crystal of sodium chloride.**  **Fig. 2-11. Growing a crystal.**

entists have developed a method of comparing the masses of different minerals by using water as a standard. In this method, the mass of a mineral sample is compared to the mass of an equal volume of water. The number that represents this comparison is called **specific gravity.**

Suppose you compare the mass of a sample of galena that has a volume of 16.4 cubic centimeters (cc) with a sample of water that also has a volume of 16.4 cc. When you determine the mass of both samples, you discover that the galena has a mass of 921.4 grams (g) and the water has a mass of 121.9 g. By dividing the mass of the water into the mass of the galena, you find that the sample of galena has a mass about 7.56 times the mass of the same volume of water:

$$\frac{\text{mass of galena}}{\text{mass of water}} = \frac{921.4 \text{ g}}{121.9 \text{ g}} = 7.56$$

The number obtained when you divide the mass of galena by the mass of water having the same volume is the specific gravity of galena, or 7.56. Specific gravity is a measure that tells you how much heavier or lighter a substance is compared to an equal volume of water. As a formula, specific gravity is written as follows:

$$\frac{\text{specific}}{\text{gravity}} = \frac{\text{mass of substance}}{\text{mass of equal volume of water}}$$

Gold is one of the heaviest minerals, with a specific gravity of 19.3. This means that a sample of gold is slightly more than 19 times as heavy as an equal volume of water. Whereas a cubic foot of water has a mass of 28.3 kilograms (kg), a cubic foot of gold has a mass of about 546 kg.

Substances that are lighter than an equal volume of water have a specific gravity of less than 1. All substances that have a specific gravity of less than 1 can float on water. For example, ice has a specific gravity of about 0.92. A solid block of pinewood has a specific gravity of 0.50.

A common method of determining the specific gravity of irregularly shaped minerals is shown in Fig. 2-12. In this method, the mass of the mineral is first determined in air, then determined in water. The spring scale shows that the mineral sample appears to weigh less when it is suspended in water. The same thing happens to you when you go swimming. Your body seems much lighter in water than in air. The weight lost by the mineral in the water is

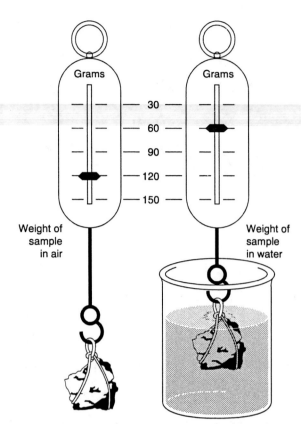

Fig. 2-12. **Determining specific gravity.**

equal to the weight of the water that it displaces.

The formula for specific gravity can be stated in another way:

$$\frac{\text{specific}}{\text{gravity}} = \frac{\text{weight of substance in air}}{\text{weight lost by substance in water}}$$

For example, the mineral sample in Fig. 2-12 weighs 120 g in air. In water, the same sample weighs only 60 g. The loss in weight is 60 g. Thus,

$$\frac{\text{weight of mineral in air}}{\text{weight lost by mineral in water}} = \frac{120 \text{ g}}{60 \text{ g}} = 2$$

The specific gravity of the mineral is 2.0. This means that the mineral's *density* (its mass, or weight, per unit volume) is 2 g per cubic centimeter.

# CHEMICAL PROPERTIES OF MINERALS

A *chemical property* of a mineral is a property that involves the way the mineral changes chemically when it reacts with other elements or compounds. For example, chemical properties may be observed when a mineral combines with oxygen or when a mineral reacts with an acid. When a mineral reacts chemically with an element or a compound, one or more new substances are formed. The properties of the new substance are entirely different from the properties of the original mineral.

**Oxidation.**   Many substances react chemically by combining with oxygen. This kind of chemical reaction is called **oxidation.** For example, when wood burns, it is combining with oxygen in the atmosphere to produce heat, light, and new substances. When a piece of iron combines chemically with oxygen and moisture in the atmosphere, a new substance called *limonite (iron oxide)* is produced. Limonite—a major ore of iron—is a brownish material that crumbles easily, unlike the iron that was originally present.

**Acid Reaction.**   Another chemical property is the reaction that occurs when a mineral comes into contact with an acid. Many minerals consist of metals that are combined with carbon and oxygen. These minerals are called *carbonates.* For example, calcite is the chemical compound calcium carbonate ($CaCO_3$).

When a drop of hydrochloric acid is placed on a piece of calcite, the acid reacts with the carbonate in the calcite. Bubbles of carbon dioxide gas appear on the surface of the calcite. (Carbon dioxide is also the gas that fizzes in a bottle of soda pop.) Thus, when hydrochloric acid is applied to an unknown mineral and bubbles appear, the mineral is probably a carbonate. If the mineral is calcite, the chemical equation for the reaction is written as follows:

$CaCO_3$ +  2 HCL $\longrightarrow$
(calcium  (hydrochloric
carbonate)  acid)

$CaCl_2$   + $H_2O$ + $CO_2$
(calcium (water) (carbon
chloride)  dioxide)

**CAUTION: Hydrochloric acid must be used carefully because it can cause serious burns to the skin and eyes.** A safe substitute for hydrochloric acid is vinegar, which is a dilute form of *acetic acid.*

## SPECIAL PROPERTIES OF MINERALS

Some minerals can be identified by their unusual properties. Some of these unusual properties include radioactivity, double refraction, magnetism, and fluorescence (see Fig. 2-13).

**Radioactivity.**   A few minerals give off invisible radiation that can be detected by a special instrument called a *Geiger counter.* These minerals, which include pitchblende and carnotite, are said to be **radioactive.** A Geiger counter converts the radiation from these minerals into electrical signals. These signals can either be read on a numerical meter or used to produce clicking sounds in the counter.

**Double Refraction.**   Some mineral crystals can split a beam of light into two beams. If you look at an object through one of these crystals, you will see two images of the object. This special property is called *double refraction.* Crystals of transparent calcite (called *Iceland spar*) exhibit this property more clearly than any other mineral.

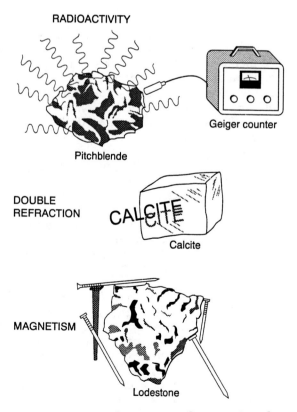

Fig. 2-13. Special properties of some minerals.

**Magnetism.** *Magnetic* minerals, such as magnetite, are attracted to magnets. In fact, the first magnetic compasses used by sailors to navigate ships were a form of magnetite called *lodestone.* Large concentrations of magnetite in Earth's crust cause compass needles to point to the body of magnetite rather than to magnetic north. Airplane pilots navigating by compass have sometimes become lost because of the influence of these large masses of magnetite.

**Fluorescence.** Certain minerals, such as fluorite and some kinds of calcite, glow with beautiful colors when they are exposed to X rays or ultraviolet rays. Both of these rays are invisible to humans. But, when they are absorbed by fluorescent minerals, the minerals give off visible light. This special property of minerals is called **fluorescence.**

## USES OF MINERALS

Traces, or very small amounts, of many minerals can be found in most rock samples. These small amounts of minerals are commonly called *impurities.* However, when a useful mineral occurs in amounts large enough to make it profitable to mine, the mineral is called an *ore.* Most metals you use every day are extracted from ores.

A metal is usually separated from its ore by chemical or electrical methods. For example, iron is extracted from its ore (hematite) by **smelting**—a chemical method of separation that requires high temperatures. In the smelting process, coke (a form of carbon) and limestone are mixed with the ore. Coke removes oxygen from the hematite, and limestone removes the sandlike impurities.

Aluminum is extracted from its ore (bauxite) by an electrical method. In this process, an electric current melts the bauxite and then separates the aluminum metal from the ore. Copper is extracted from its ores (chalcocite, azurite, and cuprite) by both chemical and electrical methods.

Some useful elements may be found in nature in their native form, that is, uncombined with other elements. These are called *native elements.* Gold, silver, copper, sulfur, graphite, and carbon (in the form of diamond) are examples of native elements.

A variety of economically valuable substances and their ores or mineral sources are listed in Table 2-3.

## TABLE 2-3. USEFUL SUBSTANCES AND THEIR MINERAL SOURCES

| Substance | Mineral Source |
|---|---|
| Abrasives: sandpaper, grinding wheels, drills | Diamond, corundum, garnet, quartz |
| Acids: sulfuric acid ($H_2SO_4$) | Sulfur, pyrite ($FeS_2$) |
| Aluminum products | Bauxite |
| Bricks | Kaolinite (clay) |
| Cement | Calcite |
| Chalk | Calcite, gypsum |
| China | Kaolinite |
| Copper products | Chalcocite, chalcopyrite, native copper |
| Gems | Diamond, ruby and sapphire (corundum), emerald (beryl), garnet, amethyst and agate (quartz) |
| Insecticides | Sulfur |
| Iron products | Hematite, magnetite |
| Lead: pipes, sinkers, storage batteries, solder (lead and tin) | Galena |
| Mercury: thermometer, barometer, fillings (teeth), mirrors | Cinnabar |
| Paints | Galena (lead), sphalerite (zinc), ilmenite (titanium) |
| Pencil lead | Graphite and kaolinite |
| Plaster of paris: wallboard | Gypsum |
| Salt: table salt, rock salt | Halite |
| Silver: coins, jewelry, fillings (teeth) | Argentite |
| Tin | Cassiterite |
| Uranium | Pitchblende, carnotite |
| Whitewash | Calcite |
| Zinc | Sphalerite |

# CHAPTER REVIEW

## Science Terms

*The following list contains all of the boldfaced words found in this chapter and the page on which each appears.*

atmosphere (p. 18)
chemical compound (p. 20)
chemical element (p. 20)
cleavage (p. 21)
crust (p. 19)
crystal (p. 22)
fluorescence (p. 26)

fracture (p. 22)
geologists (p. 19)
hardness (p. 20)
hydrosphere (p. 18)
inner core (p. 18)
lithosphere (p. 19)
luster (p. 23)
mantle (p. 19)

mineral (p. 20)
outer core (p. 19)
oxidation (p. 25)
radioactive (p. 25)
smelting (p. 26)
specific gravity (p. 24)
streak (p. 21)

## Matching Questions

*On the blank line, write the letter of the item in column B that is most closely related to the item in column A.*

*Column A*

____ 1. calcite

____ 2. corundum

____ 3. halite

____ 4. graphite

____ 5. kaolin

____ 6. feldspar

____ 7. mica

____ 8. galena

____ 9. fluorite

____ 10. lodestone

*Column B*

*a.* natural magnet
*b.* reacts with hydrochloric acid to produce carbon dioxide gas
*c.* glows under ultraviolet light
*d.* commonly called rock salt
*e.* used in electrical insulation
*f.* type of clay
*g.* radioactive
*h.* composed solely of the element carbon
*i.* harder than all other minerals except diamond
*j.* two directions of cleavage
*k.* lead mineral with a specific gravity of about 8.0

## Multiple-Choice Questions

*On the blank line, write the letter preceding the word or expression that best completes the statement or answers the question.*

**1.** The deepest distance that scientists have ever drilled into Earth's crust is about
*a.* 5 km   *b.* 13 km   *c.* 40 km   *d.* 90 km                    1 ____

**2.** Silicon and oxygen combine chemically to form
*a.* $Fe_2O_3$   *b.* $SnO_2$   *c.* $SiO_2$   *d.* $CaCO_3$                    2 ____

3. The most abundant elements in Earth's crust are
   *a.* iron and oxygen      *c.* oxygen and silicon
   *b.* iron and nickel      *d.* water and iron            3 ____

4. A mineral that consists of calcium, carbon, and oxygen is
   *a.* calcite  *b.* graphite  *c.* quartz  *d.* table salt      4 ____

5. A mineral that can be scratched easily by all other minerals is
   *a.* talc  *b.* calcite  *c.* quartz  *d.* corundum      5 ____

6. A mineral that has a single direction of cleavage is
   *a.* calcite  *b.* mica  *c.* feldspar  *d.* galena      6 ____

7. The way in which a mineral cleaves is determined by its
   *a.* hardness  *b.* streak  *c.* luster  *d.* internal structure      7 ____

8. An example of a chemical property of iron is
   *a.* magnetism  *b.* fluorescence  *c.* oxidation  *d.* radioactivity      8 ____

9. A mineral that can be detected by a Geiger counter is
   *a.* pitchblende  *b.* fluorite  *c.* magnetite  *d.* calcite      9 ____

10. A mineral that can be used as a magnet is
    *a.* lodestone  *b.* calcite  *c.* quartz  *d.* corundum      10 ____

11. Of the following elements, the only one *not* found in native form is
    *a.* gold  *b.* silver  *c.* copper  *d.* aluminum      11 ____

# Modified True-False Questions

*In some of the following statements, the italicized term makes the statement incorrect. For each incorrect statement, write the term that must be substituted for the italicized term to make the statement correct. For each correct statement, write the word "true."*

1. Three fourths of Earth's surface consists of *land*.      1 _____

2. The solid part of Earth is called the *lithosphere*.      2 _____

3. Earth's crust is *thickest* beneath the continents.      3 _____

4. Earth's crust ranges in thickness between 5 and *90* km.      4 _____

5. Geologists think that the inner and outer cores of Earth consist of *iron and nickel*.      5 _____

6. Earth's crust is made of rocks that were once in a *molten* state.      6 _____

7. The simplest kind of substance in nature is *a mineral*.      7 _____

8. Both diamond and graphite consist of *carbon*.      8 _____

9. Halite (NaCl) consists of *four* different elements that are chemically combined.      9 _____

10. Luster is an example of a *chemical* property.      10 _____

11. *Quartz*, topaz, and corundum can scratch glass.      11 _____

12. *Talc* is a mineral that can be scratched with a fingernail.      12 _____

13. *Mica* splits easily into thin sheets.      13 _____

**14.** A windowpane has *metallic* luster.                                     14 _____

**15.** The standard of comparison for determining the specific gravity of different substances is *water*.                                     15 _____

**16.** Carbon dioxide is a gas that is produced when hydrochloric acid reacts with *calcite*.                                     16 _____

**17.** The first magnetic compasses used by sailors were made from a form of *pitchblende* called lodestone.                                     17 _____

**18.** Copper can be found *uncombined* with other elements.                                     18 _____

# Testing Your Knowledge

**1.** Describe the two methods by which geologists study Earth's internal structure.

   *a.* _____

   *b.* _____

**2.** Describe two differences between the crust and the mantle.

   *a.* _____

   *b.* _____

**3.** What is the difference between a rock and a mineral?_____

   _____

   _____

**4.** How can you use a steel nail, a copper penny, a piece of glass, and your fingernail to

   determine the hardness of different minerals?_____

   _____

   _____

**5.** Why may a geologist have to break a mineral sample to determine its true color?_____

   _____

   _____

**6.** Devise a method to separate grains of gold from grains of quartz sand by using the difference

   in specific gravity between gold and quartz._____

   _____

   _____

**7.** List five properties that you can test to identify an unknown mineral.

   _____

   _____

# Rocks in Earth's Crust

## LABORATORY INVESTIGATION

### IDENTIFYING A FEW COMMON ROCKS

**A.** Examine the numbered rock samples on your desk that are in group A. These rocks are called *igneous rocks*. They are formed when hot, molten rock materials solidify.

Study the descriptions of the igneous rocks in the table below. In the space provided, write the number of the rock sample from group A that best matches each description.

**IGNEOUS ROCKS**

| Rock | Description | Sample |
|------|-------------|--------|
| Granite | Contains quartz, feldspar, and mica or hornblende. The quartz looks like chips of broken glass. The feldspar is light-colored and occurs in small, blocklike chunks. The mica and hornblende look like dark, shiny specks. | |
| Obsidian | Usually looks like a piece of dark, smooth glass. A thin slice or edge of this rock allows light to shine through. | |
| Pumice | Very light in weight. Has a rough surface that contains many tiny, spongelike pores. | |
| Basalt | Has a dark, slightly rough surface. | |

1. The rocks that do not have any visible grains or crystals in their structure are _____
_____.

2. The rock that has comparatively large crystals is _____.

3. The rock that looks as if gases had bubbled through it is _____.

**B.** Examine the numbered rock samples on your desk that are in group B. These rocks are called *sedimentary rocks*. Most sedimentary rocks are formed when rock particles are cemented together.

Study the descriptions of the sedimentary rocks in the table below. In the space provided, write the number of the rock sample from group B that best matches the description.

### SEDIMENTARY ROCKS

| Rock | Description | Sample |
|---|---|---|
| Sandstone | Contains rock particles about the size of grains of sugar. The grains can often be scraped from the rock. | |
| Conglomerate | Consists of various-sized pebbles cemented together in a cementlike mass of smaller rock particles or fragments. | |
| Shale | Resembles thin sheets of hardened clay or mud that have been pressed tightly together. Shale gives off an earthy odor when moist. | |
| Limestone | Usually light-colored and relatively smooth. A drop of acid placed on limestone causes bubbles of carbon dioxide to appear on the rock's surface. This is the *acid test*, instructions for which are given in part E. | |

**4.** The rock that looks like raisin bread or peanut brittle is _____.

**5.** The rock that looks somewhat like an irregular lump of sugar is _____.

**6.** The rock that looks like dried mud is _____.

**C.** Examine the numbered rock samples on your desk that are in group C. These rocks are called *metamorphic rocks*. They are rocks that have been subjected to great heat and pressure. Metamorphic rocks often have a "squeezed" look.

Study the descriptions of the metamorphic rocks in the table below. In the space provided, write the number of the rock sample from group C that best matches each description.

### METAMORPHIC ROCKS

| Rock | Description | Sample |
|---|---|---|
| Gneiss | Has alternating light and dark bands of different thickness. | |
| Schist | Consists of very thin bands, or layers. | |
| Quartzite | Very hard. Scratches glass more easily than does any other rock of this group. | |
| Slate | Very smooth. Consists of hard, thin sheets pressed tightly together. | |
| Marble | Surface of a freshly broken piece looks like the surface of a sugar cube. Most kinds of marble react with acid, forming bubbles of carbon dioxide. | |

**D.** Study the rock samples listed below. From your observations of the igneous and sedimentary rocks in group I, which of these rocks do you think could be transformed into the metamorphic rocks of group II? Write your answers in the spaces provided.

I. *Igneous and Sedimentary Rocks*

    *a.* Limestone

    *b.* Granite

    *c.* Shale

    *d.* Sandstone

II. *Metamorphic Rocks*

Gneiss _____

Slate _____

Quartzite _____

Marble _____

**E.** Dip a glass rod in dilute hydrochloric acid. Put a drop of acid on each of the rock samples in groups B and C.

**CAUTION: Wear goggles to protect your eyes. Acids can burn your skin and destroy your clothing. Rinse the glass rod in water after each use. Spilled acid should be diluted with water. Dry each rock sample with a tissue after this activity.**

**7.** On which rock surface does bubbling occur? _____

**8.** What does the bubbling suggest? _____

_____

# Rocks in the Crust

## HOW ROCKS WERE FORMED

Most scientists think that billions of years ago Earth was a mass of hot, liquid material. This *molten* material was similar to the fiery lava that pours out of volcanoes today. Over millions of years, the molten material at the surface of Earth cooled enough to solidify into rock. Cooling continued downward from the surface until Earth developed a relatively thick and solid crust. Rocks that form as molten material cools are called **igneous rocks.**

When Earth first formed, its thin atmosphere did not contain water and many other gases. However, the molten material within Earth contained water vapor and other gases. These gases were expelled from Earth through volcanoes and other openings in the crust, and added to Earth's original atmosphere. As Earth continued to cool, the water vapor in the atmosphere formed dense clouds, from which rain fell.

Continual rainfalls over millions of years slowly wore away the rocks of Earth's crust. The water collected in depressions in the crust, forming the oceans. At the same time, the rain washed loosened grains of rocks into rivers, lakes, and oceans, where they settled to the bottom. These particles of sand, silt, and clay that are transported and deposited by moving water are called **sediments.** Over time, the sediments piled up in deep layers. In some places, the layers of sediments became extremely thick and heavy. As a result, the top layers pressed down on the bottom layers, compressing them.

In time, the bottom layers of sediments were compressed sufficiently to form rocks. In other places, chemicals dissolved in the water were deposited between the particles of sediment. These chemicals acted as a cement, binding the rock particles together. Rocks that form when sediments are compressed or cemented together are called **sedimentary rocks.**

Both igneous and sedimentary rocks may be subjected to the heat, pressure, and chemically active fluids that are involved in mountain building and volcanic eruptions. These conditions change rocks from their original form. Rocks that have been changed from their original form by heat, pressure, or chemical action are called **metamorphic rocks.**

In this chapter, you will read about igneous, sedimentary, and metamorphic rocks. You also will learn about the minerals that make up these three types of rocks. Your laboratory investigation will help you recognize them.

## MINERALS THAT MAKE UP ROCKS

Some rocks consist of one mineral. Most rocks, however, contain a mixture of different minerals. Nearly 2000 kinds of minerals have been found in rocks. Yet, out of this variety, about 90 percent of the rocks in Earth's crust consist of only 10 minerals. Two of the most common minerals are *quartz* and *feldspar*.

When examined with a magnifying glass, quartz grains in rocks resemble shattered chunks of glass. Grains of feldspar look like small, rectangular blocks and are usually white, gray, or pink in color. Other common minerals found in rocks include mica, which appears as light-to-dark flat sheets; and *augite* and *hornblende*, which appear as dark, rectangular blocks. Augite and hornblende are usually dark green or black in color.

# Igneous Rocks

Inside Earth, molten rock is called **magma.** When molten rock pours out onto the surface of Earth, it is called **lava.** Most magma lies far below the surface where the temperatures are very high. Thus, magma takes many years to cool and solidify into rock. In contrast, lava is at the surface of Earth where temperatures are much cooler. Consequently, lava usually cools and solidifies in a much shorter time. A lava flow one meter thick may solidify into rock within a few months.

Molten magma consists of various elements and compounds, all dissolved together. As the magma cools, these elements and compounds begin to form crystals. The slower the magma cools, the larger the crystals grow. As shown in Fig. 3-1, molten materials that cool deep inside Earth form much larger crystals than do molten materials that cool closer to the surface. Rocks with large crystals are said to have a *coarse-grained* texture. (*Texture* means the size, shape, and arrangement of the particles in a rock.) Granite is an example of a coarse-grained igneous rock (see Fig. 3-2).

Fig. 3-2. Granite: a coarse-grained rock.

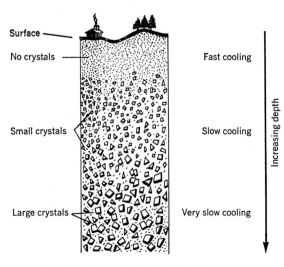

Fig. 3-1. Crystal size and cooling rate.

Rocks that solidify on or near Earth's surface cool rapidly. Rapid cooling prevents the mineral crystals in these rocks from growing very large. Crystals that form under these conditions are usually too small to be seen without magnification. Such rocks are said to have a *fine-grained* texture. Basalt is an example of an igneous rock with a fine-grained texture (see Fig. 3-3).

Some rocks cool too rapidly for any crystals to form. These rocks have a *glassy* texture. Obsidian, commonly called "volcanic glass," is an example of an igneous rock with a glassy texture (see Fig. 3-4).

Igneous rocks can be classified in a number of ways: according to their texture, their composition, the way they formed, or some other characteristic. In this book, for simplicity of identification, igneous rocks are classified according to color—whether they are *light* or *dark*.

## LIGHT-COLORED IGNEOUS ROCKS

The light-colored igneous rocks include *granite, obsidian, pumice,* and *pegmatite*. These rocks consist mainly of quartz and feldspar. The main difference among the rocks in this group is the size of their crystal grains.

**Granite.** **Granite** is usually gray or pink in color. The most abundant minerals in granite are feldspar and quartz, with smaller amounts of mica and hornblende. As shown in Fig. 3-2, the minerals in granite appear as relatively large, easily visible crystals or grains.

Feldspar makes up about half of the granite and is responsible for its color. You can identify the feldspar as relatively smooth, light-colored, blocklike grains in the rock. The quartz grains may appear as cloudy glass chips. The mica or hornblende (or both) are commonly seen as dark specks scattered throughout the granite. Granite has a coarse-grained texture because of the large size of its crystals. As you have learned, the crystals are large because the granite cooled slowly, deep within Earth's crust.

The crust, or bedrock, that makes up the continents (large land areas) consists mostly of granite. In addition, many mountain ranges, such as the Rocky Mountains and the White Mountains in New Hampshire, consist largely of granite. Because of its hardness, granite is frequently used for buildings and monuments. The Mount Rushmore National Memorial, located in the Black Hills of South Dakota, is carved out of granite (see Fig. 3-5).

**Obsidian.** A lump of **obsidian** usually is dark brown or black in color. However, a thin slice of obsidian appears colorless or lightly tinted. For this reason (and because the minerals in obsidian are usually the same as those in granite), obsidian is classified with the light-colored igneous rocks.

Obsidian is formed when molten materials cool very rapidly—too rapidly for crystals to form. Consequently, obsidian has a glassy texture. Obsidian was used by some tribes of Native Americans to make arrowheads and knives because of its hardness and sharp, glasslike edges.

Fig. 3-3. Basalt: a fine-grained rock.

Fig. 3-4. Obsidian: a rock with a glassy texture.

**Fig. 3-5. Mount Rushmore National Memorial in South Dakota.**

**Pumice.** At times, hot gases bubbling through lava cause the lava to froth up like whitecaps on an ocean. When such lava cools and solidifies, a lightweight, spongelike rock forms. This rock is called **pumice.** Many tiny air holes within the pumice usually make this rock light enough to float in water. In fact, pumice is often found floating on the ocean near an erupting volcano. Powdered pumice is a fine polishing material used in dental powders and household scouring mixtures.

**Pegmatite.** Pegmatite is similar in composition to granite, but pegmatite has a much coarser-grained texture than granite. In fact, pegmatite is noted for its huge crystals. One mica crystal found in a deposit of pegmatite weighed about 90 tons. Crystals more than 12 meters (m) long have also been found in pegmatite. Pegmatite is mined commercially because of its large crystals, especially crystals of feldspar and mica. Sometimes, emeralds, garnets, rubies, and other gemstones are found in pegmatite.

## DARK-COLORED IGNEOUS ROCKS

The minerals feldspar, augite, and olivine are abundant in the dark-colored igneous rocks, which include *basalt, diabase,* and *gabbro.* Like the light-colored igneous rocks, the main difference among dark-colored igneous rocks is their texture.

**Basalt.** Basalt is dark in color. The color of basalt is caused by dark-colored minerals such as augite, olivine, and magnetite. Augite and olivine usually are dark green; magnetite is black.

As shown in Fig. 3-3, the very small size of the mineral grains in basalt gives it a texture that is almost smooth. Basalt is the most abundant of all the igneous rocks that cooled near Earth's surface. A great lava deposit is the Columbia River Plateau, which extends through the states of Washington, Oregon, and Idaho. The plateau consists of layers of basalt that are hundreds of meters thick (see Fig. 3-6).

**Diabase.** Diabase is similar to basalt in color and mineral composition. However, diabase has a coarser texture than basalt. The Palisades, along the Hudson River opposite New York City, consist largely of diabase.

Basalt and diabase are sometimes referred to as *traprock.* Traprock is crushed basalt or diabase used as filler under railroad tracks, for the foundations of highways, and for making concrete.

**Gabbro.** The coarsest-grained member of the dark-colored igneous rocks is called **gabbro.** Gabbro contains the same minerals found in basalt and diabase. Gabbro is found in parts of Canada, in northern Michigan around Lake Superior, and in the Adirondacks of New York. Gabbro often is cut, polished, and sold as "black granite."

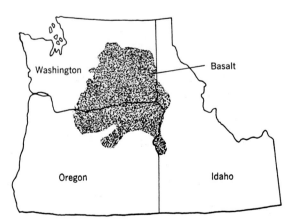

**Fig. 3-6. The Columbia River Plateau.**

# Sedimentary Rocks

The particles, or sediments, that make up sedimentary rocks originate in three different ways. First, the weathering action of water, wind, and ice can gradually wear down or crumble igneous rock or other masses of bedrock (see Fig. 3-7). The rock particles produced by weathering often form a layer of loose surface materials that hides the bedrock. Eventually, these particles may be washed away and deposited as sediments in rivers, lakes, and oceans.

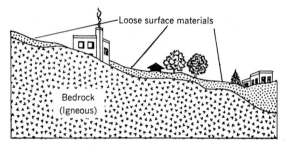

**Fig. 3-7. The weathering of bedrock.**

Second, water can dissolve and carry away the minerals in a rock. These dissolved minerals may later solidify again to form particles of sediment.

Third, the particles of sediment can be *organic*. Organic sediments consist of the skeletons and shells of small marine animals that have settled to the ocean floor.

## SEDIMENTARY ROCKS FORMED FROM ROCK FRAGMENTS

When a rock crumbles, it breaks into pieces of different sizes. Pieces of rock with a diameter larger than 25 centimeters are called *boulders*. In decreasing order of size, smaller pieces of rock are called *cobbles, pebbles, gravel, sand, silt,* and *clay*. The particles of silt and clay are about the size of dust particles.

You recall that natural agents such as rivers, glaciers, and winds often carry away the crumbled pieces of rock. The largest and heaviest pieces, such as boulders and pebbles, usually are carried only short distances by a river. The smallest and lightest particles, such as clays and silts, can be carried many kilometers by water and wind.

A stream of rainwater pouring down a street or hillside often has a muddy appearance. The muddy appearance is due to silt and clay particles suspended in and carried along by the water. When the water slows down, the suspended particles of silt and clay may sink to the bottom of the stream.

Under natural conditions, differences in size and weight cause particles suspended in a body of water to settle out at different rates. Thus, the large and heavy particles, such as gravel and sand, are the first to sink. The light-weight particles of clay and silt remain suspended in the water long after the gravel and sand have settled to the bottom (see Fig. 3-8).

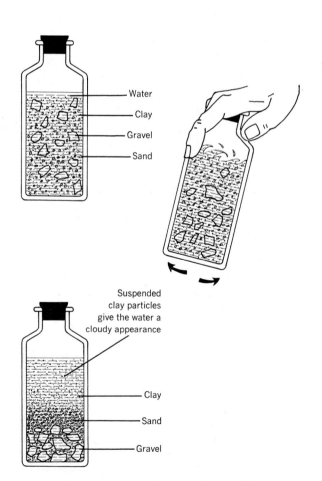

**Fig. 3-8. Demonstration showing the settling out of various-sized particles.**

# CLASSIFYING SEDIMENTARY ROCKS

Running water separates rock particles according to their sizes. The heaviest particles (boulders, pebbles, and gravels) travel the shortest distance before settling to the bottom. The lightest particles (sands, silts, and clays) travel the farthest distance (see Fig. 3-9). In this way, the rock particles are sorted out to form deposits of different types of sedimentary rock. As shown in Fig. 3-10, the most common types are *sandstone, shale,* and *conglomerate.*

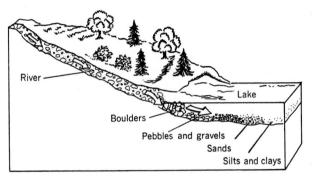

Fig. 3-9. Moving water separates particles of rock.

**Sandstone.** A sample of **sandstone** often feels gritty because it consists largely of grains of quartz sand. The grains of quartz sand are held together by natural cements. These cements consist of minerals, such as silica, cal-

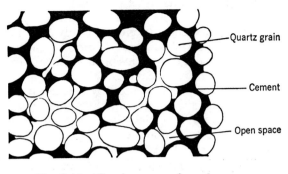

Fig. 3-11. The structure of sandstone.

cium carbonate, or iron compounds that are dissolved in the water. The dissolved minerals seep into the deposits of sand, fill in many of the open spaces, and bind the grains together (see Fig. 3-11). The color of sandstone is determined mainly by the color of its cement. Rocks cemented by silica or calcium carbonate are generally light in color; rocks cemented by iron compounds are red to reddish brown in color.

Although the grains of sand in sandstone are cemented together, relatively large spaces remain between the grains. For this reason, water and other liquids can soak through sandstone (see Fig. 3-12).

Sandstone

Conglomerate

Fig. 3-10. Types of sedimentary rocks.

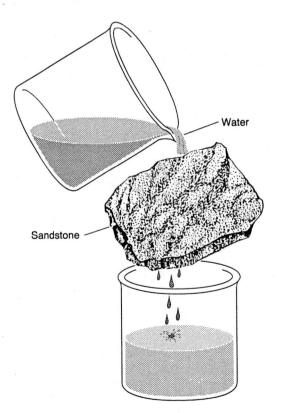

Fig. 3-12. Water soaks through sandstone.

**Shale.** **Shale** consists of clay and silt particles packed so closely together that water carrying dissolved cements cannot freely pass between the particles. Shale crumbles rather easily because the particles of clay and silt are cemented together weakly.

Shale is fairly easy to identify because it resembles hardened mud, and it splits easily into relatively flat pieces. Most shales are gray, although impurities in the sediments may give the shales other colors.

Finely ground shale is used for making bricks, cement, and tiles. Some forms of shale contain oil locked tightly between the particles of clay and silt. At present, recovering oil from these shales is difficult and expensive. In the future, oil shales may become an important source of petroleum.

**Conglomerate.** One of the easiest rocks to identify is **conglomerate.** Conglomerate is made up of rounded, pebble-sized rock fragments set in a cementlike mass of sand, silt, or clay. A common type of conglomerate consists of smooth, rounded quartz pebbles cemented together by finer materials. Conglomerate gets its name from a Latin word that means "to gather or collect together." Some conglomerates may even have boulders cemented into their mass.

# SEDIMENTARY ROCKS: FORMED BY CHEMICAL ACTION

Rain seeping into the crust dissolves minerals from the rocks. These dissolved minerals are carried by running water into oceans and other large bodies of water. When the water evaporates or becomes overloaded with minerals in solution, the dissolved minerals may separate out and become solid matter again. Solid matter formed in this way is called *chemical sediment*. The most common sedimentary rocks formed by chemical action are *limestone, gypsum,* and *rock salt*.

**Limestone.** **Limestone** consists mostly of calcium carbonate, or calcite. Limestone forms when a calcium-containing compound, dissolved in water, separates from the water and becomes solid particles of calcite.

As you recall from your laboratory investigation, limestone can be identified easily by performing the acid test. Bubbles of carbon dioxide gas appear on the surface of limestone when drops of dilute hydrochloric acid or vinegar are placed on it.

Large deposits of limestone are found in Kentucky, Indiana, and Florida. Limestone is widely used for buildings. For example, the Empire State Building in New York City and the Pentagon in Washington, D.C., are both built with large amounts of limestone.

**Gypsum and Rock Salt.** The minerals **gypsum** and **halite** (rock salt) dissolve slowly in water. During the history of Earth, enormous quantities of both of these minerals were dissolved in ancient oceans. When these ancient oceans dried up, great deposits of gypsum and halite were left behind.

Gypsum is composed of the elements calcium, sulfur, and oxygen combined with the compound water. When gypsum is heated in an oven, it forms plaster of paris. A pearly-white or pink rock made of gypsum, called *alabaster,* is often carved into statues and vases. The White Sands National Monument in New Mexico consists of deposits of beautiful white gypsum sands.

Deposits of rock salt are usually found with deposits of gypsum because rock salt also is left behind when seawater evaporates. The Great Salt Lake in Utah is a small remnant of a much larger body of water, called Lake Bonneville, that once existed there (see Fig. 3-13). The great Bonneville Salt Flats that exist

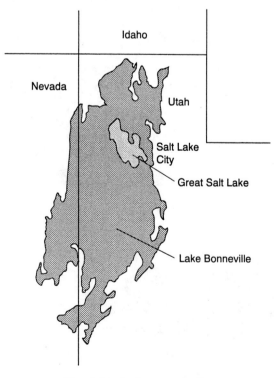

**Fig. 3-13. Lake Bonneville.**

today were formed when Lake Bonneville lost most of its water by evaporation. The extensive salt deposits and the Great Salt Lake are all that remain of ancient Lake Bonneville. The flatness of this area makes it an ideal place for testing high-speed racing cars.

# SEDIMENTARY ROCKS: FORMED FROM ORGANIC MATERIALS

Sedimentary rocks that form from the piling up of the remains of plants and animals are called *organic sedimentary rocks*. Among the most important organic sedimentary rocks are *organic limestone* and *coal*.

**Organic Limestone.** Great numbers of shell-forming animals live in the shallow waters off the coasts of continents and islands. These animals, which include clams, oysters, snails, and corals, remove calcium carbonate from seawater and use it to build the shells and other hard parts of their bodies. When the animals die, their soft, fleshy parts rot away, leaving only their hard parts and shells. As these hard materials pile up on the ocean floor and become cemented together, they form different kinds of **organic limestone**. Three different organic limestones are *coquina, chalk,* and *coral limestone*.

As shown in Fig. 3-14, *coquina* is made up of loosely cemented fragments of shells and corals. Coquina is common along the coast of Florida.

*Chalk* consists of the shells of billions of microscopic organisms that piled up in deep layers on the ocean floor over thousands of years.

Fig. 3-15. The white cliffs of Dover in England.

In some places, these tiny shells have formed chalk beds over 100 m thick. The famous white cliffs of Dover, England, were formed in this way (see Fig. 3-15). In the distant past, England and France probably were connected by a strip of land made of this chalk.

*Coral limestone* is produced by tiny animals that live together in vast colonies, inhabiting relatively warm, shallow water. Hard skeletons grow around the bodies of these animals. After the coral animals die, their skeletons remain behind. In time, other coral animals grow on top of the skeletons. As the coral colonies continue to live, reproduce, and die, the skeletons accumulate and form *coral reefs*.

Coral reefs are found today along the coasts of Bermuda, Florida, the Virgin Islands, and Hawaii. The Great Barrier Reef, a huge coral reef off the eastern coast of Australia, is about 1900 kilometers (km) long and 145 km wide.

The continual pounding of waves may gradually break down a coral reef, reducing the reef to a sedimentary deposit. When this deposit is cemented together by silica or calcium carbonate dissolved in the water, coral limestone is formed.

**Coal.** The accumulated remains of dead plants are the source of **coal**. Coal is made up mainly of the element carbon.

Fig. 3-14. Coquina.

# FORMATION OF COAL

Coal is found in many sections of the United States. The regions where coal is found today were once covered by great swamps. Giant ferns and other large plants thrived in these swamps. Over thousands of years, as the swamp plants died, their remains accumulated in great deposits and formed coal.

As shown in Fig. 3-16, the plants changed into coal slowly. At first, as the dead plants were buried, the weight of the deposits lying above them gradually compressed the plants into rocklike masses. Heat drove off elements such as hydrogen and oxygen, leaving carbon as the most abundant element remaining.

The major stages in the formation of coal are:

1. *Peat formation.* During the first stage of coal formation, the remains of plants became *peat*. Peat is a loosely packed, brownish mass of decayed leaves, twigs, and other plant parts. The carbon content of peat averages about 55 percent.

2. *Lignite formation.* Further decay and compression of peat produce *lignite*, or brown coal. Lignite is harder and more compressed than peat. The carbon content of lignite averages about 73 percent.

3. *Bituminous coal formation.* Further changes in lignite produce *bituminous*, or soft coal. Bituminous coal is black. The carbon content of bituminous coal is about 83 percent.

4. *Anthracite coal formation.* Continued heat and pressure applied to soft coal produce *anthracite*, or hard coal. The carbon content of anthracite averages about 94 percent, making it a better fuel than peat, lignite, or bituminous coal. Because of the changes it has undergone, anthracite is classified with the metamorphic rocks, which are described later in this chapter.

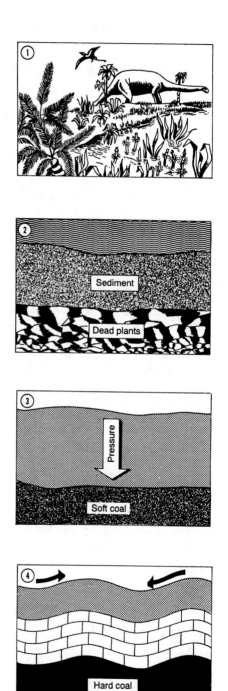

**Fig. 3-16. Formation of coal.**

## OIL AND NATURAL GASES

According to geologists, tiny sea plants and animals that lived in the distant past are the source of today's deposits of petroleum and natural gas. These deposits were formed in a manner similar to the formation of coal from plants.

Over millions of years, the remains of billions of tiny plants and animals settled on the bottom of seas and swamps. The remains of these organisms mixed with sediments such as clay and silt to form thick deposits. Certain chemical reactions changed these remains into oil and natural gas.

Eventually, the oil collected in the open spaces of sandstone or in the cracks of limestone. Today, the bodies of oil trapped in these spaces are called *reservoirs*, or pools of oil. Above and below these reservoirs of oil are layers of shale and other impermeable rocks (rocks through which liquids cannot pass). The shale prevents the oil from seeping out of its natural reservoir.

Underground structures in which oil has accumulated are called *traps* (Fig. 3-17a). One common type of trap is shown in Fig. 3-17b. Notice that natural gas is trapped above the oil. This gas is under great pressure. Beneath the oil is a reservoir of water. When an oil drill breaks into a deposit of oil, the pressure of the natural gas forces the oil up the drilled hole. If the gas pressure is great enough, the oil escapes as a *gusher*. The natural gas is pumped to distant towns and cities, where it is used as a fuel for heating and cooking.

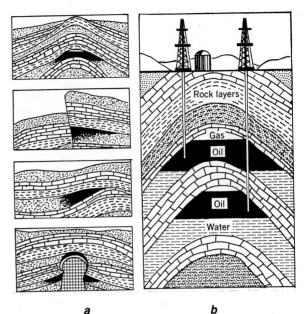

**Fig. 3-17. (a) Oil traps. (b) A detailed oil trap.**

a          b

# Metamorphic Rocks

Rocks that have been changed in form or composition as a result of being subjected to high temperatures or great pressure are called *metamorphic rocks*. The process by which the rocks undergo these changes is called *metamorphism*. One reason why rocks undergo change is that large sections of Earth's crust moving sideways may fold and break. These Earth movements exert tremendous pressure on rocks. Another reason is that molten materials may flow through breaks in the crust. The great heat of the molten materials changes the composition of the surrounding rocks.

Metamorphic rocks are classified into two groups: *foliated* (leaflike) rocks and *unfoliated* rocks. Foliated rocks consist of compressed, parallel bands of minerals, which give the rocks a striped appearance. These bands can be very thick or thinner than the pages of this book. Unfoliated rocks are not banded.

## FOLIATED METAMORPHIC ROCKS

**Slate.** When shale is compressed and heated, **slate** is formed. Because slate is highly foliated, it splits easily into thin, smooth slabs.

These slabs are used for roofs, sidewalks, billiard tables, and chalkboards. Slate occurs in many colors, including black, blue, green, and red.

**Schist.** The continued metamorphism of slate or the metamorphism of basalt produces **schist.** As shown in Fig. 3-18, schist consists of very thin, parallel bands, or flakes, of minerals. There are various types of schist. Each

**Fig. 3-18. Schist.**

**Fig. 3-19. Gneiss.**

of heat and pressure and the length of time that the shale undergoes metamorphism (see Table 3-1).

# UNFOLIATED METAMORPHIC ROCKS

**Quartzite.** When quartz sandstone undergoes metamorphism, a rock called **quartzite** is formed. Quartzite is very hard and durable and resembles sandstone. However, unlike sandstone, which is porous, there are very few open spaces between the grains in quartzite. Thus, water cannot seep through quartzite. Although quartzite is a hard rock, it is not used as a building stone because it is full of cracks. The cracks result from the great pressures that changed the sandstone to quartzite.

**Marble.** Limestone that undergoes metamorphism becomes **marble.** Marble is harder, more crystalline, and coarser than the limestone from which it is formed. The mineral grains in marble often have a somewhat rounded appearance.

Marble occurs in many colors. It often has a streaked appearance due to impurities forced into the limestone during metamorphism. Marble is widely used for sculptures and monuments. Highly polished marble is used as an ornamental stone in buildings. Large amounts of marble were used in the Lincoln and Jefferson memorials in Washington, D.C.

**Anthracite.** When large sections of the crust move, the movement forces huge masses of rock to buckle and fold. If layers of bituminous coal are present in the folding rock, the bituminous coal is changed into **anthracite.** Greater pressures may even change the anthracite coal into graphite.

type is named for the most abundant mineral in the schist. For example, mica schist contains many flakes of mica. Talc schist and graphite schist contain large amounts of talc and graphite. Schists form the bedrock in many parts of New England, as well as making up most of the bedrock of Manhattan Island.

**Gneiss.** Further metamorphism of schist or the metamorphism of granite forms **gneiss.** Like schist, the minerals in gneiss appear as parallel bands. As shown in Fig. 3-19, however, the coarse, alternating bands of light- and dark-colored minerals are much thicker in gneiss than in schist. Gneiss often is cut into blocks and used as a building stone.

Notice that the three foliated metamorphic rocks can be produced from the same rock: Shale can be the source rock for slate, schist, and gneiss. The type of metamorphic rock produced from shale depends upon the amount

### TABLE 3-1. THE METAMORPHISM OF FIVE COMMON ROCKS

| *Source Rock* | | *Metamorphic Rock* |
|---|---|---|
| Shale ——→Slate ——————→ Schist ——————————→ Gneiss | | |
| Sandstone ——————————————————————→ Quartzite | | |
| Limestone ——————————————————————→ Marble | | |
| Basalt ————————————————————————→ Schist | | |
| Granite ————————————————————————→ Gneiss | | |

——————————→ Increasing Heat, Pressure, and Time ——————————→

# The Rock Cycle

The three types of rocks in the crust are continually being created, changed, and destroyed, and created anew. This constant, gradual process is called the **rock cycle.** In the rock cycle, igneous rocks are formed from the cooling and hardening of molten materials. Sedimentary rocks are produced from sediments that are compressed and cemented in layers. And metamorphic rocks are produced when sedimentary and igneous rocks are subjected to great heat and pressure, deep inside Earth (see Fig. 3-20).

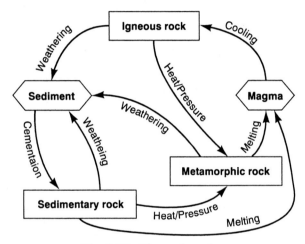

**Fig. 3-20. The rock cycle.**

As soon as an igneous, sedimentary, or metamorphic rock is exposed at Earth's surface, the rock is subjected to weathering and erosion. The actions of plants, water, ice, heat, and wind break rocks down into sediments. The sediments are transported by running water and eventually deposited in lakes and oceans.

Sediments harden into sedimentary rocks. Later, movements of Earth's crust may force the sedimentary rocks downward to great depths, where they are heated and compressed into metamorphic rock. Similar movements of the crust also can change igneous rock into metamorphic rock. Any of the three rock types may melt if pushed sufficiently far into Earth's interior. Then, upon hardening, the molten material becomes igneous rock once again—renewing the rock cycle.

## IMPORTANCE OF ROCKS

The rocks you have read about in this chapter are used in many ways to build our present-day world. Glass windows, concrete highways, building stones, fertilizers, metal objects, and thousands of other objects around us came originally from rocks. Truly, we are living in a "modern stone age."

Another important material that comes from rocks is *topsoil.* Topsoil is a mixture of rock particles and the remains of plants and animals. The thin layer of topsoil on Earth's surface provides the environment in which plants live and grow. Water and minerals that plants need for proper growth are stored in the topsoil. The plants, in turn, are used as food by people and many other animals. In fact, life on Earth would be impossible without topsoil. Soils are discussed in more detail in Chapter 4.

**TABLE 3-2. SOME COMMON ROCKS AND THEIR USES**

| Rocks | Uses |
|---|---|
| *Igneous* | |
| Basalt and diabase (traprock) | Filler for roads and railroad beds; concrete ingredient |
| Granite | Building stone, monuments, paving blocks |
| Pumice | Dental powders, scouring mixtures, hand cleansers |
| Pegmatite | Source of large sheets of mica, used as electrical insulators; gems |
| *Sedimentary* | |
| Limestone | Manufacture of steel; used as building stone; in the making of lime, cement |
| Sandstone | Building stone; glass manufacturing |
| Shale | Component of cement; making of bricks, tiles |
| *Metamorphic* | |
| Marble | Ornamental stone for monuments and table-tops |
| Slate | Roofing; chalkboards |

# CHAPTER REVIEW

## Science Terms

*The following list contains all of the boldfaced words found in this chapter and the page on which each appears.*

anthracite (p. 42)
basalt (p. 35)
coal (p. 39)
conglomerate (p. 38)
diabase (p. 35)
gabbro (p. 35)
gneiss (p. 42)
granite (p. 34)
gypsum (p. 38)
halite (p. 38)
igneous rocks (p. 32)
lava (p. 33)
limestone (p. 38)
magma (p. 33)
marble (p. 42)

metamorphic rocks (p. 33)
obsidian (p. 34)
organic limestone (p. 39)
pegmatite (p. 35)
pumice (p. 35)
quartzite (p. 42)
rock cycle (p. 43)
rock salt (p. 38)
sandstone (p. 37)
schist (p. 41)
sedimentary rocks (p. 32)
sediments (p. 32)
shale (p. 38)
slate (p. 41)

## Matching Questions

*On the blank line, write the letter of the item in column B that is most closely related to the item in column A.*

*Column A*

_____ 1. igneous

_____ 2. sedimentary

_____ 3. metamorphic

_____ 4. lava

_____ 5. gypsum

_____ 6. marble

_____ 7. sandstone

_____ 8. shale

_____ 9. granite

_____ 10. basalt

*Column B*

*a.* coarse-grained rock consisting mainly of feldspar and quartz
*b.* molten substances that appear at the surface of Earth
*c.* sedimentary rock consisting of sand particles cemented together
*d.* probably the first rocks to form on Earth
*e.* sedimentary rock composed of clay particles
*f.* rocks consisting of the broken pieces of other rocks
*g.* formed from the remains of microscopic plants
*h.* common chemical sediment rock
*i.* rocks changed by chemical action, heat, or pressure
*j.* metamorphic rock that was once limestone
*k.* substance of which coral reefs are made
*l.* dark-colored, fine-grained igneous rock

# Multiple-Choice Questions

*On the blank line, write the letter preceding the word or expression that best completes the statement.*

**1.** Rocks that consist of the cemented fragments of other rocks are called
  *a.* volcanic  *b.* igneous  *c.* metamorphic  *d.* sedimentary          1 ___

**2.** The two most abundant minerals always present in granite are
  *a.* quartz and calcite          *c.* feldspar and fluorite
  *b.* mica and gypsum          *d.* quartz and feldspar          2 ___

**3.** An example of a rock that cools very slowly from magma is
  *a.* quartz  *b.* granite  *c.* obsidian  *d.* pumice          3 ___

**4.** A lightweight rock that forms when gases bubble through cooling lava is called
  *a.* pumice  *b.* obsidian  *c.* volcanite  *d.* basalt          4 ___

**5.** The type of rock fragment that is most likely to remain suspended in slow-moving water is
  *a.* clay  *b.* sand  *c.* pebble  *d.* boulder          5 ___

**6.** Sedimentary rocks that are mainly reddish brown are probably cemented together by
  *a.* calcite  *b.* clay  *c.* iron compounds  *d.* silica          6 ___

**7.** Water has great difficulty seeping through shale because shale is
  *a.* composed of fine particles  *b.* solid  *c.* heavy  *d.* well-cemented          7 ___

**8.** Which of the following rocks is *not* a chemical sediment?
  *a.* rock salt  *b.* limestone  *c.* gypsum  *d.* shale          8 ___

**9.** The reefs along the shores of Bermuda, Florida, and Australia are made up of
  *a.* quartz sands  *b.* bones  *c.* coral  *d.* boulders          9 ___

**10.** The type of coal that has undergone the greatest change in its formation is
  *a.* peat  *b.* lignite  *c.* bituminous  *d.* anthracite          10 ___

**11.** Oil that is used as a fuel and as a lubricant was probably formed millions of years ago from
  *a.* dinosaur skeletons          *c.* giant ferns
  *b.* tiny plants and animals          *d.* sea shells          11 ___

**12.** The most important substance derived from eroded rocks is
  *a.* soil  *b.* natural gas  *c.* clay  *d.* slate          12 ___

# Modified True-False Questions

*In some of the following statements, the italicized term makes the statement incorrrect. For each incorrect statement, write the term that must be substituted for the italicized term to make the statement correct. For each correct statement, write the word "true."*

**1.** Geologists think that early in Earth's history, all the rocks that make up the crust were in the *liquid* state.          1 _____

**2.** Rocks that form when molten magma hardens are called *igneous* rocks.          2 _____

3. Two of the most common minerals that make up rocks are quartz and *talc*.

3 _____

4. Coarse-grained rocks form when magma cools *slowly*.

4 _____

5. Crystals weighing more than a *ton* have been found in pegmatites.

5 _____

6. The most abundant type of rock that forms from cooling lava is *granite*.

6 _____

7. Diabase is to basalt as pegmatite is to *granite*.

7 _____

8. Sedimentary rocks, such as sandstone and conglomerate, are held together by *pressure*.

8 _____

9. Dissolved minerals found in the ocean were once part of the *rocks* that make up Earth's crust.

9 _____

10. Because most limestones contain calcite, they can be identified by their *light color*.

10 _____

11. Although chalk consists of microscopic-sized particles, it is not unusual for deposits of chalk to be over *30 meters* thick.

11 _____

12. One type of rock that is likely to contain oil is *sandstone*.

12 _____

13. Shale is to slate as *granite* is to marble.

13 _____

14. Bituminous coal is to anthracite as anthracite is to *graphite*.

14 _____

# Testing Your Knowledge

1. Explain how each of the following rock types is formed:

   *a.* igneous _____

   _____

   *b.* sedimentary _____

   _____

   *c.* metamorphic _____

   _____

2. At the base of a very high cliff of igneous rock, you find several large fragments of rock. Some of these rocks look like masses of dark-colored glass; other rocks consist of large crystals. Assuming that the fragments of rock came from the cliff, which fragments most probably came from the top of the cliff, and which fragments came from the bottom of the cliff? Explain.

   _____

   _____

   _____

3. Name the three different kinds of sedimentary rock. Describe in one or two sentences how each kind of sedimentary rock is formed.

   a. _____

   _____

   _____

   b. _____

   _____

   _____

   c. _____

   _____

   _____

4. The great Bonneville Salt Flats are located in a region that was once the bottom of a huge freshwater lake. Explain how the salt flats formed.

   _____

   _____

   _____

5. Describe the stages that dead and decaying plants pass through as they change into anthracite

   coal. _____

   _____

   _____

6. Describe the stages that organic matter passes through in order to form oil deposits.

   _____

   _____

7. Describe briefly what each of the following discoveries tells you about Earth's history:

   a. Beds of coal have been discovered beneath the South Pole. _____

   _____

   _____

   b. Deposits of oil have been found beneath the Sahara Desert. _____

   _____

   _____

   c. Sea shells have been found in sedimentary rocks located on mountaintops. _____

   _____

   _____

*d.* The remains of a coral reef have been found in Texas in rocks deep underground. _____

_____

_____

*e.* Thick deposits of salt are located beneath the ground in Louisiana and New York. _____

_____

_____

*f.* Metamorphic rocks, such as gneiss, schist, and marble, make up the bedrock of Manhattan Island. _____

_____

_____

*g.* Diamonds are found within masses of solidified magma. _____

_____

_____

**8.** Complete the table by filling in the names of the following rocks under the correct heading: basalt, schist, granite, sandstone, slate, shale, marble, pumice, pegmatite, gabbro, limestone, conglomerate, gypsum, anthracite, gneiss.

| *Igneous* | *Sedimentary* | *Metamorphic* |
|-----------|---------------|---------------|
|           |               |               |
|           |               |               |
|           |               |               |
|           |               |               |

# *Weathering Changes the Land*

## LABORATORY INVESTIGATION

### *THE BREAKING DOWN OF ROCK*

**A.** Obtain two small pieces of limestone of about the same size. Carefully crush one of the pieces. You can do this either by grinding the piece of limestone in a mortar or by placing the piece in a plastic bag, resting the bag on a solid surface, and then tapping the bag with a hammer. Collect the crushed limestone, and place it in a test tube.

    **1.** Which sample of limestone has more surface exposed to the air? _____

    Explain. _____

    _____

**B.** As shown in Fig. 4-1, place the solid piece of limestone in another test tube. Add about 50 milliliters of dilute hydrochloric acid to each test tube, and observe the limestone for a few minutes. **CAUTION: Remember that all acids are dangerous and should be handled carefully. (Wear goggles.)**

    **2.** What happens to the crushed limestone? _____

    _____

    **3.** What happens to the solid piece of limestone? _____

    _____

    **4.** Which of the two samples reacts more rapidly with the acid? _____

    Explain. _____

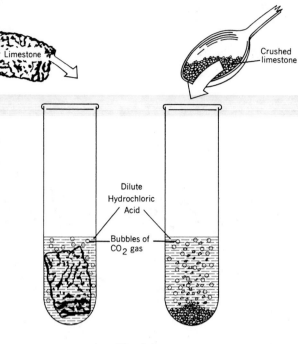

**Fig. 4-1.**

**5.** Of the two pieces of limestone you had originally, one piece may have been slightly larger than the other. Could this difference in the size of the original samples have had any effect on the results of this laboratory investigation?

Explain. _____

_____

# Weathering of Rock

You have learned that Earth's crust was formed from molten materials that cooled and solidified into rock. The crust, or bedrock, is usually covered by layers of sand, gravel, or soil. These materials come largely from bedrock that has been subjected to weathering by water, ice, and other agents. **Weathering** is the breaking down of rocks at or near Earth's surface into smaller and smaller pieces. Weathering may also result in the formation of new substances from elements in the original rocks. There are two types of weathering: *physical weathering* and *chemical weathering*.

## PHYSICAL WEATHERING

**Physical weathering** is the process by which rocks are broken down into smaller fragments without undergoing any change in chemical composition. Physical weathering is mainly caused by the freezing of water, the expansion of rock, and the activities of plants and animals.

**Frost Wedging.** In areas with temperate and cold climates, as well as in high mountainous areas, rocks are weathered by the action of freezing water. During the daytime, when the temperature is above the freezing point of water (0°C), rainwater, melted snow, or ice trickles into cracks in the rocks. At night, when the temperature falls below the freezing point of water, the water trapped inside the rocks changes into ice.

When water freezes, it increases in volume by about 9 percent. The expansion of water into ice pushes against the sides of the cracks with great force, wedging the rocks apart (see Fig. 4-2). This process, which is characterized by a cycle of daytime thawing and refreezing at night, is called **frost wedging.**

Frost wedging causes large rock masses, especially the rocks exposed on mountaintops, to be broken into smaller pieces. Frost wedging can have the same effect on pavement. During the winter, water trapped in cracks in the pavement freezes into ice. The ice often expands enough to crack the pavement and form potholes.

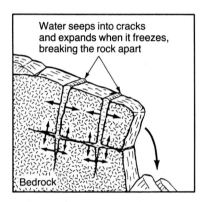

Fig. 4-2. Frost wedging of a rock.

**Exfoliation.** Weathering continually breaks down exposed bedrock into smaller fragments, which are then carried away by wind and water. Consequently, rock formed deep underground (under great pressure) becomes exposed at Earth's surface. The release of the overlying pressure causes the newly exposed bedrock to expand, forming cracks parallel to the rock's surface. Then, frost wedging causes large, curved slabs of rock to peel away from the main body of the rock. The peeling away of the outer layers from a rock is called **exfoliation.** Rounded mountaintops called *exfoliation domes* are formed in this way (see Fig. 4-3).

Miners and quarry workers have been injured and killed by exfoliation. As a mine shaft or quarry pit is dug into the bedrock, the removal of rock causes a rapid decrease in pressure on the surrounding bedrock. The reduced pressure can result in sudden exfoliation, sending dangerous missiles of rock flying through the air.

**Animals and Plants.** Insects, earthworms, rabbits, woodchucks, and many other animals burrow through the soil. The tunnels these or-

Fig. 4-3. Exfoliation dome.

ganisms make often expose parts of the bedrock to the weathering action of the air and water. Frost wedging or chemical action can then break down the bedrock. In addition, the act of burrowing through soil helps to break down rock particles into smaller and smaller pieces.

You may have seen trees that appear to be growing out of solid rock. Trees occasionally grow in soil that has collected in small cracks in the bedrock. In other cases, trees grow in soil upon which a cracked boulder is resting. In either case, as the tree continues to grow, it exerts great pressure against the cracks in the rock. The pressure causes the rock to split apart even wider (see Fig. 4-4).

Fig. 4-4. Growing trees can split rocks.

# CHEMICAL WEATHERING

Exposure to air, water, and organisms can change the minerals in a rock into new substances that have different chemical compositions. These new substances are generally softer or weaker than the original materials, so they tend to cause the rock to crumble and fall apart. The breaking down of rocks through changes in their chemical composition is called **chemical weathering.** An example of chemical weathering is the change of feldspar in granite to clay. When acted upon by water, the feldspar changes into powdery clay minerals. Feldspar and clay have different chemical compositions.

Water, oxygen, and carbon dioxide are the main agents of chemical weathering. When water and carbon dioxide combine chemically, they produce a weak acid that can break down rocks. Rocks also are weathered chemically by the action of acids produced by plants and animals. These acids are formed by living plants and by the decomposition of plant and animal materials.

**Action of Acids.** You may have noticed that the exposed surfaces of bedrock sometimes appear to be covered by a grayish or bluish-green crust. A closer look reveals that this colored crust actually consists of tiny plants called *lichens*. Lichens "eat" rocks by releasing acids that slowly dissolve the minerals in the rocks.

In addition, the action of bacteria and other microscopic organisms on dead plants and animals changes the composition of their remains. As these changes take place, acids are produced. The acids then dissolve the minerals in the rocks, and the rocks crumble apart.

**Water.** When water comes into contact with certain minerals, the water combines chemically with these minerals and changes them. This kind of chemical weathering is called **hydrolysis.** You learned that when water reacts with the feldspar in granite, the feldspar changes into clay. Clay is much softer than the original feldspar. Thus, the process of hydrolysis weakens granite, making it more susceptible to physical weathering.

**Oxygen.** When an iron nail is exposed to the atmosphere, rusting takes place. That is because iron, in the presence of moisture, combines chemically with oxygen in the atmosphere. The combining of oxygen with another substance is called **oxidation.** The oxidation of an iron nail produces a powdery reddish-brown substance called *rust.* Rust is composed mainly of a chemical compound called *iron oxide.*

Oxidation also takes place in rocks that contain iron-bearing minerals. When these rocks are exposed to oxygen in the atmosphere, the minerals slowly change into softer, more crumbly substances. Like an iron nail that rusts when exposed to oxygen and moisture, the iron-bearing minerals in these rocks also rust when exposed to the atmosphere.

The chemical weathering of rocks that contain iron often produces reddish or brownish soils. Many of the world's largest iron ore deposits were formed by chemical weathering. For example, the great deposits of iron ore in Minnesota and upper Michigan were formed over a vast period of time by the chemical weathering of basalt, an igneous rock that contains iron.

**Carbon Dioxide.** When carbon dioxide and water combine chemically, they form a weak acid called **carbonic acid.** (A common name for carbonic acid is *soda water,* which is made by dissolving carbon dioxide gas, under pressure, in water.) As you learned in Chapter 2, when you place a drop of hydrochloric acid on a rock containing calcite, bubbles of carbon dioxide gas appear on the rock. At the same time that the hydrochloric acid is dissolving the calcite, carbon dioxide gas is being released.

Like hydrochloric acid, carbonic acid dissolves calcite from rocks, but much more slowly. The world's largest underground caverns were formed by this type of chemical weathering. The original limestone bedrock was gradually dissolved by water containing carbonic acid. The water then carried the dissolved limestone away, leaving the hollow spaces called *caverns.* (Caverns are described more fully in Chapter 5.)

# RESISTANCE TO WEATHERING

The ability of a rock to resist weathering depends mainly on the mineral composition of the rock and on the number of cracks in the rock. Rocks that consist mostly of quartz are more resistant to either physical or chemical weathering than most other kinds of rocks. The chemical weathering of quartz is very slow because quartz does not combine readily with other substances. Unlike rocks that contain abundant quartz, rocks that consist mainly of feldspar weather rapidly, especially in warm, moist parts of the world. In these

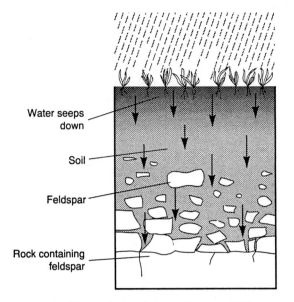

**Fig. 4-5. Chemical weathering changes feldspar into clay.**

areas, chemical weathering rapidly changes feldspar into clay (see Fig. 4-5).

Granite consists mainly of quartz and feldspar. Although granite is a relatively hard rock, the chemical weathering of feldspar eventually causes granite to crumble. Quartz weathers also, but at a much slower rate than feldspar. Eventually, physical weathering breaks down the quartz into small grains of sand.

In general, sedimentary rocks weather more rapidly than other rock types because sedimentary rocks consist of many small grains that are cemented together. Poorly cemented sedimentary rocks, such as some shales and sandstones, contain small air spaces between the grains. Consequently, water easily penetrates the rocks. If the water freezes within these rocks, the rocks are broken apart by frost wedging.

If water dissolves the cement holding the grains of a sedimentary rock together, the grains will separate. Cements that consist of iron compounds or calcite dissolve more rapidly than do cements composed of silica compounds. In moist climates, therefore, sedimentary rocks cemented by calcite and iron compounds weather rapidly, whereas those cemented by silica are more resistant to weathering.

Cracks in a rock also cause the rock to weather more rapidly. The more cracks in a rock, the faster that weathering will split the rock into fragments. Once a large rock has been split into smaller fragments, the smaller fragments can then be broken down more rapidly by the action of air and water.

You observed this effect in your laboratory investigation. In this investigation, the crushed limestone in one test tube provided more exposed surfaces for the acid to act upon than the unbroken piece. Thus, the crushed limestone weathered more rapidly than the large piece of limestone in the other test tube.

# Soil

The **soil** consists of particles formed by physical and chemical weathering. Soil usually contains particles of sand, clay, various minerals, tiny living organisms, and humus. **Humus** is the decayed remains of plants and animals. In addition, some types of soil have large numbers of air spaces between their particles.

## SOIL TYPES

Soils are divided into three main classes, according to texture. These classes are *sandy soils, clay soils,* and *loamy soils.* You can determine the texture of each soil type by squeezing and rubbing a small amount of moist soil between your fingers.

Sandy soils feel gritty, and their particles do not bind together firmly. Sandy soils are porous, which means that water passes through them rapidly. Consequently, sandy soils do not hold much water. Adding a large amount of humus to a sandy soil permits the soil to hold more water.

Clay soils feel smooth and greasy, and their particles bind together firmly. Clay soils are usually moist, but they do not permit water to pass through easily.

Loamy soils feel somewhat like velvet, and their particles clump together. Loamy soils consist of a mixture of sand, clay, and silt. A loamy soil holds water well and permits some water to pass through.

## SOIL LAYERS

As shown in Fig. 4-6, a typical soil consists of three different layers. The top layer (the soil layer in which most plants grow) is called the **topsoil.** Only the topsoil contains humus. Because the topsoil contains humus, it tends to be darker than the underlying soil layers. Topsoil ranges in thickness from about 10 to 60 centimeters (cm). Topsoils contain different amounts of sand, clay, and humus. Loamy soil is the best kind of topsoil for growing crops, such as corn, oats, and wheat.

Underneath the topsoil is the second layer of soil, or **subsoil.** The subsoil contains more clay and various minerals than the topsoil. The subsoil is usually reddish, brownish, or yellowish. Larger plants, especially trees, extend their roots into the subsoil.

The bottom layer of soil, which may extend several meters downward, consists mostly of rock fragments produced by the physical and chemical weathering of the bedrock. The color of the bottom layer of soil depends largely on the color of the underlying bedrock.

Although the bottom two layers of the soil are relatively thick, the topsoil is always a thin layer. Even fertile farmlands have a relatively thin layer of topsoil. Because topsoil is a loose material, it is easily carried away by wind and water. In the 1930s, winds blew away billions of tons of topsoil from the farmlands of the Plains states. The dust was carried in great clouds more than 3000 kilometers east to the Atlantic Ocean. The United States still loses large amounts of topsoil each year because of the destructive action of wind and water. Topsoil conservation is discussed in Chapter 9.

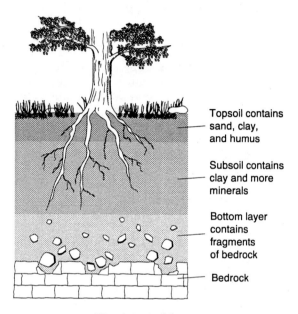

Topsoil contains sand, clay, and humus

Subsoil contains clay and more minerals

Bottom layer contains fragments of bedrock

Bedrock

**Fig. 4-6. Soil layers.**

# CHAPTER REVIEW

## Science Terms

*The following list contains all of the boldfaced words found in this chapter and the page on which each appears.*

carbonic acid (p. 52)
chemical weathering (p. 52)
exfoliation (p. 51)
frost wedging (p. 51)
humus (p. 53)
hydrolysis (p. 52)

oxidation (p. 52)
physical weathering (p. 50)
soil (p. 53)
subsoil (p. 54)
topsoil (p. 54)
weathering (p. 50)

# Matching Questions

*On the blank line, write the letter of the item in column B that is most closely related to the item in column A.*

<table>
<tr><td colspan="2">Column A</td><td>Column B</td></tr>
<tr><td>____ 1.</td><td>breakdown of rocks into smaller pieces</td><td>a. hydrolysis</td></tr>
<tr><td>____ 2.</td><td>breaking apart of rocks by ice</td><td>b. topsoil<br>c. carbonic acid</td></tr>
<tr><td>____ 3.</td><td>peeling away of outer rock layers</td><td>d. subsoil</td></tr>
<tr><td>____ 4.</td><td>chemical weathering by water</td><td>e. humus<br>f. physical weathering</td></tr>
<tr><td>____ 5.</td><td>chemical combination of $CO_2$ and $H_2O$</td><td>g. frost wedging</td></tr>
<tr><td>____ 6.</td><td>remains of plants and animals in soil</td><td>h. exfoliation<br>i. oxidation</td></tr>
<tr><td>____ 7.</td><td>uppermost soil layer, having humus</td><td></td></tr>
<tr><td>____ 8.</td><td>second layer of soil, having more clay</td><td></td></tr>
</table>

# Multiple-Choice Questions

*On the blank line, write the letter preceding the word or expression that best completes the statement.*

**1.** The destructive effects of frost wedging are most likely to take place on
*a.* city streets   *b.* mountaintops   *c.* country roads   *d.* lake shores                    1 ____

**2.** A mass of rock from which tons of overlying material are being removed should
*a.* contract   *b.* expand   *c.* change chemically   *d.* crumble                    2 ____

**3.** Ants, worms, and burrowing animals help weather bedrock mostly by
*a.* carrying particles of rock to the surface
*b.* secreting chemicals
*c.* crushing small particles
*d.* digging tunnels that expose the bedrock to air and water                    3 ____

**4.** The substances that are mainly responsible for chemical weathering in nature are water, carbon dioxide, and
*a.* oxygen   *b.* nitrogen   *c.* hydrochloric acid   *d.* frost                    4 ____

**5.** When carbon dioxide and water combine chemically, they form
*a.* calcium bicarbonate   *b.* clay   *c.* limestone   *d.* carbonic acid                    5 ____

**6.** Ordinarily, the rock that is most resistant to weathering is
*a.* limestone   *b.* shale   *c.* granite   *d.* sandstone                    6 ____

**7.** Humus refers to
*a.* minerals in the soil                      *c.* sandy soil
*b.* decayed plant and animal matter   *d.* silt                    7 ____

8. The soils in which most plants grow best are
   *a.* loamy soils   *b.* sandy soils   *c.* clay soils   *d.* silt soils          8 ____

9. Topsoil usually ranges in depth from several centimeters to about
   *a.* 60 cm   *b.* 100 cm   *c.* 160 cm   *d.* 200 cm          9 ____

10. The bottom layer of soil is usually composed of
    *a.* clay          *c.* broken bedrock
    *b.* sand          *d.* decayed animal and plant remains          10 ____

# Modified True-False Questions

*In some of the following statements, the italicized term makes the statement incorrect. For each incorrect statement, write the term that must be substituted for the italicized term to make the statement correct. For each correct statement, write the work "true."*

1. *Weathering* is the physical and chemical breakdown of rock.          1 _____

2. The expansion of water that occurs during freezing is an example of a *chemical* action.          2 _____

3. The separation of the outer layers from a mass of rock is called *exfoliation*.          3 _____

4. Clay can be produced by the *chemical* breakdown of feldspar.          4 _____

5. A coating of bluish-green material on a mass of rock is an indication that the rock is probably being broken down by the *chemical* action of lichens.          5 _____

6. Hydrolysis is to feldspar and water as rusting is to iron and *carbonic acid*.          6 _____

7. Rocks containing large amounts of quartz usually weather *slowly*.          7 _____

8. In moist regions, rocks cemented by calcite weather at a *slower* rate than do rocks cemented by silica compounds.          8 _____

9. The greater the number of cracks in a rock, the *faster* the rock will weather.          9 _____

10. Fertile soils consist of a mixture of sand, clay, and *pebbles*.          10 _____

# Testing Your Knowledge

1. Name two agents of physical weathering and two agents of chemical weathering. Briefly describe the effects of any one of these agents on the following rock types:

    *a.* granite _____

    *b.* limestone _____

    *c.* sandstone _____

2. Describe the differences in the ability of the following soils to hold water:

    *a.* sandy soil _____

    *b.* clay soil _____

    *c.* loamy soil _____

3. If you were to dig a deep hole in a nearby park, what three layers of soil would you probably expose? Briefly describe each layer.

    *a.* _____

    *b.* _____

    *c.* _____

# Sand Rights: Do They Belong to Us or to Beaches?

In late October of 1991, a howling storm whipped down the coast of New England all the way south to Virginia. This "northeaster," as such storms are called, packed winds of 74 miles per hour and unleashed waves 25 feet high. Together, the screeching winds and pounding waves swept away the sands of entire beaches, destroyed hundreds of homes, and were responsible for killing at least four people.

Whereas some homes could not be rebuilt, and the four human victims could not be brought back to life, geologists say many of the beaches *would* grow again—provided the activities of people did not prevent it.

Beaches, like other landforms, are changed by *erosion*, or wearing away by wind and moving water. Some of the changes are triggered by natural events like storms, the ebb and flow of tides, and the steady lapping of ordinary waves. These forces often strip away sand in one place, only to deposit it in another.

In most cases, the stripping away and replacing processes tend to preserve beaches over time. However, when people interfere with these natural processes, an unnatural erosion can occur that destroys a beach, perhaps forever.

What kinds of human activities speed up erosion of beaches? Anything that interferes with the movement of sand along a coast.

This happens, for instance, when people build rock jetties that jut into the sea. Although the jetties produce relatively calm water for harbors, they block the movement of sand. Picture a coastline where sand moves from north to south. The beaches to the north lose their sand, but those to the south that would ordinarily gain this sand are blocked from getting it by the jetties.

Similarly, any kind of construction—a home, parking lot, or other structure—that is built too close to the water of a beach may keep the sands from shifting naturally.

Sometimes sand robbers are located far from beaches. For example, some beaches get their sand from rivers and creeks that constantly sweep *silt*, fine grains of eroded rock, toward the sea. But when such rivers are dammed to produce hydroelectricity, or the bottoms and sides of creeks are paved with concrete to inhibit flooding, the flow of silt is interrupted. The beach sand that is carried away by natural forces is not replenished, and the beaches shrink.

Certainly, we do need electricity and we don't need flooding rivers and creeks. And people should have the right to build homes near the shoreline. But what about a beach's right to *its* sand? Should that be protected too? Put another way, don't the beaches belong to all of us and, if so, shouldn't something be done to protect our rights to have beaches?

A beach's right to its sand and our right to enjoy unspoiled beaches is a position supported by Los Angeles lawyer Katherine E. Stone. Ms. Stone suggests that "certain kinds of resources are held in common for everyone under a trust. Those resources include the air, the sea, fish, swimming in the sea, and the shore of the sea as well."

In recent years, local, state, and federal governments have recognized such rights by passing laws or instituting regulations that require builders not to interfere with the natural flow of sand or, if they do, to replace the sand at their own expense. This, of course, increases the cost of construction, electricity, and flood control, which people have to pay for when they buy products or pay taxes.

What's more, some people say such laws and regulations are unconstitutional. What happens, they ask, if a person buys a piece of ocean-front land and then is not permitted to build on it? The value of the land plummets and the person loses a lot of money. Isn't this the same as the government taking land and not compensating the owner for it, which is a violation of the Fifth Amendment to the U.S. Constitution?

No, say supporters of beach erosion control. The government isn't taking the land. It is only regulating how the land can be used.

Clearly, this is a complex issue with reasonable arguments on both sides. Which arguments make more sense? What do you think?

1. *A rock jetty causes erosion by*

   a. wearing away a beach.
   b. protecting a harbor.
   c. controlling floods.
   d. preventing the movement of sand.

2. *One source of sand for a beach is silt from*

   a. rivers.          c. jetties.
   b. dams.           d. storms.

3. *Lawyer Katherine E. Stone believes that*

   a. people should build homes where they want.
   b. a beach has a right to its sand.
   c. builders should not have to replace sand.
   d. the Fifth Amendment is unconstitutional.

4. *With regard to beaches, what are some causes of natural and unnatural erosion?*

   _____

   _____

   _____

   _____

5. *Look up the Fifth Amendment to the U.S. Constitution. On a separate sheet of paper, identify the part that applies to the protection of beaches. In your library, look up the Supreme Court decision of 1992 in the case of Lucas v. South Carolina Coastal Commission. Explain why you do or do not agree with the Supreme Court's decision.*

# *Groundwater*

## LABORATORY INVESTIGATION

### *HARD WATER*

**A.** Pour tap water into a test tube to a height of 3 centimeters (cm). Using a medicine dropper, add 4 drops of liquid soap to the test tube. Note if a white, curdlike substance, or *precipitate*, appears in the water. Record your observations under *Precipitate* in the table.

Place your thumb over the mouth of the test tube, and shake the test tube vigorously for 15 seconds. Measure the height of the suds in the test tube with a metric ruler. Record your observations under *Suds* in the table. Repeat the procedure using distilled water, and then record your observations in the table.

**B.** Pour hard water into a test tube to a height of about 1 cm. Heat the test tube over a burner until almost all of the water has evaporated. The heat from the test tube will evaporate the remaining water.

**CAUTION: Attach the test tube to the clamp as shown in Fig. 5-1. Before heating the water, make sure that the mouth of the test tube is not pointed toward anybody. (Wear goggles.)**

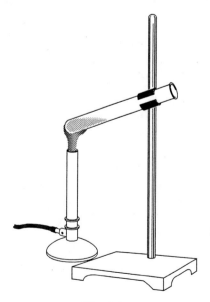

**Fig. 5-1.**

Next, pour tap water into a test tube to a height of 1 cm. Heat the test tube until almost all the water has evaporated. The heat from the test tube will evaporate the remaining water. Examine the two test tubes, and record your observations in the table under *Appearance after Evaporation*. Repeat the procedure using distilled water. Record your observations in the table.

| | *Precipitate* | *Suds* <br> *(in centimeters)* | *Appearance* <br> *after* <br> *Evaporation* |
|---|---|---|---|
| *Tap water* | | | |
| *Distilled water* | | | |
| *Hard water* | | | |

1. The water sample in which the largest amount of precipitate appeared is _____.

2. The water sample in which the least amount of precipitate appeared is _____.

3. The water sample that produced the least amount of suds is _____.

4. The water sample that produced the largest amount of suds is _____.

5. After evaporation, the test tube that showed the heaviest deposit was the test tube that contained _____ water.

6. Based on this investigation, what appears to be an important factor that determines the amount of suds that will be produced in a given water sample?

_____ .

# ☐ Water Enters the Ground ☐

A huge amount of water falls on Earth's surface each year. The water that falls to Earth in the form of rain and snow is called *precipitation*. A large amount of the precipitation either drains off the land and returns to the ocean or evaporates into the atmosphere and later falls again as rain or snow. Some of the precipitation is absorbed by plants and the soil. The remaining precipitation soaks into the ground. The continuous process in which water at Earth's surface evaporates, condenses into clouds, and returns to Earth as precipitation is called the **water cycle** (see Fig. 5-2, page 62).

## GROUNDWATER

There is many times more fresh water located below Earth's surface than in all of the world's lakes, rivers, and reservoirs combined. All of the water located below the surface is called **groundwater.** In Fig. 5-3, on page 62, groundwater is located in the spaces between grains of soil or rock fragments and in cracks in the bedrock.

Not all of the water that enters the topsoil soaks completely into the spaces between soil particles. A small amount of the water clings

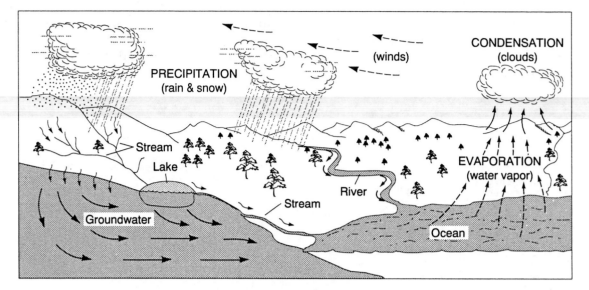

**Fig. 5-2. The water cycle.**

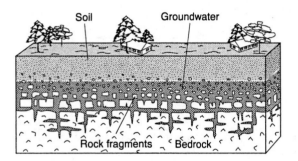

**Fig. 5-3. Groundwater.**

**Amount of Precipitation.** The amount and rate of precipitation vary throughout the world. These two factors greatly influence the amount of water that soaks into the ground and collects as groundwater. During periods of heavy precipitation, the ground cannot absorb all of the water that falls. Consequently, the excess water runs off the surface of the land into streams, lakes, or the ocean. This excess water is called *runoff*. In contrast, during a period of light precipitation, rainwater soaks into the ground as quickly as it falls—even if the rain continues for several days.

**Slope of the Land.** The slope of the land is another factor that determines how much precipitation soaks into the ground. For example, on a steep slope, rainwater runs downhill too rapidly for the ground to absorb much of the water. On a gentle slope, rainwater runs off at a much slower rate. Consequently, the rainwater has more time to soak into the ground. In general, as the land surface becomes flatter, the amount of water absorbed by the ground increases.

**Amount and Types of Plant Cover.** The presence of plants that cover the ground, or *plant cover*, is another important factor in reducing the amount of runoff. The leaves and stems of plants such as grasses, shrubs, and trees slow down falling rain before it reaches the ground. Plant stems and roots slow down the water as it runs off sloping land. This gives the soil and the roots of plants more time to absorb the water. Plant cover is especially important on

to soil particles near the surface. Thus, most topsoil is slightly damp. Farther underground, the spaces between soil particles, rock fragments, and cracks in the bedrock may be completely filled with water. In fact, groundwater has been found at depths of more than five kilometers in fractured bedrock.

Groundwater can be found underground throughout most of the world—even beneath deserts. However, the amount of groundwater and the depth at which it is located vary greatly in different climates. In a tropical rain forest, a large volume of groundwater lies near the surface. In contrast, a desert contains very little groundwater, and it is usually located deep below the surface.

The amount of groundwater in a particular region depends on four main factors: (1) the *amount of precipitation,* (2) the *slope of the land,* (3) the *amount and types of plants* that cover the ground, and (4) the *types of soil and bedrock* present.

steep slopes. (Chapter 9 describes how farmers use plants to prevent water from washing away the soil.)

**Surface Materials.** The amount of precipitation that soaks into the ground is also affected by the types of soil and rock materials at the surface. Very little precipitation soaks into the ground where the surface materials have few air spaces or cracks. The more cracks and air spaces there are, the more water soaks into the ground. For example, rain falling on a hillside of unbroken granite rapidly runs off the granite surface. In contrast, rain falling on a sandy beach immediately soaks into the sand.

**Sand.** Sand grains do not pack closely together because of their relatively large size and their irregular shapes. As you can see in Fig. 5-4, the spaces between sand grains are large enough for water to flow easily through them. Consequently, water falling or flowing on sandy soil is rapidly absorbed and moves downward to become groundwater. Soil that consists mainly of sand and gravel also lets water pass through readily.

**Clay.** Clay consists of flakelike particles that are much smaller than sand grains. When clay particles accumulate, they pack together very closely. Thus, the spaces between the clay particles are tiny.

Fig. 5-5 shows how the size of the spaces between rock particles affects the passage of water. Notice that the closer two fingers are placed together, the more tightly water will cling to the space between them. In the same

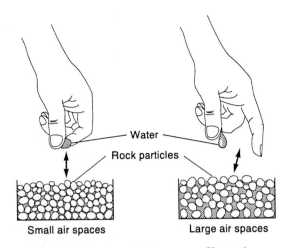

**Fig. 5-5. The size of air spaces affects the passage of water through rocks.**

way, the first drops of water falling or flowing on a clay soil cling tightly to the clay particles near the surface. This forms a barrier that blocks the downward movement of additional water. Therefore, precipitation usually remains on the surface of a clay soil for a long time.

**Rocks.** Water can pass easily through rocks that have many large, interconnected spaces and cracks. Such rocks are said to be *permeable* (able to transmit water). For example, most sandstones have relatively large surfaces between sand grains and thus allow water to pass through easily.

Water usually will not pass through a solid rock such as granite or limestone. Thus, a solid rock is said to be *impermeable*. However, weathering processes and movements of Earth's crust often cause many cracks to develop in solid rock. These cracks then act as passageways for the flow of water, so that the rock becomes permeable.

## SOURCES OF GROUNDWATER

Groundwater tends to move downward through the ground until stopped by an impermeable layer of material. Then, water collects in the permeable layers above until all of the air spaces in these layers are filled with water. The zone of soil and rock within which all spaces and cracks are completely filled with water is called the *zone of saturation*. The upper surface of the zone of saturation is called the **water table** (see Fig. 5-6, page 64).

**Springs.** A **spring** occurs wherever the water table meets the surface of the ground (see Fig.

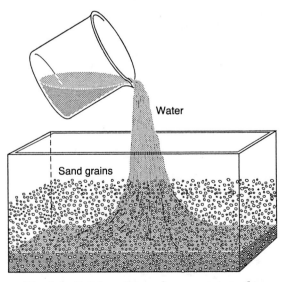

**Fig. 5-4. Sand grains soak up water easily.**

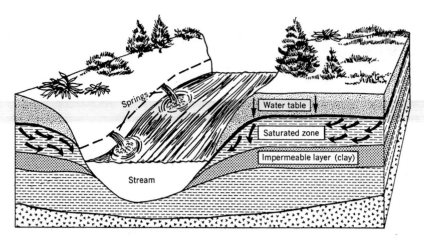

**Fig. 5-6. Groundwater emerges from springs.**

5-6). In many places, the water table meets the ground surface on the side of a hill. In hilly country, the top of the water table often is higher than the valleys between the hills. Consequently, the water table emerges as a spring from the hillside and flows into the valley below.

Water emerging from a spring is usually cool. Groundwater found down to about 15 meters (m) underground has about the same temperature as the average yearly air temperature of the region in which the groundwater is found. The average yearly air temperature in most parts of the United States ranges between 5°C and 15°C. Below 15 m, the temperature of groundwater gradually increases. Thus, water obtained from deep wells usually is much warmer than water from springs fed by groundwater near the surface.

In some parts of the United States, large masses of magma are located near Earth's surface. The magma heats any nearby groundwater. The hot springs found in western states such as Wyoming, Utah, and California result from the heating of groundwater by magma located in the upper region of Earth's crust.

**Geysers.** Magma is also responsible for geysers. A **geyser** is a type of hot spring that periodically shoots a fountain of hot water and steam into the air. A geyser is fed by a network of channels that extend down to a region of hot bedrock deep underground (see Fig. 5-7). Groundwater that collects in these channels gradually is heated. When the groundwater reaches the boiling point, it explodes upward in an eruption. In some cases, the ejected column of steam and hot water may rise more than 50 meters into the air.

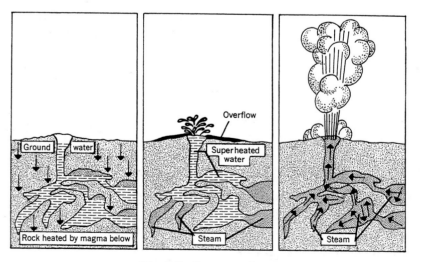

**Fig. 5-7. Geyser action.**

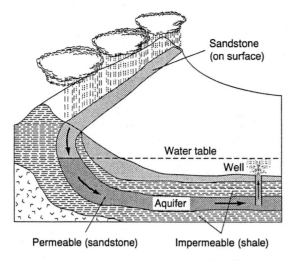

**Fig. 5-8. Artesian system and well.**

After an eruption drains a geyser's underground network, a new supply of groundwater begins seeping into the channels. Eventually, this water is heated to the boiling point, and the eruption is repeated. The most famous geyser in the United States is "Old Faithful" in Yellowstone National Park, Wyoming. Old Faithful erupts, on average, about once every 70 minutes. In fact, Old Faithful has not missed an eruption in more than 80 years of being observed.

**Wells.** People often obtain groundwater by drilling or digging holes from the surface into the zone of saturation. These holes are called **wells.** Groundwater from wells may be brought to the surface by pumps, or it may rise to the surface under its own pressure.

A well in which water rises to the surface under its own pressure is called an *artesian well.* Artesian wells are found on the Great Plains and along the eastern and southern coasts of the United States.

An artesian well can be drilled only where the underground rock structure forms an **artesian system** (see Fig. 5-8). Artesian systems may be hundreds of kilometers long, several kilometers wide, and over 30 m thick. Some artesian systems are located over 500 m below the surface. The Great Plains artesian system has been tapped for the abundant water it could supply under its own pressure.

The "walls" of an artesian system consist of layers of impermeable rocks or rock materials, such as shale or clay. The impermeable layers enclose a layer of permeable rock or rock fragments, such as sandstone or loose sand, forming a natural pipe. One end of this pipe is tilted up so that the permeable sand or sandstone is exposed on the surface at a high elevation. Precipitation falling on the exposed sand or sandstone soaks downward, eventually filling spaces in the permeable materials. The water-bearing layer of permeable rocks or rock fragments is called an **aquifer.** Groundwater in the upper levels of the aquifer presses down on the water confined in the lower levels. Thus, water in the lower part of the aquifer is under greater pressure.

Groundwater will emerge from a hole drilled into such an aquifer due to the high water pressure on the lower part of the aquifer. Water may ooze slowly out of the hole or may gush out with great force. The force with which the water emerges depends on the difference in height between the water in the upper part of the aquifer and the level at which the hole pierces the aquifer. The greater the difference in height between these two points, the greater is the water pressure at the lower point.

Water pressure in an artesian system can be demonstrated by attaching a rubber tube to a funnel (see Fig. 5-9). First, the funnel is filled with water. Then, the height of the funnel is changed in relation to the end of the rubber tube. The higher the funnel is raised above the end of the rubber tube, the greater is the force with which water comes out of the tube, and the higher the stream of water rises.

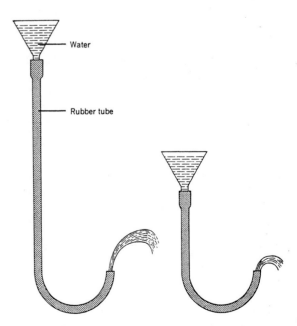

**Fig. 5-9. Demonstrating water pressure in an artesian system.**

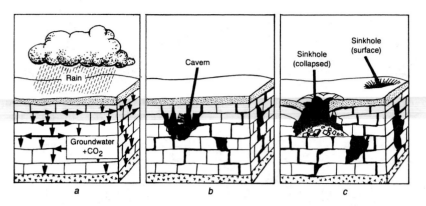

**Fig. 5-10. Formation of caverns and sinkholes.**

In the Great Plains artesian system, the higher part, or *source region*, of the aquifer is located along the eastern side of the Rocky Mountains. This region is at a much higher elevation and hundreds of kilometers away from the farms and communities that drill artesian wells into the aquifer.

## CAVERNS

Large underground chambers and passages are called **caverns.** Caverns are mostly found in areas where there is limestone bedrock. They are formed by groundwater dissolving the limestone (see Fig. 5-10a).

As groundwater flows through cracks in limestone bedrock, the groundwater (which contains carbonic acid) dissolves the limestone and enlarges the cracks. In time, this process forms a network of tunnels and large chambers (see Fig. 5-10b). Caverns that have formed as the result of chemical weathering by groundwater include Carlsbad Caverns in New Mexico, Mammoth Cave in Kentucky, and Howe Caverns in New York.

**Stalactites and Stalagmites.** In many caves, icicle-shaped deposits of calcite hang from the ceiling. These calcite deposits are called **stalactites.** Stalactites are formed when limestone, which is composed mainly of calcite, is dissolved by groundwater.

As the groundwater slowly drips from a cave ceiling, some of the water evaporates. Evaporation of the water releases carbon dioxide dissolved in the groundwater. As a result, a small amount of calcite is deposited on the cave ceiling. As the dripping continues, the calcite deposit extends downward, forming a stalactite.

If the dripping groundwater falls on the cavern floor, a small amount of calcite is depos-

ited on the floor as well. This deposit grows upward and forms a blunt, cone-shaped mass called a **stalagmite.** You can see stalactites and stalagmites in Fig. 5-11.

Sometimes, a stalactite and a stalagmite meet and form a column that extends from the floor of the cavern to the ceiling. In time, enough of these columns may form to fill in parts of a cavern.

## SINKHOLES

In regions that have limestone bedrock, large, round depressions in the ground called sinks, or **sinkholes,** are common (see Fig. 5-10c). Sinkholes form in two ways. One way is when acidic groundwater dissolves the edges of small openings in exposed limestone bedrock, enlarging them. In time, these openings become sinkholes.

**Fig. 5-11. Stalagmites and stalactites in Carlsbad Caverns National Park, New Mexico.**

**Fig. 5-12. A sinkhole in Florida.**

Another way that a sinkhole may form is when the roof of a cavern collapses, leaving a large, round hole in the ground (see Fig. 5-12). Sinkholes often fill with water, forming a pond or a lake. Sinkholes containing ponds and lakes are common in areas of Florida and Kentucky that have limestone bedrock.

## PETRIFIED WOOD

Sometimes, groundwater containing dissolved minerals seeps into wood that lies buried in sediments. The wood gradually is replaced by minerals that precipitate out of the groundwater. In time, the wood may be completely replaced by mineral materials. Because rocks are made up of minerals, this type of wood is referred to as **petrified wood**—the wood having turned to stone.

Pieces of trees that became petrified in this way are shown in Fig. 5-13. Petrified trees can

**Fig. 5-13. Petrified wood in Arizona.**

be seen in Petrified Forest National Park in Arizona and in Yellowstone National Park in Wyoming. These petrified trees consist of minerals such as *opal* and *chalcedony*, which are varieties of quartz (silica).

## HARD WATER

Groundwater that contains a large amount of dissolved calcium and magnesium is called **hard water.** Groundwater that is relatively free of these elements is called *soft water.*

As you observed in the laboratory investigation at the beginning of this chapter, when a sample of hard water evaporates, a solid deposit is left behind. You also observed that hard water prevents the formation of abundant soapsuds. Instead, the dissolved minerals in the hard water react with the soap, forming a **precipitate,** or curdlike deposit.

When the dissolved minerals in hard water react with enough soap, the water becomes soft. If more soap is then added to this water, abundant soapsuds will form. In fact, a sufficient amount of soap can soften hard water. However, the addition of extra soap is costly. Thus, cheaper methods of softening water have been developed.

The water in precipitation is soft because raindrops and snowflakes usually do not contain dissolved minerals. However, as rain or snow falls through the atmosphere, it may pick up carbon dioxide and become acidic. When the acidic water soaks into the ground, it can dissolve minerals in the bedrock and become hard water.

The factors that determine the mineral content of groundwater, and thus its hardness, include (1) the *solubility of the rock* the groundwater comes into contact with, (2) the *temperature of the groundwater,* and (3) the *length of time* the groundwater is in contact with the rock.

For example, in regions with limestone bedrock, the groundwater is hard because limestone is very *soluble,* or easily dissolved by groundwater. The higher the temperature of the groundwater, the more rapidly the limestone is dissolved. If the layers of limestone bedrock are thick, the groundwater stays in contact with the rock for a long time after penetrating cracks in the surface. Thus, the thicker the layers of limestone, the harder the groundwater.

Because of the great size and depth of most artesian systems, artesian groundwater usually stays in contact with the minerals in the aquifer for a long time. Consequently, water from artesian wells is harder than water from most shallow wells or from lakes and rivers.

Hard water is a costly nuisance in homes and in factories. Oftentimes, hard water has an unpleasant mineral taste. In addition, a water softener or extra soap may have to be added to hard water that is used for washing and doing laundry.

Hard water also creates a problem in hot-water pipes. The minerals dissolved in the water are continually deposited on the inner walls of the pipes. These mineral deposits, called *boiler scale*, can grow thick enough to block the passage of water through the pipe and cause the pipe to burst. Consequently, many communities add water softeners to groundwater before it enters the pipes.

# CHAPTER REVIEW

## Science Terms

*The following list contains all of the boldfaced words found in this chapter and the page on which each appears.*

aquifer (p. 65)
artesian system (p. 65)
caverns (p. 66)
geyser (p. 64)
groundwater (p. 61)
hard water (p. 67)
petrified wood (p. 67)
precipitate (p. 67)

sinkholes (p. 66)
spring (p. 63)
stalactites (p. 66)
stalagmites (p. 66)
water cycle (p. 61)
water table (p. 63)
wells (p. 65)

## Matching Questions

*On the blank line, write the letter of the item in column B that is most closely related to the item in column A.*

Column A

_____ 1. permeable

_____ 2. aquifer

_____ 3. stalactite

_____ 4. sinkhole

_____ 5. zone of saturation

Column B

a. section of soil and bedrock in which all spaces are completely filled with water
b. icicle-shaped deposit in a limestone cavern
c. a circular depression formed by the dissolving of limestone
d. the water-bearing layer of an artesian system
e. capable of transmitting water
f. large underground chamber or passage

# Multiple-Choice Questions

*On the blank line, write the letter preceding the word or expression that best completes the statement.*

1. The water that is found in a layer of sand is mainly located
   a. within the sand grains
   b. on the surface of the sand grains
   c. in the spaces between the sand grains
   d. on the surface of the ground                                             1 ____

2. The type of surface material through which water usually passes most rapidly is
   a. clay  b. sand  c. shale  d. granite                                      2 ____

3. Solid granite can become permeable if it
   a. develops large numbers of cracks
   b. is exposed to the action of groundwater
   c. becomes molten and then solidifies again
   d. is under great pressure                                                  3 ____

4. The temperature of groundwater does *not* depend on the
   a. depth below the surface          c. permeability of rock
   b. presence of molten rock          d. temperature at the surface          4 ____

5. A natural groundwater outlet through which boiling water and steam explodes into the air is called a (an)
   a. geyser  b. artesian system  c. sinkhole  d. mud volcano                 5 ____

6. The water pressure in a well drilled into an artesian system is mainly produced by
   a. expanding gases
   b. heating of the water
   c. the difference in height between the well and the water in the upper part of the system
   d. the distance that the water travels through the system                  6 ____

7. Which of the following rocks would make the best aquifer?
   a. granite  b. shale  c. sandstone  d. basalt                              7 ____

8. The source region of the Great Plains artesian system is located in
   a. Canada              c. the Appalachians
   b. the Great Lakes     d. the Rocky Mountains                              8 ____

9. Underground caverns usually form in regions underlain by
   a. shale  b. sandstone  c. limestone  d. granite                           9 ____

10. Common surface features in regions where the bedrock has been dissolved and carried away by groundwater include
    a. sinkholes  b. kettles  c. stalagmites  d. canyons                      10 ____

11. Petrified trees usually consist of minerals such as
    a. talc  b. silicon  c. silica  d. soapstone                             11 ____

12. Minerals usually found in hard water include compounds of magnesium and
    a. calcium  b. silicon  c. sodium  d. carbon                             12 ____

13. Washing clothes is difficult in hard water because the minerals dissolved in the water react with the soap and form a
    a. solution  b. precipitate  c. gas  d. crystal                          13 ____

**14.** The hardest water would probably be found in a region where the ground consists largely of
*a.* limestone   *b.* quartz sands   *c.* granite   *d.* clay          14 ____

**15.** The *least* important factor in determining the mineral content of groundwater is the
*a.* solubility of the rocks
*b.* temperature of the water
*c.* amount of time the water is in contact with the rocks
*d.* age of the bedrock          15 ____

**16.** The hardest water usually comes from
*a.* artesian wells          *c.* lakes
*b.* rivers          *d.* melted ice and snow          16 ____

# Modified True-False Questions

*In some of the following statements, the italicized term makes the statement incorrect. For each incorrect statement, write the term that must be substituted for the italicized term to make the statement correct. For each correct statement, write the word "true."*

**1.** Steep slopes usually absorb *less* runoff water than do gentle slopes.          1 _____

**2.** *More* rainwater usually soaks into grassy slopes than into bare slopes.          2 _____

**3.** Water usually soaks into sandy soils rapidly because the grains of sand have *small* spaces between them.          3 _____

**4.** Clay soils are *impermeable* because the particles of clay are too small and closely packed to allow water to pass between them.          4 _____

**5.** The top surface of the saturated zone of groundwater is called the *permeable layer*.          5 _____

**6.** Water brought to the surface from hundreds of meters underground is usually *colder* than water from near the surface.          6 _____

**7.** Wells in which water from a distant source rises to the surface under its own pressure are called *artesian* wells.          7 _____

**8.** Caverns are usually found in regions where the bedrock consists of *limestone*.          8 _____

**9.** The icicle-shaped deposits that hang from the ceilings of caverns are called *stalagmites*.          9 _____

**10.** Most of the elements found in a sample of petrified wood are *unlike* the elements found in the original wood.          10 _____

# Testing Your Knowledge

1. How does the type of surface material determine the amount of rainfall that soaks into the ground and becomes groundwater? _____

_____

2. How might you explain the fact that swampy places are sometimes located on hilltops?

_____

_____

3. Draw a diagram of a simple artesian system, and label its parts. How can an artesian system be made to supply well water without the use of pumps? _____

_____

_____

4. How do caverns, such as Carlsbad Caverns or Howe Caverns, form in a limestone region?

_____

_____

5. A topographic map of an area in Florida shows numerous depressions in the ground. The bedrock in the area is limestone. Describe two ways in which the depressions may have been formed. _____

_____

_____

6. How is petrified wood formed? _____

_____

7. If you were given six different water samples, how could you determine which sample is best for washing clothes? _____

_____

_____

# *Running Water Changes the Land*

## LABORATORY INVESTIGATION

### *THE EFFECT OF RUNNING WATER*

**NOTE:** This experiment is a teacher demonstration.

**A.** Set up an inclined trough and a large collecting pan, as shown in Fig. 6-1. Mix together 1000 milliliters (ml) of sand, 250 ml of powdered clay, and 100 ml of gravel. Add a few pebbles to the mixture. Prepare another identical mixture, and set it aside for later use.

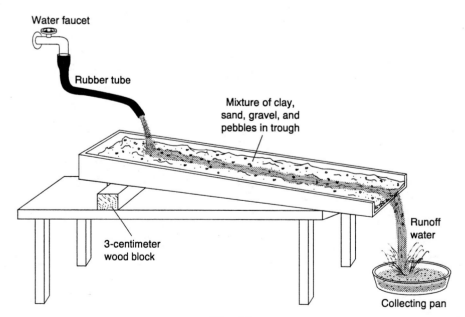

**Fig. 6-1.**

Pour one of the mixtures of sand, clay, gravel, and pebbles into the inclined trough, and spread the mixture evenly throughout the trough.

Attach the rubber tube to a water faucet, and let a small, gentle stream of water flow down the trough. Collect a sample of the runoff water in a small beaker held beneath the lower end of the trough. Then, let the runoff pour into the collecting pan for a minute or two. Turn off the faucet, and examine the trough and collecting pan to see which materials, or *sediments,* were most easily moved by the running water.

**1.** What effect did the running water have on each type of sediment in the mixture?

Clay _____

Sand _____

Gravel _____

Pebbles _____

**2.** Describe the appearance of the water sample collected in the beaker. _____

_____

**3.** Describe the sediments you collected in the beaker. _____

_____

**4.** Describe the groove, or gully, produced in the trough by the running water. _____

_____

**5.** Use a ruler to measure the depth and width of the gully made by the running water.

　　*a.* The depth of the gully is _____ centimeters.

　　*b.* The width of the gully is _____ centimeters.

**B.** Place the second mixture you prepared in a second trough. Set up the second trough so that the raised end of the trough is twice as high as the first trough. Repeat the procedure you followed in part A.

**6.** What effect does increasing the slope of the trough have on the speed of the running water?

_____

_____

**7.** What effect does increasing the slope of the trough have on the depth and width of the

gully? _____

Explain. _____

_____

**8.** Describe the sediments in the beaker. _____

_____

**9.** What is the relation between the speed of the running water and its ability to carry sed-

iments? _____

_____

# Weathering and Erosion

You learned in Chapter 4 that bedrock is broken down into small fragments by physical and chemical weathering. For example, an outcrop of granite bedrock may be broken down by frost wedging and by carbonic acid dissolved in rain and snow. Then, natural agents such as running water, moving ice (glaciers), and wind can remove the weathered materials from their original location. The process by which weathered rock material is removed and carried away by running water, glaciers, and wind is called **erosion.**

## IMPORTANCE OF GRAVITY

The fundamental force responsible for the erosion of the land is *Earth's gravity.* Rain and snow fall to the ground because of Earth's gravitational attraction. Earth's gravity also causes streams and rivers to flow downhill to the ocean. As streams flow, they wear down the land by scraping bedrock and removing loosened rock materials and carrying them downstream.

The removal of loose materials by erosion exposes fresh bedrock surfaces to the agents of weathering and erosion. From time to time, large pieces of bedrock break loose from a cliff or mountainside, and move downslope under the pull of gravity. The mound of rock fragments that collects at the base of a cliff or steep slope is called **talus.** Fig. 6-2 shows talus at the base of the Palisades—basalt cliffs along the west bank of the Hudson River. Occasionally, a huge mass of rock and soil breaks loose from the side of a cliff or mountain and rushes downhill as a unit. This kind of sudden mass movement is called a **landslide.**

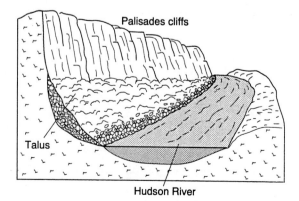

**Fig. 6-2. Talus at the base of the Palisades.**

## RUNNING WATER AND EROSION

The most important agent of erosion is *running water.* Brooks, creeks, streams, and rivers are constantly at work removing weathered fragments of bedrock and carrying them away from their original location.

Almost all of the water that erodes the land originates in the ocean. Water evaporates from the ocean and becomes *water vapor.* The water vapor cools and condenses into huge masses of tiny water droplets called *clouds.* In time, *precipitation* falls from clouds in the form of rain or snow.

You learned in Chapter 5 that only a small amount of the precipitation that falls on Earth's surface soaks into the ground. Most of the water that falls as rain or snow quickly evaporates back into the atmosphere or returns to the ocean by way of streams and rivers.

Water that seeps into the ground moves slowly downward and becomes groundwater. When groundwater meets an impermeable layer of rock, the groundwater flows sideways along the impermeable layer until it meets openings in the surface. At these surface openings, lakes, ponds, and springs appear. Lakes, ponds, and springs provide streams with a steady supply of water.

## RIVER SYSTEMS

When rain falls on sloping land, the water flowing downhill carries away loose rock fragments and soil. The deep channels carved into Earth's surface by running water are called **gullies.** In the laboratory investigation, you noticed that a gully was formed when the running water carried off particles of clay and sand as it flowed through the trough.

As shown in Fig. 6-3, water flowing downhill from repeated rainfalls deepens and widens a gully. Eventually, the bed of the gully may reach the water table. If there is sufficient groundwater throughout the year, a *permanent stream* forms in the gully. Otherwise, the gully contains flowing water only when it rains a lot or when snow melts. For example, many gullies in the deserts of the southwestern United States are dry for much of the year. They fill with water after a heavy rain and then dry up. Because water does not flow continuously in such a gully, the stream is called an *intermittent stream.*

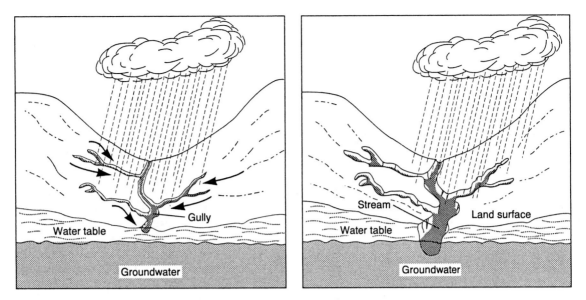

**Fig. 6-3. Formation of a permanent stream.**

Small streams often join to form larger streams. In turn, these streams combine and form a still larger stream called a **mainstream.** All of the streams that supply a mainstream are called **tributaries** of that mainstream.

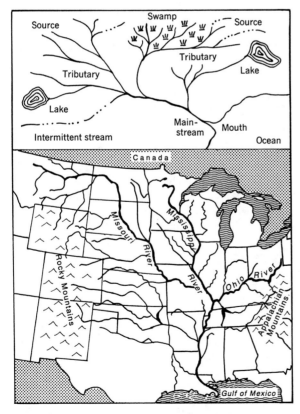

**Fig. 6-4. Top: Components of a river system. Bottom: The Mississippi River system.**

A mainstream and all of its tributaries form a **river system** (see Fig. 6-4). The Missouri River and Ohio River are examples of river systems; both rivers are themselves tributaries of the Mississippi River. The Mississippi River and its thousands of tributaries form a huge river system that extends from the Rocky Mountains to the Appalachian Mountains and from Canada to the Gulf of Mexico.

The area of land drained by a river system is called its **watershed.** River systems usually empty into a large body of water such as the ocean. The place where a river empties into a larger body of water is called the *mouth* of the river.

## EROSION BY STREAMS AND RIVERS

You have seen that running water carries away loose rock materials. Streams and rivers can also wear down solid bedrock, gradually eroding high mountains into low hills and flat plains. Two ways that streams and rivers erode bedrock are by *abrasion* and by *solution* of minerals in the bedrock.

**Abrasion.** A stream erodes bedrock by the grinding action of the sand, pebbles, and other rock fragments it carries. Like sandpaper against wood, these materials scrape against the streambed and banks. Particle by particle, the bedrock is worn away. The grinding action of rock fragments scraping against the streambed and against each other is called **abrasion.**

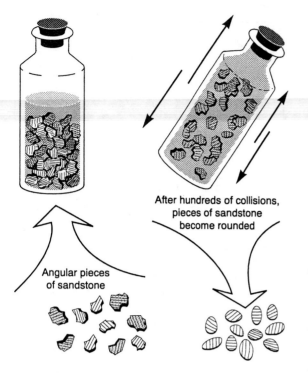

After hundreds of collisions, pieces of sandstone become rounded

Angular pieces of sandstone

**Fig. 6-5. Demonstrating abrasion.**

You can observe the effect of abrasion on rock fragments by placing some angular pieces of sandstone in a bottle half-filled with water and then shaking the jar for several minutes. As shown in Fig. 6-5, after hundreds of collisions, the edges of some of the sandstone fragments become more rounded.

Similarly, when large rock fragments first enter a stream, the fragments are usually angular with sharp edges. These sharp edges act as cutting tools. As the sharp-edged, angular rocks bounce and scrape along the riverbed, they chip off pieces of the bedrock. In time, the angular rocks lose their sharp edges and become rounded. Thus, the roundness of a stone is an indication of how long it has been carried along by a stream.

Streams may carry a large volume of fine-grained silt and clay (see Fig. 6-6). The silt and clay are produced when coarse-grained rock

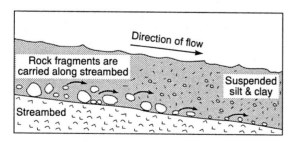

**Fig. 6-6. Abrasion in a stream.**

particles collide with each other or with the streambed, or when loose materials are eroded from the banks of a stream. In the laboratory investigation, you noticed that fine particles of silt and clay remained suspended in the water sample you collected while there was any movement of the water in the beaker. Likewise, silt and clay can remain suspended in a stream of water for a long time. The motion of the stream prevents the silt and clay particles from settling to the streambed.

Rivers that carry abundant silt and clay have a muddy appearance. The Mississippi River and the Missouri River are two examples. In fact, the Missouri River has been nicknamed the "Big Muddy." It has been estimated that the rivers of the United States carry about 700 million tons of sediments into the oceans every year.

**Solution.** Besides abrasion, streams also erode rocks by the dissolving, or **solution,** of their minerals. For example, limestone and marble are readily dissolved by streams. Running water also causes sedimentary rocks cemented together by calcite or iron compounds to crumble, because these cementing agents are easily dissolved.

The dissolved mineral substances in water are invisible. However, they can be detected by chemical analysis. Geologists estimate that the rivers of the United States carry about 250 million tons of dissolved minerals into the oceans each year.

## HOW A STREAM ERODES ITS BED

How quickly a stream erodes its bed depends on several factors. These factors include (1) the *velocity* of the stream, (2) the *volume of water* carried by the stream, (3) the *size and shape of sediments* in the stream, and (4) the *type of bedrock* beneath the stream.

**Velocity.** Rapidly flowing streams erode their beds faster than slow-moving streams. A rapidly flowing stream has great cutting power because of its speed and because it can carry large, coarse-grained sediments such as gravel, pebbles, and even small boulders. These coarse-grained materials scrape with great force against the streambed.

In contrast, a slow-moving stream has little cutting power because it can carry only fine-grained sediments such as silt and clay. Such fine-grained sediments have almost no effect on the streambed.

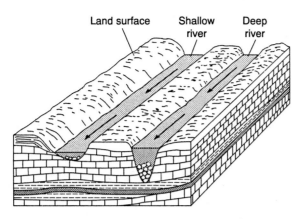

**Fig. 6-7. The stream that carries the larger volume of water cuts its bed more rapidly.**

A stream moving about 1 kilometer (km) per hour can carry silt, clay, and small sand grains. A stream moving at about 8 km per hour can move small boulders. You recall that in the laboratory investigation when you raised the trough, the stream of water flowed faster, and larger rock particles were moved farther down the trough.

**Volume.** A deep, wide stream can transport a greater volume of water than a narrow, shallow stream can. Suppose two streams move with the same velocity but carry different volumes of water. The two streams also flow over the same type of bedrock. In this case, the stream with the greater volume of water will erode its bed faster (see Fig. 6-7). This stream erodes its bed faster because it carries more sediments.

**Size and Shape of Sediments.** Most stream erosion is produced by the abrasion of coarse-grained, angular sediments. Large rock fragments with sharp edges make much better cutting tools than do small, rounded particles. Consequently, large coarse-grained sediments such as gravel and pebbles enable a stream to erode its bed most effectively.

**Type of Bedrock.** Streams readily erode bedrock that is soft or soluble in water. For example, most streams will easily erode bedrock made of limestone and shale, which are relatively soft rocks. Streams also readily erode sedimentary rocks cemented together by calcite or iron compounds, which are soluble in water. In contrast, bedrock made of granite, slate, or quartzite is much harder and more resistant to chemical weathering than are limestone and shale. Consequently, streams erode bedrock made of granite, slate, and quartzite very slowly.

## RUNNING WATER AND DEPOSITION

You already know that the ability of a stream to carry sediments depends partly on its velocity. As shown in Fig. 6-8, a stream's velocity decreases when the stream (a) *enters a larger body of water*, (b) *overflows its banks*, (c) *changes direction*, or (d) *flows from a steep slope to a gentle slope.*

As the velocity of a stream decreases, its ability to carry rock particles also decreases. Thus, when a stream slows down, the coarse-grained sediments settle out first, followed by

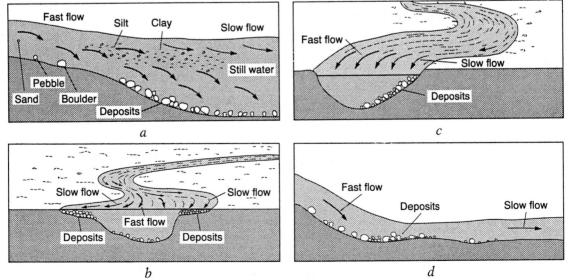

**Fig. 6-8. Factors that cause streams to slow down and deposit sediments.**

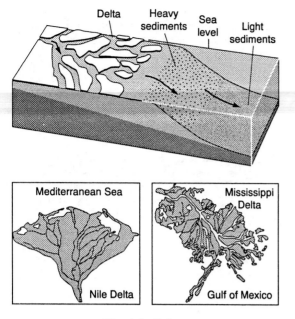

**Fig. 6-9. Deltas.**

However, not all deltas have this shape. Another type of delta looks like long toes extending into a larger body of water. This type is called a *bird-foot delta*. In the United States, the Mississippi River has built a bird-foot delta out into the Gulf of Mexico.

**Alluvial Fans.** A stream that flows from steep mountains out onto a broad, flat plain abruptly slows down and deposits sediments on the plain. The resulting fan-shaped deposit is called an **alluvial fan** (see Fig. 6-10). Alluvial fans are most common in dry regions, such as the southwestern United States, where they are often seen at the base of mountain passes.

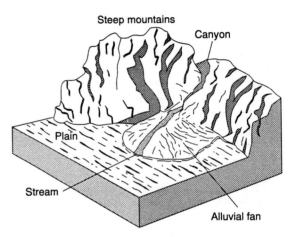

**Fig. 6-10. Alluvial fan.**

the fine-grained sediments. An accumulation of sediments left by a stream or river is called a **deposit.** As a deposit of sediment grows, it may produce landscape features such as *deltas, alluvial fans, levees,* and *sandbars.*

**Deltas.** When a river enters a large, still body of water (such as a lake or bay), the river immediately slows down. This sudden decrease in speed causes sediments to be deposited on the bottom of the larger body of water. In time, these deposits extend far into a lake or bay, producing a landform called a **delta** (see Fig. 6-9).

The name *delta* comes from the fact that many of these deposits have a triangular shape that resembles the Greek letter *delta* ($\Delta$). For example, the Nile River in Egypt has built a triangular delta where it empties into the Mediterranean Sea.

**Levees.** When a river overflows its banks, the water that leaves the river channel slows down. As the water loses speed, it deposits sediments on the banks of the river. Each time the river overflows its banks, more sediments are deposited on top of the previous deposit. Layer by layer, the river builds up long, low walls of sediments along its banks. These built-up riverbanks are called **levees** (see Fig. 6-11).

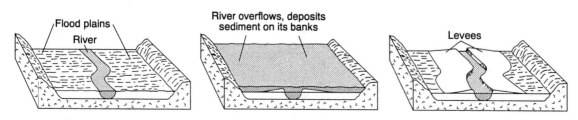

**Fig. 6-11. Levee formation.**

Cross sections

Top view

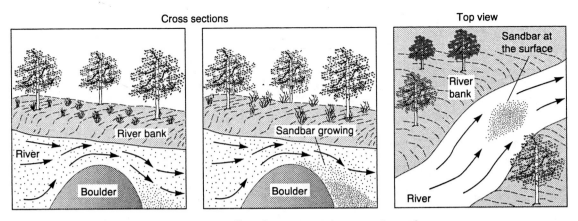

**Fig. 6-12. Sandbar formation in a river channel.**

Towns located near a flood-prone river often build levees along the riverbanks. These artificial levees usually prevent the river from overflowing and flooding the nearby towns. Many levees, both natural and artificial, are found along the banks of the Mississippi River.

**Sandbars.**   When a river flows around a curve or bend, the velocity of the river changes. The water in the river slows down on the inside of a curve and speeds up on the outside of the curve. The loss of speed on the inside of the curve causes sediments to settle out of the water and to be deposited along the inside bank of the river. This type of deposit is called a **sandbar.** A sandbar also may form when a boulder or some other obstruction, such as a fallen tree, partly blocks a river channel. As shown in Fig. 6-12, a large boulder in a river channel causes the river to slow down temporarily at that spot and to deposit sediments.

## LIFE CYCLE OF A RIVER

As a river flows and erodes the land, the river undergoes changes, passing through different stages of development. These stages make up the river's *life cycle*. There are three main stages in the life cycle of a river: *youth, maturity,* and *old age*. Each stage has its own characteristic features.

**Youth.**   During youth, a river flows rapidly down mountains or hills. A young river tumbles noisily over falls and races through rapids, enclosed within a narrow, rock-walled valley. Because a young river moves swiftly, it carries large quantities of coarse-grained sediments. Consequently, the river cuts rapidly downward into the underlying bedrock, eroding a deep, narrow, V-shaped channel (see Fig. 6-13a). The Colorado River, which has produced the Grand Canyon, is an example of a young river.

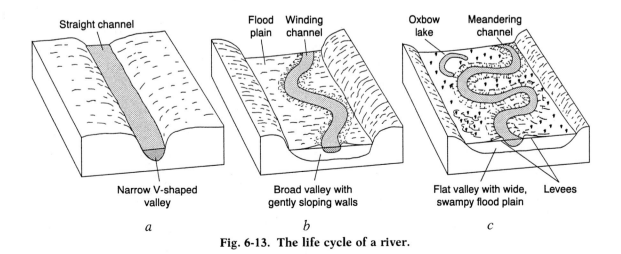

**Fig. 6-13. The life cycle of a river.**

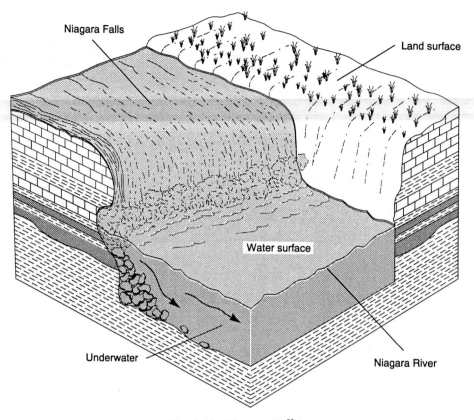

**Fig. 6-14. Niagara Falls.**

A **waterfall** is perhaps the most spectacular feature of a young river. A waterfall occurs wherever the elevation of a riverbed drops suddenly. For example, as the Niagara River flows from Lake Erie to Lake Ontario, the elevation of the riverbed drops about 50 meters at one spot (see Fig. 6-14). The sudden drop of the river's bed in this location produced the famous Niagara Falls.

A waterfall also may form where a river travels over different types of bedrock. For example, a river flowing over a layer of basalt followed by shale erodes the softer shale faster than the harder basalt. In time, the uneven erosion produces a basalt cliff in the river channel. A waterfall is formed as the river flows over this cliff (see Fig. 6-15). Yellowstone Falls in Yellowstone National Park is an example of a waterfall produced in this way.

**Maturity.**   As a young river continues to erode and smooth out its channel, the river loses its waterfalls and rapids. Erosion also decreases the slope of its channel. As a consequence, the velocity of the river decreases as well. The river is then said to be in its mature stage (see Fig. 6-13b). The Missouri River is an example of a mature river.

A mature river flows more slowly and has less cutting power than a young river. Con-

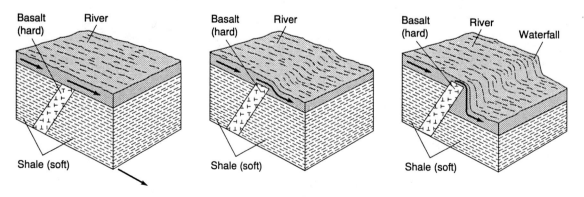

**Fig. 6-15. Formation of a waterfall by uneven erosion.**

sequently, when a mature river encounters an obstruction, the river flows around the obstruction. In this way, a mature river begins to form a curved, or S-shaped, channel. In time, the curves become larger, and the river channel becomes more winding. These looping, S-shaped curves in the river's course are called **meanders.**

During maturity, a river's once-steep valley walls are worn down by the agents of physical and chemical weathering. Instead of being V-shaped, the valley walls of a mature river are somewhat rounded and slope gently toward the river.

As a mature river flows back and forth across its valley, the river gradually widens the valley floor. When flooding causes the river to overflow its channel, coarse-grained sediments, such as gravel and pebbles, are deposited along the banks to form levees. Finer-grained sediments, such as silt and clay, are carried beyond the riverbanks and deposited on the valley floor. These sediments level out the valley floor to form a relatively flat area called a **floodplain.** Swamps often form on a floodplain because the deposits of fine silt and clay do not readily allow the floodwaters to soak into the ground.

**Old Age.** When the slope of a river channel becomes almost flat, the water flows very slowly. Then, a river is said to be in old age (see Fig. 6-13c). The lower part of the Mississippi River is in this stage of development.

Instead of causing erosion, an old river mainly affects the land by depositing sediments. Consequently, the levees found along an old river are higher than the levees along a mature river. An old river also has a wider floodplain than a mature river because the old river flows back and forth over a much wider valley.

Sometimes, an old river cuts through the narrow part of one of its meanders. This causes a sharp decrease in the velocity of water flowing through the meander. As a result, sediments accumulate where the ends of the meander meet the new river channel. In time, sediments completely cut off the meander from the river. The cutoff meander is called an **oxbow lake** (see Fig. 6-13c). When sediments fill in an oxbow lake, a swampy area, or **bayou,** is formed.

You should be aware that a river may not be in the same stage of its life cycle along its entire course. In fact, many rivers show several different stages of development along their length, depending on the steepness of the land. For example, a single river may exhibit signs of youth in its upper reaches, near its source; characteristics of a mature river in its middle stretch; and features of old age near its mouth, where it empties into the sea (see Fig. 6-16).

Sometimes, an old or mature river becomes youthful again. For example, forces inside Earth may raise the crust, causing the slope of an old, meandering river channel to become steeper. The steeper slope increases the water velocity, enabling the river to cut through the underlying bedrock again.

With its renewed cutting power, the river begins to deepen its old meanders. A deepened meander looks like a curved trench with steep walls. These deeply eroded curves of a river are called *entrenched meanders.* You can see them along the Susquehanna River in New York and Pennsylvania, and along the San Juan River in Utah and New Mexico.

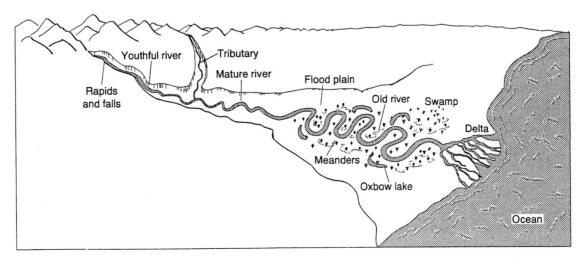

**Fig. 6-16. A single river may show different stages of development along its length.**

# CHAPTER REVIEW

## Science Terms

*The following list contains all of the boldfaced words found in this chapter and the page on which each appears.*

abrasion (p. 75)  gullies (p. 74)  sandbar (p. 79)
alluvial fan (p. 78)  landslide (p. 74)  solution (p. 76)
bayou (p. 81)  levees (p. 78)  talus (p. 74)
delta (p. 78)  mainstream (p. 75)  tributaries (p. 75)
deposit (p. 78)  meanders (p. 81)  waterfall (p. 80)
erosion (p. 74)  oxbow lake (p. 81)  watershed (p. 75)
floodplain (p. 81)  river system (p. 75)

## Matching Questions

*On the blank line, write the letter of the item in column B that is most closely related to the item in column A.*

### Column A

_____ 1. how wind and water remove rock material

_____ 2. sudden movement of rock mass downhill

_____ 3. huge masses of tiny water droplets

_____ 4. channels carved into Earth by running water

_____ 5. all the streams that supply a mainstream

_____ 6. grinding of rock fragments in a streambed

_____ 7. river deposit extending into lake or bay

_____ 8. stream deposit on a broad, flat plain

_____ 9. long walls of sediment along riverbanks

_____ 10. swampy sediment-filled oxbow lake

### Column B

*a.* gullies
*b.* abrasion
*c.* alluvial fan
*d.* delta
*e.* levees
*f.* erosion
*g.* bayou
*h.* landslide
*i.* watershed
*j.* tributaries
*k.* clouds

## Multiple-Choice Questions

*On the blank line, write the letter preceding the word or expression that best completes the statement.*

**1.** Most of the abrasion performed by rivers is done by
*a.* moving rock fragments          *c.* waterfalls
*b.* dissolved gases                 *d.* rapids                                    1 _____

**2.** The *least* effective cutting tools carried by a stream are
*a.* particles of clay   *b.* particles of sand   *c.* pebbles   *d.* boulders          2 _____

3. Which of the following is the *least* important factor in determining how swiftly a river will erode the land?
   a. volume of water in the river
   b. velocity of the river
   c. thickness of the bedrock being eroded
   d. types of rocks being eroded

   3 _____

4. The rock that is *most* resistant to erosion by a river is
   a. quartzite  b. shale  c. sandstone  d. limestone

   4 _____

5. A river becomes able to carry large rock fragments when it
   a. overflows its banks
   b. enters a larger body of water
   c. decreases in volume
   d. increases in velocity

   5 _____

6. A landscape feature that is *not* the result of deposition by a river is a
   a. V-shaped valley  b. floodplain  c. delta  d. sandbar

   6 _____

7. A feature that is *not* characteristic of a young river is a
   a. floodplain  b. V-shaped channel  c. waterfall  d. rapid

   7 _____

8. When a river overflows its banks, the river may build a natural wall called a
   a. cliff  b. ridge  c. levee  d. rampart

   8 _____

9. Old rivers are characterized by
   a. small floodplains  b. steep canyons  c. oxbow lakes  d. waterfalls

   9 _____

10. An old river that meanders slowly over the land may begin to cut downward again like a young river if
    a. it reaches rock layers that are difficult to erode
    b. the slope of the land is increased
    c. the velocity of the river decreases
    d. the amount of sediment in the river decreases

    10 _____

# Modified True-False Questions

*In some of the following statements, the italicized term makes the statement incorrect. For each incorrect statement, write the term that must be substituted for the italicized term to make the statement correct. For each correct statement, write the word "true."*

1. *Most* of the rain that falls on Earth's surface sinks deep into the ground.

   1 _____

2. Water does not flow continuously throughout the year in *intermittent* streams.

   2 _____

3. A *well-rounded* pebble was probably worn down by the action of moving water in a river.

   3 _____

4. The Missouri River is called the "Big Muddy" because it carries large quantities of rock material in *solution*.

   4 _____

5. Most of the rock material carried by rivers has been *dissolved* by the water.

   5 _____

6. A river that has a velocity of 8 km per hour can move *small boulders*.

   6 _____

7. Sediments that settle out from rivers are called *deposits*.

   7 _____

8. When a river overflows its banks, the river usually *erodes* materials along the banks.

   8 _____

9. Sandbars usually form along the *inside* curves of a winding river.     9 _____

10. The main work performed by a *young* river is cutting downward into its bed.     10 _____

11. The largest floodplains are found along the banks of *young* rivers.     11 _____

# Testing Your Knowledge

1. Explain how repeated rainfalls may eventually result in the development of a river.

   _____

   _____

2. Describe two ways in which a river may erode its bed.

   *a.* _____

   *b.* _____

3. The erosion of a riverbed depends upon four factors. Name these factors, and briefly describe how each one affects the rate of erosion of a riverbed.

   *a.* _____

   _____

   *b.* _____

   _____

   *c.* _____

   _____

   *d.* _____

   _____

4. Describe two landforms that rivers produce when they deposit sediments.

   *a.* _____

   *b.* _____

5. Name two features that are typical of each of the following stages in the life cycle of a river.

   *a.* Youth _____

   _____

   *b.* Maturity _____

   _____

   *c.* Old Age _____

   _____

# Watersheds: Should We Pay the Price to Protect Them?

Fresh water "may well be the most precious resource the earth provides to humankind," said J. W. Maurits la Rivière, one of the world's leading environmental scientists. Yet in many parts of the world, including the United States, humankind seems to be doing a great deal to waste or pollute that natural resource.

Waste and pollution can occur in many places, but there is one place where its impact can be most striking: a watershed. A *watershed* is an area of land that receives water from rain or snow and feeds it into countless creeks, streams, and rivers.

Some watersheds are truly huge, like the Mississippi River watershed that stretches from the Rocky Mountains in the west to the Appalachian Mountains in the east and from the Canadian border to the Gulf of Mexico. Others may be as small as your local forest.

But whether large or small, an undamaged watershed preserves clean fresh water. It does so by keeping creeks, streams, and rivers freely flowing and by allowing water to slowly seep into the ground.

Now let's explore where these natural processes may go wrong. Let's start in the great forests of a watershed. Here, where thick stands of trees spring from the slopes of mountain ranges and rolling hills, is the first stop in fresh water's journey to farms, towns, and cities.

Tree trunks and roots, shrubs, bushes, and blankets of leaves on the ground slow the downhill rush of water. This reduces the chance of floods at lower altitudes. It also reduces the amount of fine dirt, called *silt*, that will be carried into waterways.

Silt can clog streams and rivers, sometimes making them little more than muddy marshes. Such marshes accumulate dead vegetation that, as it decays, robs the water of oxygen. Then fish and other living things die. The water becomes polluted and not fit for use by people.

So the forests are a kind of first line of defense against the loss of clean, available fresh water. According to environmentalists, this "line" is being systematically broken down by logging companies. Many such companies practice a form of logging called "clear-cutting." In clear-cutting, an entire section of forest is cut down. The land becomes bare, and water rushes over it freely, picking up dust and dirt as it goes. The amount of water that soaks into the soil falls, while the amount of water and dirt that pours into streams and rivers rises.

The logs, however, are collected very efficiently. This increases the profits of logging companies and tends to decrease the cost of lumber for construction. So there is an economic benefit to many people, including consumers. But is the benefit worth the cost in fresh water?

Now let's move beyond the forests to the place where our food is grown: our country's farmland. Crops need water to grow and ripen. However, in many farm areas the amount of rainfall is not enough to keep crops healthy. To overcome this problem, farmers have turned to *irrigation*, which is the channeling or piping of water to fields. Much of the water used in the United States goes for irrigation. For example, Utah uses 90 percent of its fresh water for farming.

Now, there's nothing wrong with irrigation unless it wastes water, and a good many conservationists say that this often happens. They further argue that there aren't enough regulations—and those that do exist are poorly enforced—to make farmers cut down on wasteful irrigation practices, like plowing fields up and down the slopes of hills instead of across them.

What's more, critics point out that in some states the path of a stream or river may be changed, or its waters tapped, for irrigation. In many instances, these rivers and streams dry up or become filled with muddy sediments. Another valuable source of fresh water is then lost.

All this happens very slowly, of course—too slowly to trouble most people. But we do need food from farms now! So, can we satisfy our need for food and still have clean fresh water?

Conservationists say yes. But we will have to have tough water preservation regulations that are strictly enforced, they argue. Other people say such regulations interfere with the rights of farmers, loggers, and others to earn a fair living or profit. Some people insist that our country has too many regulations as it is. What do you think?

**1.** *Scientist J. W. Maurits la Rivière said the most precious resource on Earth is*

   a. lumber.     c. water.
   b. food.      d. mountains.

**2.** *The cutting down of entire sections of a forest is called*

   a. clean-cutting.
   b. clear-cutting.
   c. watershedding.
   d. conservation.

**3.** *Irrigation is used*

   a. where not enough rain falls.
   b. where too much rain falls.
   c. in forests.
   d. in rivers.

**4.** *How might the cutting down of a forest affect a stream or river?* _____

_____

_____

_____

_____

**5.** *On a separate sheet of paper, discuss the implications of strict regulations to preserve watersheds.*

# Glaciers and Wind Change the Land

## LABORATORY INVESTIGATION

### *THE EFFECT OF PRESSURE ON ICE*

**A.** Support an ice cube on two small blocks of wood, as shown in Fig. 7-1. Tie one 250-gram (g) weight to each end of a piece of thin, strong thread or wire. Place the string or wire over the piece of ice as shown in the figure. Allow the setup to remain undisturbed until you have completed part B.

**B.** Obtain two small pieces of cardboard. Push four nails through each piece of cardboard in the shape of a square, as shown in Fig. 7-2. Place one large ice cube on top of each group of four nails.

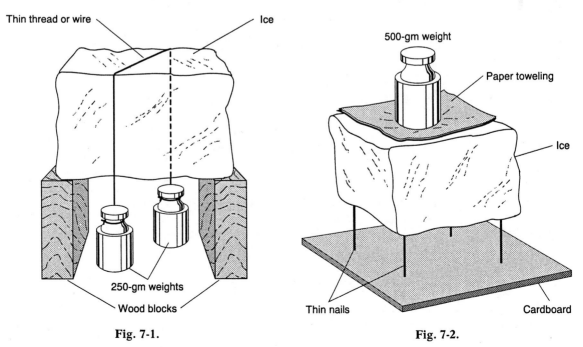

**Fig. 7-1.**

**Fig. 7-2.**

Place a piece of paper toweling about 2.5 centimeters square on top of one of the ice cubes. Place a 500-g weight on top of the toweling. Do not place anything on top of the second ice cube.

During a 10-minute period, carefully observe each ice cube every 2 minutes where the ice cube is supported by the nails. In the chart below, record the following information:

*a.* The amount of melting that occurs at the points of the nails. Compare the difference between the two ice cubes, using the words "more," "less," or "same."

*b.* How deep the nails penetrate into each ice cube. With a metric ruler, measure the distance between the bottom of each ice cube and the top of the cardboard.

| Minutes | Melting (more, less, same) | | Distance Remaining (centimeters) | |
|---|---|---|---|---|
| | Weighted ice cube | Unweighted ice cube | Weighted ice cube | Unweighted ice cube |
| 0 | | | | |
| 2 | | | | |
| 4 | | | | |
| 6 | | | | |
| 8 | | | | |
| 10 | | | | |

**1.** Into which ice cube have the nails penetrated deepest after 10 minutes? _____

_____

**2.** How might you explain the fact that the nails penetrated the ice cubes without cracking or breaking them? _____

_____

**3.** Which ice cube exerted the greater pressure on the nails? _____

_____

**4.** Which ice cube showed the greater amount of melting at its base? _____

_____

**5.** What caused the difference in melting between the two ice cubes? _____

_____

**6.** What is the relationship between the amount of pressure on the ice and the rate at which the ice melted? _____

_____

**C.** Examine the ice cube and weighted thread you prepared in part A.

**7.** What was the position of the thread when you began the experiment? _____

_____

**8.** What is the position of the thread now? _____

_____

**9.** How did the thread reach its present position? _____

_____

**10.** What happened to the thin slit in the ice when the thread reached the bottom of the ice?

Explain. _____

_____

# Glaciers

About 12,000 years ago, a vast sheet of ice covered a large part of the northern United States. In some places, the ice was more than 3 kilometers (km) thick. This huge, frozen mass had moved southward from the northern regions of Canada as several large bodies of slow-moving ice, or **glaciers.**

Geologists believe that glaciers have advanced southward from northern Canada and retreated, or melted back, northward four times in the past one million years. A time period in which glaciers advance over a large part of a continent is called an **ice age.** During the most recent ice age, the southern edge of the glaciers extended westward from Long Island, New York, along the courses followed by the Ohio and Missouri rivers and then northwest to Montana (see Fig. 7-3).

## FORMATION OF GLACIERS

A glacier is a large mass of ice that moves, or flows, over the land in response to gravity. Glaciers form among high mountains and in other cold regions, where winter snowfalls do not melt completely during the following summer. Year after year, layers of snow accumulate. In time, the snowflakes buried in the deepest layers of snow are compressed by the weight of the upper layers of snow. The pressure of the upper layers causes melting of the tightly packed snowflakes deep within the mass. The melted snow, or *meltwater*, trickles down toward the bottom of the snow mass and refreezes in the spaces between snowflakes.

This sequence of events produces a grainy form of ice similar to the ice under an old pile of snow. The grainy ice continues to undergo change, and is gradually converted into *glacier ice*. Glacier ice is not as clear as ordinary ice, and often appears bluish gray because of tiny air bubbles and dust particles trapped in the ice.

## MOVEMENT OF GLACIERS

The movement of a glacier is caused in part by the great pressure exerted by the surface ice on the ice located deep within the glacier.

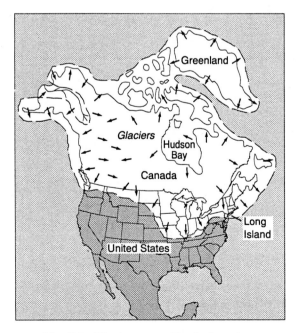

**Fig. 7-3. Glaciers over North America.**

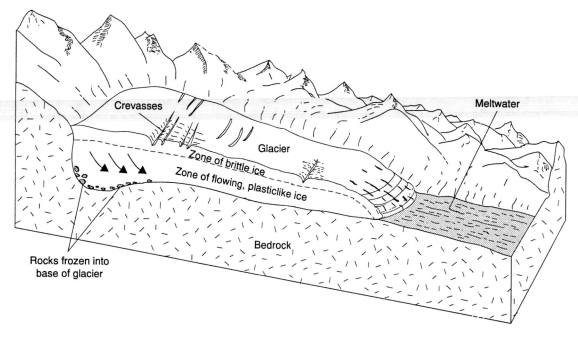

**Fig. 7-4. Movement of a glacier.**

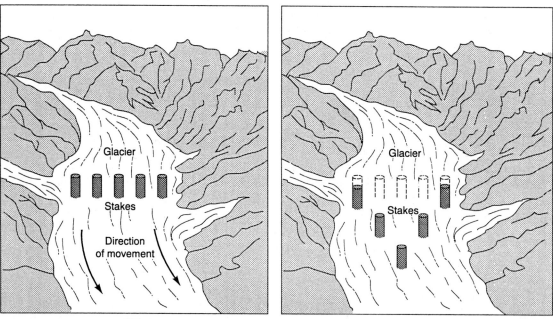

*a. Before*          *b. After*

**Fig. 7-5.** (*a*) **A row of stakes is driven into the glacier's surface in a straight line.** (*b*) **The stakes in the center have moved further than the stakes on the sides because the center of a glacier moves faster than its sides.**

This pressure causes the deep ice, which would ordinarily be solid and brittle, to become *plastic*. In other words, the deep ice behaves like dough or taffy, materials that look solid but that can flow slowly and change shape.

You may have noticed how the asphalt along the edge of a busy highway seems to have been pushed up against the curb. That is because the weight of passing cars and trucks causes the solid asphalt to become plastic and flow slowly toward the curb. In a similar way, deep ice in a glacier becomes plastic and flows outward because of the enormous weight of the overlying snow and ice (see Fig. 7-4).

Glacial movement also involves the melting of ice at the base of a glacier. Deep ice in a glacier melts under great pressure, forming a thin layer of water beneath the glacier. The meltwater acts as a lubricant, causing the entire glacier to slip a small distance downhill. This movement decreases the pressure at the base of the glacier, allowing the meltwater to refreeze. In time, the weight of the ice causes the deep ice to melt again, and the process is repeated.

A row of stakes driven into the surface of a glacier reveals that the center of the glacier moves faster than its sides (see Fig. 7-5). The sides of a glacier move more slowly than the center because of friction between the glacier and the valley walls. Some glaciers move as far as 20 meters (m) in a day. Other glaciers move very slowly, advancing only a few centimeters in a day. Because of additional weight, glaciers move faster after winters with heavy snowfall than after winters with light snowfall. Also, the steeper the slope, the faster a glacier moves.

## TYPES OF GLACIERS

There are two main types of glaciers: *valley glaciers* and *continental glaciers*.

**Valley Glaciers.** Valley glaciers are found in stream-eroded valleys in high mountain ranges, such as the Alps, the Rocky Mountains, and the Himalayas. A valley glacier forms high in the mountains and flows downhill to lower elevations, following the course of the valley it occupies. The largest valley glacier is Beardmore, in Antarctica, which is 190 km long, 40 km wide, and more than 300 m thick. The smallest valley glaciers are about 2 km long, 300 m wide, and less than 100 m thick.

**Continental Glaciers.** A glacier that covers a vast area of land is called a **continental glacier,** or an ice sheet. Unlike a valley glacier, which moves downhill confined within its valley, a continental glacier spreads outward in all directions from a central area where the ice is thickest.

During past ice ages, continental ice sheets were far more extensive than they are today. Currently, continental glaciers are confined to landmasses in Earth's polar regions. Greenland, which lies mostly within the Arctic Circle, is almost completely covered by an ice sheet. The world's largest continental glacier covers most of Antarctica. The Antarctic ice sheet is nearly 5 km thick in places.

Ice sheets store huge amounts of water. As a result, the advance and retreat of continental glaciers can change the level of the world's oceans. During the last ice age, so much of Earth's water was locked up in glaciers that the sea level was about 100 m lower than it is today. When the glaciers melted at the end of the ice age, the water was released, raising the sea level to its current position. Geologists estimate that if the Antarctic and Greenland ice sheets were to melt, the sea level would rise an additional 60 meters.

## GLACIAL EROSION

Glaciers wear down Earth's surface and create distinctive landscape features. You saw in your laboratory investigation that pressure can force solid objects to penetrate an ice cube. In the same way, the weight of a glacier causes rock fragments to penetrate into its base and become embedded in the ice (see Fig. 7-4). As a glacier moves, the rock material frozen into its base helps the glacier erode the land. A glacier also erodes the land by freezing to the bedrock and then tearing, or "plucking" away, chunks of rock as the glacier moves downhill.

**Erosion by Valley Glaciers.** A valley glacier scrapes away the walls and floor of its valley as it moves downhill. As shown in Fig. 7-6, valleys that have been eroded by valley glaciers have a characteristic U-shape, rather than the V-shape typical of stream erosion.

Valley glaciers also produce sharp, pyramid-shaped mountain peaks called **horns.** One famous example is the Matterhorn in Switzerland (see Fig. 7-7). A horn is produced when glaciers erode a mountain from several sides. First, frost wedging and glacial plucking work

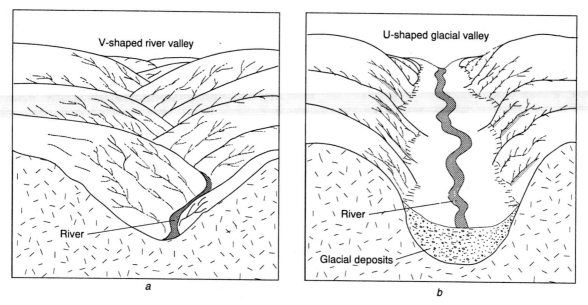

Fig. 7-6. Glaciers carve U-shaped valleys.

Fig. 7-7. The Matterhorn: a glacier-carved mountain in Switzerland.

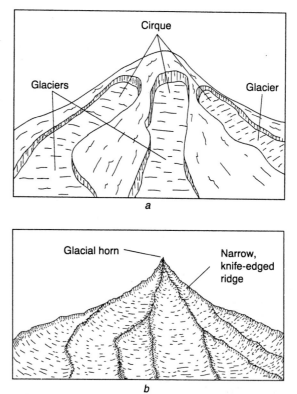

Fig. 7-8. (a) Valley glaciers carve bowl-shaped cirques into the sides of a mountain. (b) Edges of the cirques meet at the mountaintop, forming a pyramid-shaped horn.

together to carve bowl-shaped depressions called **cirques** (pronounced *serks*), into the sides of a mountain (see Fig. 7-8*a*). As weathering and erosion continue, the cirques become deeper and wider until their edges meet at the mountaintop, forming a horn. As shown in Fig. 7-8*b*, a horn looks like a mountain that has been chiseled into a pyramidlike shape.

**Erosion by Continental Glaciers.** The erosional effects of continental glaciers differ in some ways from those of valley glaciers. Valley glaciers in mountainous regions carve sharp, horn-shaped peaks and narrow, jagged ridges. In contrast, continental glaciers are so thick that they often ride completely over mountains in their path, leaving smoothed, rounded mountaintops and ridges.

When a continental ice sheet moves over an area of weak, easily eroded bedrock, the ice may gouge out a deep basin that later fills with glacial meltwater to produce a large lake. The Great Lakes occupy basins that were scooped out and widened by continental glaciers.

Both valley glaciers and continental glaciers contain many rock fragments frozen into their bases (see Fig. 7-9). As a glacier drags these rock fragments along, the embedded rocks act as cutting tools, carving deep scratches and large grooves into the bedrock (see Fig. 7-10).

Glacial movement also grinds up the rock fragments trapped in the ice into a powdery material called *rock flour*. As a glacier moves, the rock flour acts like sandpaper, smoothing and polishing the underlying bedrock.

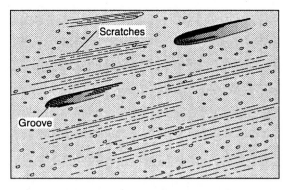

**Fig. 7-10. Glacial scratches and grooves in bedrock.**

## GLACIAL DEPOSITS

When a glacier melts, the rock materials carried in the ice and on its surface are deposited to form a variety of landscape features. Large ridges and mounds of rock debris carried by a moving glacier or left behind by a melting glacier are called **moraines**. The rock materials carried in different parts of the glacier form different types of moraines. The four kinds are *lateral moraines*, *medial moraines*, *terminal moraines*, and *ground moraines*.

**Lateral Moraines.** As shown in Fig. 7-11, **lateral moraines** form along the sides of a glacier.

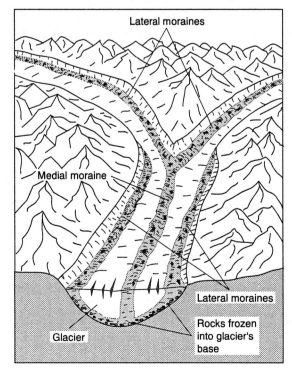

**Fig. 7-11. Glacial moraines.**

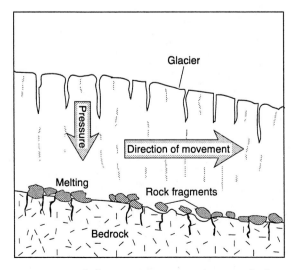

**Fig. 7-9. Rock fragments can penetrate a glacier.**

A lateral moraine consists of rock fragments that the glacier plucked from the valley walls and rocks that tumbled down from the valley walls onto the glacier's surface. When a glacier melts, the lateral moraines are left behind as ridges along the sides of the valley.

**Medial Moraines.**   When two glaciers merge and form a single, larger glacier, the inner lateral moraines of the glaciers combine in the center of the larger glacier to form a **medial moraine** (*medial* means "in the middle of"). When the glacier melts, the torrents of meltwater dismantle the medial moraine so that it is usually not preserved as a recognizable landform.

**Terminal Moraines.**   A glacier advances until its leading edge melts as fast as ice comes from behind to replace it. The front of the glacier then remains stationary. However, the ice within the glacier keeps moving forward. Consequently, rock material in the ice is carried forward as though on a conveyer belt and deposited along the front edge of the glacier. In this way, the glacier gradually builds up a long ridge of rock debris called a **terminal moraine** (see Fig. 7-12). This ridge marks the glacier's farthest advance.

Terminal moraines may be over one hundred meters high, several kilometers wide, and hundreds of kilometers long. For example, much of Long Island, New York, consists of two large terminal moraines. As shown in Fig.

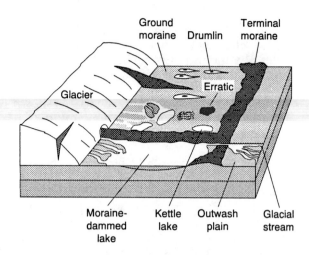

**Fig. 7-12. Some features formed by glacial deposition.**

7-13, the Harbor Mill Moraine and the Ronkonkoma Moraine cover most of central and northern Long Island.

**Ground Moraines.**   When the main body of a glacier melts, all the rock materials within and on top of the glacier are released and fall to the ground. This blankets the land surface with an irregular deposit of rock debris called a **ground moraine.** The melting of an entire continental glacier leaves behind an extensive ground moraine.

**Outwash Plains.**   Meltwater pouring from the front of a glacier carries sediments away from

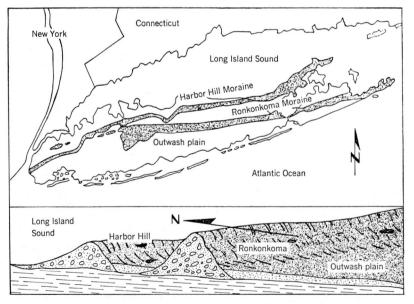

**Fig. 7-13. Terminal moraines of Long Island, New York.**

the terminal moraine. The coarser sediments are deposited close to the terminal moraine, forming a sloping mound of boulders, pebbles, gravel, and coarse sands. Finer sediments, such as sand and silt, are carried farther from the moraine by the meltwater.

Large quantities of these fine sediments form relatively flat areas called **outwash plains** (see Fig. 7-12). An outwash plain may extend for many kilometers in front of a terminal moraine. The southern half of Long Island is an outwash plain that consists mainly of sand (see Fig. 7-13).

**Erratics.** Large boulders, some as big as a house, are often carried great distances by a glacier. When the glacier melts, these boulders may come to rest on bedrock of much different composition than the boulders. These glacially transported boulders are called **erratics.** In New York's Central Park, many erratics made of basalt can be found atop bedrock made of schist. These basalt boulders were carried by moving ice from a stretch of the Palisades Cliffs northwest of New York City.

**Drumlins.** Small, rounded, oval-shaped hills that consist of glacial deposits are called **drumlins.** Most drumlins are about 25 m high, 100 m wide, and 400 m long. Drumlins are fairly common in southeastern Wisconsin, western New York State, parts of Minnesota, and near Boston, Massachusetts. Bunker Hill, a famous landmark of the American Revolution in Boston, is a drumlin.

**Kettles.** When a glacier melts back, it sometimes leaves large blocks of ice behind. The blocks of ice are often buried by outwash material or ground moraine. When the ice block melts, the space it occupied remains as a depression in the ground. These depressions are called **kettles,** or *kettle holes* (see Fig. 7-12).

## *LAKES FORMED BY GLACIERS*

A melting glacier leaves behind many depressions in the ground in which water can collect. Depending on their size and depth, these depressions fill with water to form swamps, ponds, or lakes. The largest of these bodies of water are called *kettle lakes, moraine-dammed lakes,* and *cirque lakes.*

**Kettle Lakes.** If a kettle hole fills with water, a **kettle lake** is formed. Kettle lakes are common in New England, New York, Michigan, Minnesota, Wisconsin, and Canada.

**Moraine-dammed Lakes.** Sometimes a melting glacier deposits a moraine across a valley. The moraine acts as a dam, blocking the flow of streams or glacial meltwater through the valley. In time, the dammed valley fills with water to form a lake. The Finger Lakes and Lake George in New York, and Long Lake in Maine, are examples of **moraine-dammed lakes** (see Fig. 7-14).

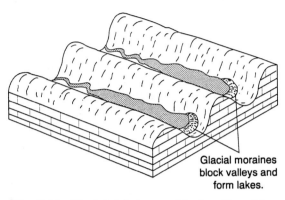

Glacial moraines block valleys and form lakes.

**Fig. 7-14. Glacial debris may block valleys like dams, forming long "finger" lakes.**

**Cirque Lakes.** You recall that a cirque is a large bowl-shaped basin carved by a glacier into the side of a mountain. Eventually, the glacier melts, and the cirque fills with water, producing a **cirque lake.**

## *ICE AGES*

As you learned at the beginning of this chapter, a continental glacier covered a large part of North America during the most recent ice age (refer back to Fig. 7-3). Geologists think that there have been at least four ice ages, separated by periods of warmth, during the past one million years.

The theory that ice ages alternated with periods of warmth is supported by the remains of plants and animals found in moraines and outwash plains. Many of the remains are those of plants and animals that lived only in warm climates. Thus, there must have been long periods of warm weather between the ice ages.

Dating of glacial deposits indicates that the last ice age ended about 12,000 years ago. Some geologists think that Earth is now in one of the periods of warmth that occur between ice ages. At some future time, our climate may gradually grow colder again as yet another ice age begins.

# Wind

In dry environments, such as deserts, very few plants cover the ground. For example, in parts of the southwestern United States, deposits of loose sand, silt, and clay lie exposed at the surface. Wherever these fine materials lie uncovered at the surface, the wind is an important agent of erosion.

## WIND EROSION

The wind changes the landscape by moving sand, silt, and clay from one place to another. As shown in Fig. 7-15, the lightweight silt and clay particles can be lifted and carried by light breezes. However, only strong winds can carry and move the heavier sand grains.

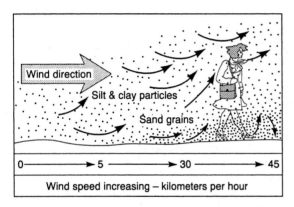

**Fig. 7-15. Transportation of loose sediments by the wind.**

Strong winds lift silt and clay particles high into the atmosphere and carry them long distances. Heavier sand grains, however, rarely are lifted more than one meter off the ground and travel relatively short distances. On a windy day at the beach, you may have noticed while standing that most sand grains strike you below knee level. Even during violent desert windstorms, most of the flying sand grains do not rise more than one or two meters above the ground.

**Wind Abrasion.**  Flying sand grains erode pebbles, boulders, and even large rock outcrops by **wind abrasion.** During wind abrasion, flying sand grains sandblast, or chip away, any solid surface that they are blown against. In some windy areas of the United States, the bases of wooden telephone poles have been completely sandblasted away by wind abrasion. On a longer time scale, isolated rock outcrops have been carved into strange, mushroomlike shapes by windblown sand.

**Loss of Topsoil.**  You learned in Chapter 4 that a long drought in the 1930s led to wind erosion of the fertile topsoil of the Plains states. Crops withered and died, leaving bare fields unprotected from the wind. Moist topsoil turned into powdery dust, and farmers were helpless as the wind stripped the precious topsoil from their fields. On several occasions, strong winds carried thick clouds of dustlike topsoil from the Great Plains to the Atlantic Ocean—over 3000 km away. Because of the disastrous loss of topsoil during such dust storms, the region from Texas to South Dakota was called the "Dust Bowl."

## DEPOSITS OF WINDBLOWN MATERIAL

Sand, silt, and clay particles are deposited when the wind slows down or is blocked by an object. Two deposits often formed by windblown sediments are *sand dunes* and *loess*.

**Sand Dunes.**  Boulders, rock outcrops, and isolated shrubs are often found in regions that have a great deal of loose sand. Such objects obstruct the wind, allowing windblown sand to be deposited. For example, a boulder blocks the wind, slowing the wind velocity in the immediate area. Consequently, sand grains suddenly fall to the ground and pile up around the boulder (see Fig. 7-16a). In time, the sand grains form a sand hill, or **sand dune.**

As shown in Fig. 7-16b, the side of a sand dune facing into the wind is called the *windward* side. The windward side of a sand dune has a gentler slope than the side facing away from the wind, or the *leeward* side. For example, winds blowing from the north will form dunes with gentle north slopes and steep south slopes. In places where the wind blows steadily from one direction, crescent-shaped sand dunes form (see Fig. 7-16c).

Sand dunes range in height from one m to more than 100 m. In the Sahara Desert there are sand dunes more than 200 m tall. Tall dunes also are found in the Great Sand Dunes National Monument in Colorado.

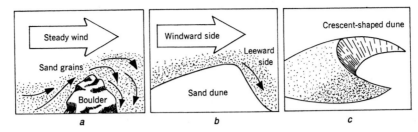

**Fig. 7-16. Formation of sand dunes.**

The wind not only forms sand dunes but also moves dunes from place to place. As the wind blows against the windward side of a sand dune, some sand is blown over the top, falling on the leeward side. Gradually, the entire dune is moved in the leeward direction. Entire farms, towns, and forests have been buried by advancing sand dunes. However, in regions where grasses and shrubs grow on the dunes, the movement of sand dunes is checked.

Most dunes are made of quartz sand. A notable exception is the White Sands National Monument in New Mexico, which has beautiful white dunes made of gypsum sand. Another exception is coral islands, such as Bermuda, which have dunes made of calcite sand. The calcite sand is produced by the weathering of coral reefs that make up the islands.

**Loess.** Unlayered deposits of windblown silt are called **loess** (pronounced *less*). Loess particles are angular and may be derived from many types of rocks. Loess weathers to form very fertile topsoil. The topsoil of several Midwestern states, including Kansas, Iowa, Missouri, and Illinois, consists largely of loess deposits. On some Illinois farmlands, fertile loess deposits are more than 90 m thick.

Loess often consists of glacial rock flour. When glaciers melt and the meltwater drains away, rock fragments of various sizes are left exposed on the ground. The wind picks up the lightweight particles of rock flour, leaving behind the heavier materials. The windblown rock flour eventually settles to the ground, forming deposits of loess.

Loess is an unusual kind of sediment. Most deposits of loose sediment collapse to form slopes when they are cut into by streams or by road builders. Loess deposits, however, form steep cliffs when they are cut into, because the angular loess particles stick together. Loess cliffs are often found along the banks of the upper Mississippi and Missouri rivers.

# CHAPTER REVIEW

## *Science Terms*

*The following list contains all of the boldfaced words found in this chapter and the page on which each appears.*

cirques (p. 93)
cirque lake (p. 95)
continental glaciers (p. 91)
drumlins (p. 95)
erratics (p. 95)
glaciers (p. 89)
ground moraine (p. 94)
horns (p. 91)
ice age (p. 89)
kettles (p. 95)
kettle lake (p. 95)

lateral moraines (p. 93)
loess (p. 97)
medial moraine (p. 94)
moraines (p. 93)
moraine-dammed lake (p. 95)
outwash plains (p. 95)
sand dune (p. 96)
terminal moraine (p. 94)
valley glaciers (p. 91)
wind abrasion (p. 96)

# Matching Questions

*On the blank line, write the letter of the item in column B that is most closely related to the item in column A.*

| Column A | Column B |
|---|---|

_____ 1. terminal moraine

_____ 2. erratic

_____ 3. outwash plain

_____ 4. horn

_____ 5. drumlin

_____ 6. kettle

*a.* boulder dropped by a glacier
*b.* pyramid-shaped mountain formed by glacial erosion
*c.* hole in the ground left by a melting block of ice
*d.* small, oval-shaped hill formed by a glacier
*e.* large scratches in bedrock formed by boulders dragged along by a glacier
*f.* deposit of rock fragments along the front edge of a glacier
*g.* relatively flat deposit composed of rock debris carried away from a glacier by meltwaters

# Multiple-Choice Questions

*On the blank line, write the letter preceding the word or expression that best completes the statement.*

**1.** Valley glaciers move slowest
   *a.* after winters of heavy snowfall      *c.* in the summer
   *b.* in the center of the glacier          *d.* along their sides                    1 _____

**2.** If the continental glaciers covering Greenland and Antarctica melted completely, the sea level would rise about
   *a.* 6 m   *b.* 60 m   *c.* 600 m   *d.* 6000 m                                      2 _____

**3.** The rock fragments carried along the sides of a glacier are called
   *a.* ground moraine   *b.* lateral moraine   *c.* medial moraine   *d.* drumlins     3 _____

**4.** The long, deep grooves left in bedrock that a glacier passed over were probably produced by
   *a.* sharp, knifelike chunks of ice
   *b.* boulders trapped at the bottom of the ice
   *c.* sand grains frozen together at the bottom of the ice
   *d.* swift rivers produced by meltwater pouring out of a glacier                     4 _____

**5.** A mountain carved into a pyramidlike shape by the combined action of frost wedging and glaciers is called a
   *a.* frost peak   *b.* single pyramid   *c.* drumlin   *d.* horn                     5 _____

**6.** Long ridges produced along the leading edges of glaciers are called
   *a.* terminal moraines      *c.* sand and rock tunnels
   *b.* erratic ridges          *d.* glacial dunes                                      6 _____

**7.** Large "out-of-place" boulders deposited by melting glaciers are called
   *a.* erratics   *b.* drumlins   *c.* ground boulders   *d.* outwash boulders         7 _____

**8.** A hole that remains in the ground after a block of glacier ice melts is called a
   *a.* pothole   *b.* sinkhole   *c.* kettle   *d.* glacier lake                       8 _____

**9.** Most sand dunes are composed of
   *a.* calcite sands   *b.* gypsum sands   *c.* quartz sands   *d.* feldspar sands      9 _____

# Modified True-False Questions

*In some of the following statements, the italicized term makes the statement incorrect. For each incorrect statement, write the term that must be substituted for the italicized term to make the statement correct. For each correct statement, write the word "true."*

1. The southern edge of the continental glacier that covered the United States about *12,000* years ago extended about as far south as the Ohio and Missouri rivers.

   1 _____

2. *Valley* glaciers form when huge portions of Earth's surface are covered with ice.

   2 _____

3. The pressure of a thick mass of snow causes the underlying snow to *freeze.*

   3 _____

4. Glaciers usually move *faster* over steep slopes than over gentle slopes.

   4 _____

5. The farthest that a glacier can advance in one day is about *3 m.*

   5 _____

6. The rock debris deposited by a glacier is called *moraine.*

   6 _____

7. Rocks become locked in the bottom of a glacier by *melting* their way into the mass of ice.

   7 _____

8. Part of the work performed by a glacier is *polishing* the underlying bedrock.

   8 _____

9. Valleys that have been carved by glaciers are *U-shaped.*

   9 _____

10. Rock debris deposited at the front of a glacier is called a *terminal moraine.*

    10 _____

11. The relatively flat plain formed from sediments deposited by meltwater in front of a glacier is called an *outwash plain.*

    11 _____

12. Wind erosion is most effective in *grassland* regions.

    12 _____

13. During a windstorm, sand grains are lifted as high as *7 m* into the air.

    13 _____

14. The region extending north from Texas to South Dakota (the Great Plains) was once known as the Dust Bowl because of the disastrous loss of *topsoil* caused by wind erosion.

    14 _____

15. Most sand dunes consist of either quartz, calcite, or *gypsum* particles.

    15 _____

# Testing Your Knowledge

1. How does a continental glacier form? _____

   _____

2. How can a glacier resting on a flat surface move across the land? _____

   _____

**3.** What causes a layer of water to form underneath a glacier? _____

_____

**4.** Why would a glacier sometimes be called "frozen sandpaper"? _____

_____

_____

**5.** Name and describe two landforms produced by glacial erosion. Name and describe two landforms produced by glacial deposition.
*Glacial erosion:*

a. _____

b. _____

*Glacial deposition:*

a. _____

b. _____

**6.** Geologists think that there have been at least four different advances and retreats of continental glaciers during the last one million years. What evidence do geologists have to support

this hypothesis? _____

_____

**7.** What conditions led to the creation of the Dust Bowl in the Plains states during the 1930s?

_____

_____

**8.** Draw three diagrams that show how wind erosion and deposition form sand dunes and cause
the dunes to "migrate."

# Movements of Earth's Crust

## LABORATORY INVESTIGATION

### *LOCATING EARTHQUAKE ZONES*

**A.** Some places where earthquakes have occurred frequently are listed below. These places are shown on the map (see Fig. 8-1, on the next page). Locate each place on the map. Then, shade the circle near each place.

### PLACES HAVING FREQUENT EARTHQUAKES

| | | | |
|---|---|---|---|
| Acapulco | Concepción | Java | Portland |
| Aleutian Islands | Costa Rica | Juneau | San Francisco |
| Anchorage | Ecuador | Kamchatka | Shanghai |
| Arica | Fiji Islands | Magellan Straits | Sumatra |
| Auckland | Hokkaido | New Guinea | Yokohama |

**B.** Draw a line connecting the shaded circles you have filled in.

    **1.** Describe any pattern or patterns you see in this region of the world. _____

_____

_____

**C.** Use a map of the world or world atlas to locate the volcanoes listed below. Place an "X" on the map in Fig. 8-1 at the approximate location of each volcano.

| | |
|---|---|
| Mount Fuji (Japan) | Cotopaxi (Ecuador) |
| Osorno (Chile) | Krakatoa (Java) |
| El Misti (Peru) | Parícutin (Mexico) |
| Mount St. Helens (Washington) | Popocatépetl (Mexico) |
| Mount Pinatubo (Philippines) | Mount Shasta (California) |
| Mount Ruapehu (New Zealand) | Mount Redoubt (Alaska) |

**Fig. 8-1. Partial world map.**

**2.** Do you see any relationship between the location of these volcanoes and earthquake zones?

Explain. _____

_____

_____

**3.** What can you conclude about the condition of Earth's crust in regions where both earth-

quakes and volcanic eruptions are occurring frequently? _____

_____

_____

In the preceding chapters, you learned that external agents such as water, wind, and ice cause weathering and erosion, wearing down the land. If these were the only forces that shape Earth's surface, the land would long ago have been reduced to a low, flat plain. How-ever, Earth is a dynamic planet; powerful forces at work inside Earth cause earthquakes, volcanic eruptions, and other movements of the crust. These crustal movements produce changes at the surface that, in general, tend to raise the level of the land.

# Earthquakes

Earth's crust is continually moving—up, down, and sideways. Most of these movements are so slight that people are unaware of them. Sometimes, however, there is a sudden move-ment of the crust, accompanied by noticeable shaking of the ground. This event is called an earthquake. Strong earthquakes have destroyed entire cities in minutes. In this century alone, more than one million people have died as a result of earthquakes.

## WHAT CAUSES EARTHQUAKES?

Geologists have discovered the cause of earthquakes. They also have found a way to predict where earthquakes are most likely to occur.

Forces within Earth continually push, pull, and twist the rocks that make up the crust. When these forces become too great, the crust breaks, or fractures (see Fig. 8-2). For comparison, consider what happens when you try to break a twig. First, the twig begins to bend. If you bend the twig far enough, it snaps, and the bent ends suddenly straighten out. The ends of the twig vibrate rapidly as they become straight, producing a snapping noise.

Likewise, when Earth's crust is bent until it suddenly fractures, the broken sections of crust vibrate as they become straight again. This vibrating of the crust is an earthquake.

**Faults.** A fracture in Earth's crust along which sections of the crust slide past each

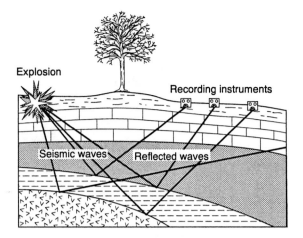

**Fig. 8-3. Seismic waves can reveal the structure of the crust.**

other is called a **fault.** The movement of the rocks along a fault is called *slippage.*

Slippage along a fault temporarily relieves stress in the crust. The rocks along the fault may then remain locked together until stress again builds up to the breaking point, causing another earthquake. Segments of an active fault that have been locked together for a long time are thus prime candidates for future earthquakes. Efforts to predict earthquakes are based largely on this idea.

Most earthquakes occur along faults, although some earthquakes are caused by erupting volcanoes. In addition, researchers intentionally create small earthquakes by setting off underground explosions or by dropping large weights on the ground.

The explosions and the falling weights generate vibrations in the crust called *seismic waves* (see Fig. 8-3). Instruments on the surface record the seismic waves when they are reflected back upward by different rock layers. By studying these records, geologists can determine the structure of the crust.

## EARTHQUAKE REGIONS

Earthquakes can occur anywhere in the world. However, most earthquakes take place in regions called *earthquake belts.* You determined the location of a major earthquake belt around the Pacific Ocean in your laboratory investigation. Another major earthquake belt runs through the Mediterranean Sea, across southern Asia to Indonesia (see Fig. 8-4). Many active volcanoes and young mountain ranges are located within these earthquake belts.

The states of Alaska and California lie in the Pacific earthquake belt. Both states have had and will continue to have strong earthquakes.

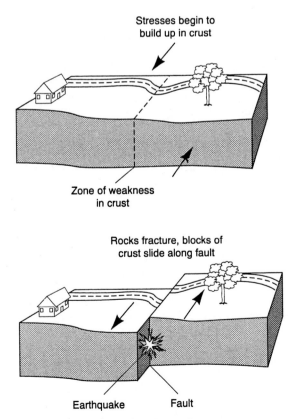

**Fig. 8-2. The breaking of Earth's crust causes an earthquake.**

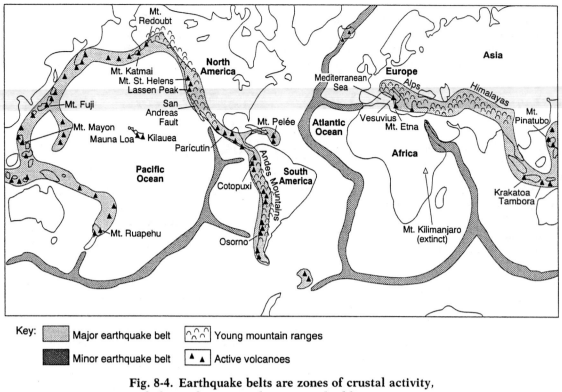

Key:  ▨ Major earthquake belt    ⌒⌒⌒ Young mountain ranges

▰ Minor earthquake belt    ▲ ▲ Active volcanoes

**Fig. 8-4. Earthquake belts are zones of crustal activity, including vulcanism and mountain-building.**

Many of California's earthquakes are caused by slippage of the crust along the *San Andreas Fault* (Fig. 8-5). This major fault extends for over 1000 kilometers (km) through the state. Alaska's earthquakes are caused by movement of the Pacific seafloor as it slides beneath the crust of Alaska.

**Fig. 8-5. The San Andreas Fault in California.**

## EFFECTS OF EARTHQUAKES

**Earthquakes on Land.**   The crustal movements that occur during strong earthquakes can sometimes be observed on land. For example, the San Francisco earthquake of 1906 produced a tear in Earth's surface more than 400 km long. The displacement of fences and roads that crossed the San Andreas Fault showed that the crust had shifted horizontally by about 6 meters (m).

Earthquakes may also involve vertical crustal movements. During the Alaskan earthquake of 1964, large areas of land were uplifted or dropped down several meters. In fact, some sections of the seafloor along the coast were raised above sea level, becoming dry land.

Strong earthquakes can be deadly. Crustal movements in large quakes can destroy buildings and trigger landslides, causing devastating loss of life. A single earthquake in China in 1976 is believed to have killed more than half a million people.

You can see some effects of a large earthquake in Fig. 8-6. In the photograph, a wrecked building rests on a section of ground

**Fig. 8-6. Earthquake damage in Alaska.**

that dropped about 4 m below the ground at the left. Earthquake damage to buildings and other structures depends largely on the *intensity* of the earthquake and the *type of material* on which the structures are built.

A structure built on solid bedrock is less likely to be damaged by an earthquake than one built on loose sediments. This is because seismic waves are amplified as they pass through loose sediments, causing the sediments to shake more violently than solid bedrock. Many of the structures that collapsed during the San Francisco earthquake of 1989, including the double-decked Nimitz Freeway, were built on loose sediments.

You can easily simulate how an earthquake affects loose sediments. Spread a thin layer of sand on top of a heavy table, and then tap the tabletop sharply with a hammer. The loose sand grains will jump about violently, while the tabletop will only vibrate slightly.

**Earthquakes Under Water.** Earthquakes that occur on the seafloor may produce fast-moving ocean waves called **tsunamis.** Tsunamis are also caused by underwater volcanic eruptions and landslides. A tsunami can cross an entire ocean, traveling at speeds of up to 800 km per hour.

On the open ocean, tsunamis are usually no higher than ordinary waves, so they often go unnoticed. However, when such waves reach shallow coastal waters, they pile up before striking the shore. Some tsunamis are more than 20 m high when they strike land. The largest of these waves on record was about 85 m high!

A tsunami is often more destructive than the geological event that produced it. In 1896, a tsunami nearly 30 m high hit the coast of Japan. Thousands of houses were swept away and more than 25,000 people were killed. In 1946, a tsunami that struck Hawaii tore railroad tracks from their beds, washed houses out to sea, and took 159 lives.

A tsunami usually gives warning before it strikes. Minutes before a large tsunami arrives, the water near the shore suddenly drains seaward, exposing parts of the beach that are normally submerged. Then, with a loud hissing sound, the wave rushes across the waterless beach and strikes the shore with a great roar. Following waves may pound the shore for several hours.

## EARTHQUAKES AND EARTH'S INTERIOR

When rocks fracture and slip along a fault, Earth's crust vibrates. The vibrations travel through Earth as seismic waves (see Fig. 8-7). Some seismic waves travel along the surface. Others travel through Earth's interior. Both types of seismic waves move outward in all directions from the source of the earthquake.

The different types of seismic waves travel

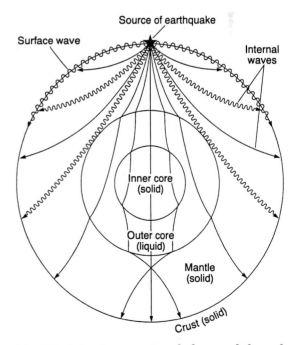

**Fig. 8-7. Seismic waves travel along and through Earth.**

at different speeds. In addition, the speed of each wave type varies with the density of the material through which it travels. The more dense the material, the faster a seismic wave moves through it. The path that a seismic wave follows through Earth depends on how rock layers of different densities are arranged within Earth.

Geologists have gained their knowledge of Earth's internal structure and composition by studying how seismic waves travel through Earth's interior. As you have learned, these studies indicate that Earth has a solid inner core, a liquid outer core, a thick mantle of dense rock, and a thin crust of lighter rocks.

**Seismograph.** Seismic waves from earthquakes can be detected and recorded by using an instrument called a **seismograph** (see Fig. 8-8). The main part of a seismograph is a heavy, suspended weight, with a pen attached to it. The tip of the pen touches a chart wrapped around a drum that slowly rotates. The seismograph is firmly anchored to the bedrock. When an earthquake takes place, vibrations in the bedrock cause the drum to vibrate against the pen, which remains stationary. Thus, the pen traces a record of the earthquake, called a *seismogram,* on the chart (see Fig. 8-9).

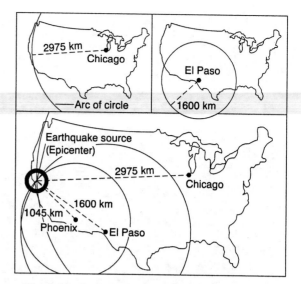

**Fig. 8-10. Locating an earthquake's epicenter.**

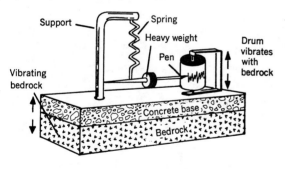

**Fig. 8-8. A seismograph.**

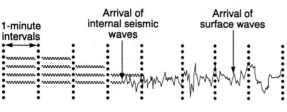

**Fig. 8-9. A seismogram.**

A seismograph can detect an earthquake that occurs thousands of kilometers away. The point where rocks first fracture is the source of seismic waves. This point, located underground, is called the **focus** of the earthquake. The point on Earth's surface directly above the focus is the **epicenter.**

Because different types of seismic waves travel at different speeds, some waves from an earthquake reach a seismographic station before others do. *Seismologists* (scientists who study earthquakes) can use this fact to locate the earthquake's epicenter.

For example, by timing the arrivals of different seismic waves, seismologists in Chicago can determine the *distance* of an earthquake (but *not* its *direction*) from their recording station. This distance is then used as the radius of a circle drawn around the station on a map. The earthquake's epicenter lies somewhere on this circle.

Seismologists at recording stations in El Paso, Texas, and Phoenix, Arizona, can also calculate their distances from the earthquake and draw circles around their stations on a map. When all three circles are combined on one map, the circles intersect at a single point (see Fig. 8-10). That point is the earthquake's epicenter.

The focus of an earthquake can be determined in a similar way. Most earthquakes occur within 50 km of Earth's surface. However, some deep earthquakes occur nearly 700 km underground.

# Vulcanism

**Vulcanism** is the movement of magma through the crust or its emergence as lava onto Earth's surface. An opening in the crust through which molten materials and rock particles reach the surface and pile up is called a **volcano.** A *volcanic mountain*, which is built up by successive deposits of volcanic materials, is also commonly called a volcano.

A volcano may be *active* (presently erupting), *dormant* (between eruptions), or *extinct* (no longer capable of erupting). Most of the world's active volcanoes are found along the rim of the Pacific Ocean, which is also a major earthquake zone. This curving belt of active faults and volcanoes is often called the *Ring of Fire.*

## VOLCANIC ERUPTIONS

At the top of a volcano there is a bowl-shaped depression called a **crater.** Within the crater is an opening called a **vent.** The vent of a dormant volcano is usually blocked by a thin layer of solidified lava. When the molten lava inside the volcano builds up enough pressure, the molten rock breaks through the thin crust of hardened lava, and the volcano *erupts.*

There are different types of *volcanic eruptions.* During a *quiet eruption,* lava flows freely from the volcano's crater or from fissures in the volcano's sides. The lava may flow for many kilometers down the slopes of the volcano. In an *explosive eruption,* molten rock explodes forcefully from the volcano in billowing clouds of volcanic ash and cinders.

A volcanic eruption can be both spectacular and devastating. Lava flows and glowing clouds of ash and toxic gases from volcanoes have destroyed entire cities. In 1902, the eruption of Mount Pelée on the Caribbean island of Martinique killed 30,000 people. In 1883, one of the most violent eruptions ever recorded took place on an Indonesian island called Krakatoa. The sound of this explosive eruption was heard more than 4800 km away. Nearly two thirds of the island was blown away by the blast.

## VOLCANIC MOUNTAINS

Vulcanism has produced some of the world's best-known mountains. Mount Etna

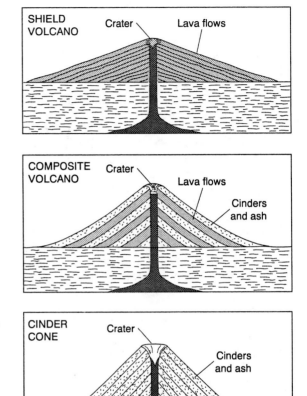

Fig. 8-11. Types of volcanoes.

in Sicily and Mount Kilimanjaro in Africa are volcanic mountains. The Hawaiian islands are actually the tops of a chain of volcanic mountains that rise from the ocean floor.

There are three types of *volcanic mountains: shield volcanoes, cinder cones,* and *composite volcanoes* (see Fig. 8-11). These volcanoes differ in size, shape, and the way they are formed.

**Shield Volcanoes.** When lava emerges from a volcanic vent or fissure in a quiet eruption, the freely flowing lava spreads out over Earth's surface until it cools and hardens into a layer of igneous rock. In time, repeated lava flows build a broad, massive mountain with gentle slopes, called a **shield volcano.** Shield volcanoes are the largest type of volcanic mountain.

Mauna Loa and Mauna Kea are the largest of several huge shield volcanoes that make up the island of Hawaii. Both of these volcanoes rise almost 10 km from the ocean floor, reaching more than 4 km above sea level. Mauna Loa is the largest volcano on Earth.

**Cinder Cones.** In explosive eruptions, lava is hurled high into the air in a spray of droplets of various sizes. These droplets cool and harden into cinders and particles of ash before falling to the ground. The ashes and cinders pile up around the volcanic vent to form a steep, cone-shaped hill called a **cinder cone.** Cinder cones are relatively small volcanoes, rarely rising more than 500 m high.

Cinder cones may form quite rapidly. In February 1943, ashes and cinders began to spew from a fissure in a cornfield near the village of Parícutin, Mexico. In just six days, a cinder cone over 150 m high had replaced the cornfield. Seven months later, the cinder cone was about 450 m high, and the surrounding region was buried beneath a layer of ash nearly 1 m thick.

**Composite Volcanoes.** A volcanic mountain built up by both lava flows and layers of ash and cinders is called a **composite volcano.** Although not as big as shield volcanoes, composite volcanoes are generally much larger than cinder cones.

Many composite volcanoes are quite famous, either for their classic, cone-shaped peaks, or their powerful, explosive eruptions. Some well-known composite volcanoes are Mount Fuji in Japan, Mount Vesuvius in Italy, and Mount St. Helens in the United States.

## DOME MOUNTAINS

Sometimes a large quantity of magma pushes up through the crust but fails to break through to the surface. Instead, the magma forces the overlying rock layers to arch up-

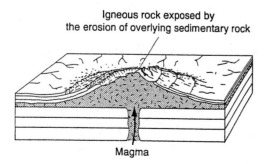

Igneous rock exposed by
the erosion of overlying sedimentary rock

Magma

**Fig. 8-12. A dome mountain.**

ward, forming a large **dome mountain** (see Fig. 8-12). A dome mountain resembles a huge blister on Earth's surface. In time, the top layers of rock are worn away, exposing the solidified magma. The Henry Mountains of Utah and the Orange Mountains of New Jersey are dome mountains.

## VULCANISM INSIDE THE CRUST

In addition to surface features like volcanoes and dome mountains, vulcanism also produces underground geologic features. Magma that remains deep within the crust eventually hardens into igneous rock, forming batholiths, sills, and dikes (see Fig. 8-13). These underground structures may be seen if they become exposed at the surface by uplift and erosion of the overlying rocks.

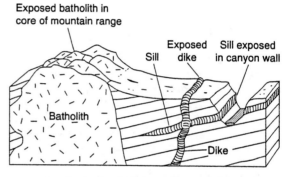

Exposed batholith in
core of mountain range

Exposed
Sill dike

Sill exposed
in canyon wall

Batholith

Dike

**Fig. 8-13. Batholiths, sills, and dikes.**

**Batholiths.** A **batholith** is a huge mass of coarse-grained igneous rock that forms deep in the core of a mountain range. When exposed by uplift and erosion, a batholith may cover thousands of square kilometers. The largest batholith in North America, located in the Coast Ranges of western Canada, is about 2000 km long and 290 km wide. Other extensive batholiths underlie the Sierra Nevada in California and the Rocky Mountains in Idaho.

**Sills.** A **sill** is formed when magma flows between layers of sedimentary rock. The magma hardens into a sheet of igneous rock that is *parallel to* the adjacent rock layers. The Palisades—cliffs of diabase that line the west bank of the Hudson River, opposite New York City—are the exposed remnants of a sill.

**Dikes.** A **dike** forms when magma forces its way into a fracture that *cuts across* sedimentary rock layers and then hardens into igneous rock.

# Slow Movements of the Crust

Not all crustal movements are sudden and violent, like earthquakes and volcanic eruptions. Forces within Earth also cause the crust to move slowly for long periods of time. In fact, most crustal movements are so slow they are barely noticeable even to the most sensitive instruments. Nonetheless, these gradual movements of the crust eventually produce dramatic changes at Earth's surface.

For example, careful surveys indicate that the Adirondack Mountains in northern New York State are currently getting taller by about 3 millimeters per year. This movement is far too slow to be noticed by a human ob-

server. However, if this rate of uplift continues for the next one million years, the mountains will have risen 3 km above their present height.

Some other mountain ranges, such as the Himalayas, also are slowly rising higher. Geologists believe that slow movements of large sections of the crust have produced all the great mountain ranges on Earth. Furthermore, these slow crustal movements also result in the sudden, violent earthquakes and volcanic eruptions that occur in the belts of crustal activity you have learned about. We will return to this subject later in the chapter.

# Landforms Produced by Crustal Movements

Movements of the crust, whether sudden or gradual, cause major changes in Earth's surface features over time. Slow crustal movements may continue relentlessly for millions of years, and sudden movements may recur at fairly regular intervals over a vast stretch of time. In both cases, major landforms are produced. These landforms include mountains, plateaus, and plains.

## MOUNTAINS

A **mountain** is a part of Earth's crust that has been raised high above the surrounding landscape. The tallest mountain in the world, Mount Everest in the Himalayas, rises almost 9 km above sea level. Mountains may be composed of igneous, metamorphic, or sedimentary rocks; some may have rock layers that are tilted or distorted.

There are different types of mountains, produced by different mountain-building processes. You have already learned how volcanic mountains and dome mountains are created. However, most major mountain ranges are formed by the processes of *folding* and *faulting*.

**Fold Mountains.** Crustal movements may press horizontal layers of sedimentary rock together from the sides, squeezing them into wavelike folds. These folds may be compared to wrinkles in a carpet that has been pushed together from opposite ends. Mountains produced by the folding of rock layers are called **fold mountains.**

Fold mountains contain upfolded sections of rock, called *anticlines,* and downfolded sections of rock, called *synclines* (see Fig. 8-14). The long ridges and valleys of the Appalachian Mountains are a series of anticlines and synclines formed by folded rock layers.

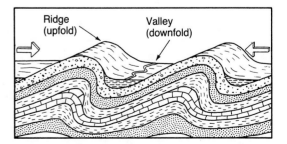

**Fig. 8-14. Fold mountains.**

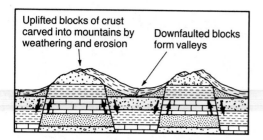

**Fig. 8-15. Fault-block mountains.**

**Fault-Block Mountains.** Some crustal movements cause the crust to fracture into huge blocks. These blocks of crust shift against each other, moving up or down along the faults that separate them. Crustal blocks that are pushed or tilted upward form **fault-block mountains,** whereas blocks that sink downward form *valleys* (see Fig. 8-15).

The Sierra Nevada in California is a large chain of fault-block mountains. Some parts of the uptilted Sierra Nevada fault block rise more than 3 km above the adjacent valleys. Smaller ranges of fault-block mountains are found in the *Great Basin* region of Utah and Nevada.

## PLATEAUS

A **plateau** is a broad, flat-topped region, underlain by horizontal layers of rock, that rises somewhat abruptly above the surrounding landscape. A plateau forms when a large section of the crust is uplifted without much folding or faulting of its rocks. The *Appalachian Plateau*, which lies west of the Appalachian Mountains along much of their length, was produced in this way.

In the western United States, the *Colorado Plateau* was formed when crustal movements gradually raised a large area of land about 2 km above sea level. As the land slowly rose, the Colorado River carved the Grand Canyon deep into the rock layers that make up the plateau (see Fig. 8-16).

A plateau that has undergone prolonged erosion may resemble a mountain range. For example, the Catskill Mountains in New York State are the remains of a plateau that has been deeply eroded by streams. However, mountains carved from a plateau can be distinguished from true mountains because the rock layers in a plateau are essentially horizontal, whereas the rock layers in a true mountain range are folded or steeply tilted.

## PLAINS

Like a plateau, a **plain** is a broad, flat region, composed of horizontal layers of rock or sediment. However, a plain lies lower than the land around it. Different types of plains are named for the kinds of sediments or rocks of which they are made. For example, there are marine plains, lake plains, and glacial plains.

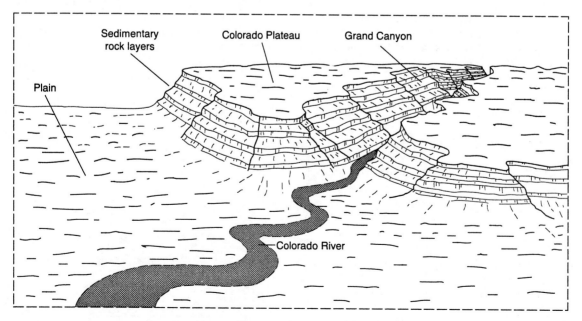

**Fig. 8-16. The Grand Canyon in Arizona.**

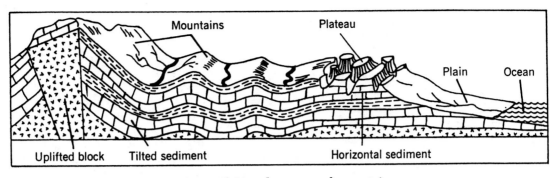

**Fig. 8-17. Plains, plateaus, and mountains.**

**Marine Plains.** Flat layers of sedimentary rocks formed on the seafloor and then slowly raised above sea level (without being folded or tilted) become *marine plains*. The *Atlantic Coastal Plain* is a marine plain that extends along the east coast of the United States, from New Jersey to Florida. The *Great Plains,* located in the interior of the United States and Canada, are marine plains that rose from a shallow inland sea. This sea once stretched from Hudson Bay in Canada down to the Gulf of Mexico.

*Lake plains* are the dried-up beds of lakes that have drained away. *Glacial plains,* also called *outwash plains* (see Chapter 7), are deposits of sediment left behind by melting glaciers. These two types of plains are not formed by crustal movements, but result mainly from climate changes and their effects on bodies of water and ice.

Fig. 8-17 shows the relationships between the major landforms produced by uplift of the crust and shaped by weathering and erosion.

# Why the Crust Moves

Geologists have long known that Earth's crust moves. The folded, crumpled rocks often seen exposed in mountainsides and cliffs suggest that these once flat rock layers were distorted by movements of the crust. Sedimentary rocks found in many mountaintops frequently contain fossils of marine organisms, indicating that these rocks were formed on the seafloor and later uplifted by crustal movements. The surface changes caused by earthquakes provide direct proof that the crust moves.

However, the underlying cause of crustal movements was long a mystery to scientists. Only in the past few decades have geologists come to understand why the crust moves.

## CONTINENTAL DRIFT

In 1912, a German scientist named *Alfred Wegener* (1880–1930) proposed that crustal movements have taken place on a global scale. Several scientists had previously noticed that

the continents bordering the Atlantic Ocean seem to fit together like pieces of a jigsaw puzzle. Wegener suggested that all the continents had once been joined in a huge "supercontinent," which he called *Pangaea* (meaning "all earth"). Eventually, Pangaea split up, and the continents drifted apart, forming the Atlantic Ocean. As evidence for this idea, Wegener pointed out that, in addition to similar coastlines, features on the different continents such as rock formations and mountain ranges would match up if the continents were reassembled (see Fig. 8-18).

Wegener's hypothesis, called **continental drift,** was rejected by most scientists because it did not adequately explain how the continents could move across the solid rock of the ocean floor. However, in the 1960s scientists studying the ocean floor made several discoveries that supported the idea of moving continents.

**Seafloor Spreading.** Surveys of the ocean floor had revealed a long mountain chain run-

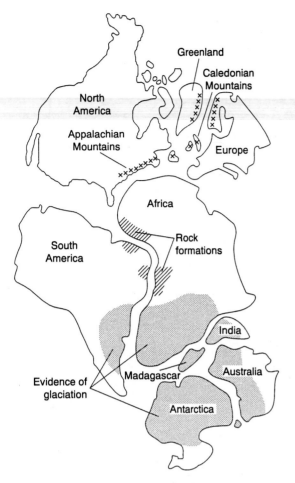

Fig. 8-18. **Evidence that the continents were once joined together.**

ning down the center of the Atlantic Ocean, called the *Mid-Atlantic Ridge.* A narrow, steep-walled valley extends along the crest of the ridge. Using deep-sea diving vehicles, researchers collected samples of ocean crust from the ridge and its surroundings. They found that the crest of the ridge is made of very young volcanic rock, and that the ocean crust becomes progressively older as its distance from the ridge crest increases. In fact, the ages of the crustal rocks on either side of the ridge form a symmetrical, mirror-image pattern (see Fig. 8-19).

These findings led geologists to realize that the seafloor is continually spreading apart at the Mid-Atlantic Ridge, opening a deep, narrow *rift valley* along the crest of the ridge. Molten rock rises into this rift and cools to form new ocean crust.

The discovery of this ongoing process, called **seafloor spreading,** helped solve the problem of how the continents move. The continents, it turns out, do not plow across a stationary ocean crust; rather, the ocean crust itself moves, pushing the continents along.

The Mid-Atlantic Ridge was found to be part of a global system of mid-ocean ridges that winds around Earth like the seams on a baseball. All along this ridge system, new ocean crust is being created in the slow process of seafloor spreading.

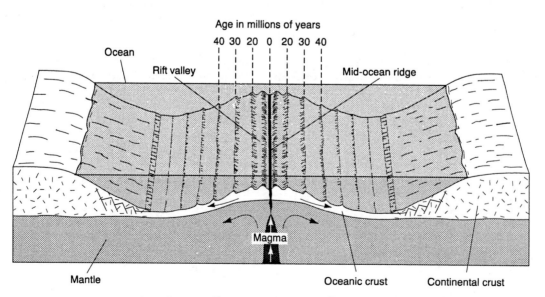

Fig. 8-19. **Seafloor spreading: age of ocean crust increases with distance from mid-ocean ridge.**

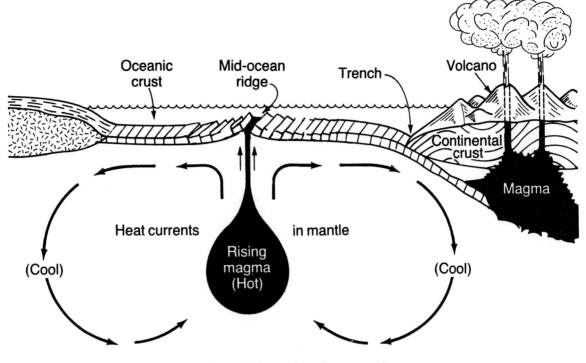

**Fig. 8-20. Effects of seafloor spreading.**

## PLATE TECTONICS

The ideas of continental drift and seafloor spreading have been combined in the modern theory of **plate tectonics.** This theory states that Earth's crust is broken into a number of large, moving slabs, called *plates.* Some plates consist entirely of oceanic crust, whereas others contain both oceanic crust and continental crust.

The plates move over a layer of hot, plasticlike rock in the upper mantle. Geologists believe that heat currents circulating within the mantle cause this plastic zone of rock to slowly flow, carrying along the overlying crustal plates (see Fig. 8-20). Where the currents of hot mantle rock are rising and separating, the crust is pulled apart at a mid-ocean ridge. Where cold mantle rock is sinking, the crust is dragged down at an ocean *trench.* The heat currents in the mantle are thought to be the ultimate cause of most crustal movements.

## PLATE MOTIONS

The movements of crustal plates are very slow, averaging only a few centimeters a year.

Over time, however, these movements produce many major features of Earth's surface, such as mountain ranges, volcanoes, and earthquake zones. Most of these features are located at *plate boundaries,* where the plates interact by spreading apart, pressing together, or sliding past each other.

**Rifting.** Boundaries between spreading plates form where the crust is forced apart in a process called *rifting.* This generally occurs at mid-ocean ridges. Sometimes, rifting takes place within a continent, splitting the continent into smaller landmasses that drift away from each other, thereby forming an ocean basin between them. As seafloor spreading takes place, new material is added to the inner edges of the separating plates. In this way, the plates grow larger, and the ocean basin widens.

This is the process that broke up the supercontinent Pangaea and created the Atlantic Ocean, which is still getting wider at a rate of about 2 centimeters (cm) per year. The early stages of this process may be occurring today in northeast Africa, where the Arabian Plate is splitting away from the African Plate, opening up the Red Sea.

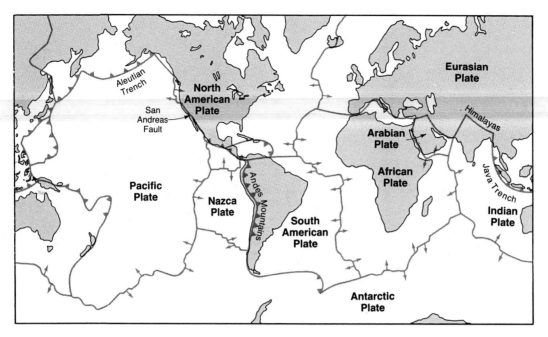

**Fig. 8-21. Earth's major crustal plates.**

**Plate Collisions.** Boundaries between plates that are colliding are zones of intense crustal activity. When a plate of ocean crust collides with a plate of continental crust, the denser oceanic plate slides under the lighter continental plate and plunges into the mantle. This process is called **subduction,** and the site where it takes place is called a **subduction zone.**

A subduction zone is usually seen on the sea-floor as a deep depression called a **trench.** Over time, large sections of ocean crust are consumed and recycled into the mantle by subduction. This makes room for the new ocean crust that is continually being created at mid-ocean ridges.

The collision of plates at a subduction zone squeezes the edge of the upper plate, causing earthquakes and mountain building. The descending plate melts from the intense heat of the mantle, causing vulcanism along the edge of the overriding plate (see Fig. 8-20). For example, the subduction of an oceanic plate beneath the west coast of South America has produced the mountains and volcanoes of the Andes, as well as a deep ocean trench offshore.

When two continents are brought together by colliding plates, the lightweight continental crust cannot be dragged down into the mantle. Instead, the edges of the colliding continents are crumpled and uplifted to form a lofty mountain range. For instance, the Himalayas were formed when plate motions caused what is now India to ram into southern Asia. This slow-motion collision continues today.

In addition to separating and colliding, crustal plates may interact by sliding sideways past each other. This motion produces a plate boundary characterized by major faults that are capable of unleashing powerful earthquakes. The San Andreas Fault forms such a boundary between the Pacific Plate and the North American Plate. The Pacific Plate is moving northwest in relation to the North American Plate at a rate of about 5 cm a year. Along the fault, the plates may creep steadily past each other at this snail's pace; or the edges of the plates may become locked together across the fault, storing energy that is released suddenly during earthquakes.

The theory of plate tectonics explains why most crustal activity occurs in long belts. These zones of earthquakes, vulcanism, and mountain-building mark the boundaries between the crustal plates. Fig. 8-21 shows Earth's major crustal plates. If you compare this figure with Fig. 8-4 on page 104, you will be able to see that the belts of earthquakes and active volcanoes coincide with the plate boundaries.

Although there is still much to learn about the workings of Earth's internal forces, the theory of plate tectonics has provided a powerful framework for understanding crustal movements and the landforms they produce.

# CHAPTER REVIEW

## *Science Terms*

*The following list contains all of the boldfaced words found in this chapter and the page on which each appears.*

batholith (p. 108)
cinder cone (p. 108)
composite volcano (p. 108)
continental drift (p. 111)
crater (p. 107)
dike (p. 108)
dome mountain (p. 108)
earthquake (p. 103)
epicenter (p. 106)
fault (p. 103)
fault-block mountains (p. 110)
focus (p. 106)
fold mountains (p. 109)
mountain (p. 109)

plain (p. 110)
plate tectonics (p. 113)
plateau (p. 110)
seafloor spreading (p. 112)
seismograph (p. 106)
shield volcano (p. 107)
sill (p. 108)
subduction (p. 114)
subduction zone (p. 114)
trench (p. 114)
tsunamis (p. 105)
vent (p. 107)
volcano (p. 107)
vulcanism (p. 107)

## *Matching Questions*

*On the blank line, write the letter of the item in column B that is most closely related to the item in column A.*

*Column A*

_____ 1. earthquake

_____ 2. batholiths

_____ 3. plate tectonics

_____ 4. fault

_____ 5. volcano

_____ 6. tsunamis

_____ 7. slippage

_____ 8. crater

_____ 9. epicenter

_____ 10. subduction zone

*Column B*

*a.* point on Earth's surface above focus of earthquake
*b.* fast-moving ocean waves
*c.* movement of rocks along a fault
*d.* bowl-shaped depression on top of volcano
*e.* noticeable shaking of the ground
*f.* Earth's crust is composed of several large, moving slabs
*g.* opening in crust through which molten rock pours out and accumulates
*h.* a fracture in Earth's crust
*i.* where oceanic plate slides under continental plate
*j.* huge masses of coarse-grained, igneous rock
*k.* deep depression in the ocean floor

# Multiple-Choice Questions

*On the blank line, write the letter preceding the word or expression that best completes the statement.*

1. As seafloor spreading occurs, Earth's crust is pulled apart in a process called
   *a.* rifting  *b.* slippage  *c.* vulcanism  *d.* drifting                                    1 ____

2. Earthquakes usually occur when huge blocks of rock slip along breaks in the bedrock called
   *a.* joints  *b.* faults  *c.* fractures  *d.* zones                                           2 ____

3. The region that is *least* likely to experience strong earthquakes is the
   *a.* Atlantic Ocean                    *c.* Mediterranean Sea
   *b.* Pacific Ocean                     *d.* Asian continent                                    3 ____

4. Geologists can discover the structure of Earth's crust by means of
   *a.* soil samples  *b.* fault lines  *c.* seismic waves  *d.* volcanic eruptions              4 ____

5. The length of the San Andreas Fault in California is about
   *a.* 10 km  *b.* 100 km  *c.* 1000 km  *d.* 2000 km                                            5 ____

6. Tsunamis reach their greatest height when they
   *a.* reach their highest speeds
   *b.* are in the middle of the ocean
   *c.* move onto the land
   *d.* travel across a calm part of the ocean                                                   6 ____

7. The first sign that a tsunami is approaching a shore is
   *a.* a hissing and roaring noise
   *b.* the movement of water away from the shore
   *c.* a sudden flattening of the waves
   *d.* the appearance of a wall of water on the horizon                                         7 ____

8. The study of earthquake seismic waves has helped geologists learn about the structure of Earth's
   *a.* surface  *b.* core  *c.* mantle  *d.* crust                                              8 ____

9. In order to function properly, a seismograph must be
   *a.* attached to bedrock
   *b.* located in the basement of a large building
   *c.* attached to the foundation of a building
   *d.* resting on loose earth                                                                   9 ____

10. The number of circles needed to locate the epicenter of an earthquake is
    *a.* one  *b.* two  *c.* three  *d.* four                                                     10 ____

11. The maximum depth at which earthquakes occur is about
    *a.* 10 km  *b.* 50 km  *c.* 100 km  *d.* 700 km                                              11 ____

12. Evidence that the continents were once joined is found partly in the
    *a.* location of reefs off the continents' coasts
    *b.* direction of river flow along their coasts
    *c.* eruptions of volcanoes along the Ring of Fire
    *d.* shapes of the coastlines of opposing continents                                         12 ____

13. Mountains in which layers of sedimentary rock have been squeezed into wavelike patterns are called
    *a.* fold mountains                    *c.* fault-block mountains
    *b.* dome mountains                    *d.* volcanic mountains                                13 ____

14. An example of a volcanic mountain is
    *a.* Mount Washington       *c.* Mount Kilimanjaro
    *b.* the Matterhorn         *d.* the Sierra Nevada                    14 _____

15. According to the theory of plate tectonics, most crustal movements are caused
    by heat currents within the
    *a.* crust   *b.* core   *c.* mantle   *d.* ocean                    15 _____

# Modified True-False Questions

*In some of the following statements, the italicized term makes the statement incorrect. For each incorrect statement, write the term that must be substituted for the italicized term to make the statement correct. For each correct statement, write the word "true."*

1. The strongest earthquakes are caused by *landslides*.                   1 _____

2. In the United States, earthquakes are most likely to occur near the
   *Pacific* Ocean.                                                        2 _____

3. Regions in which there are both *active volcanoes* and high mountains
   are most likely to experience earthquakes.                             3 _____

4. The Earth's *crust* may move more than 15 m as the result of an earthquake.                                                                 4 _____

5. Destructive waves produced by underwater earthquakes are called
   *tidal waves*.                                                         5 _____

6. The seismic waves generated by earthquakes *cannot* travel through
   parts of Earth that are below the crust.                              6 _____

7. The different types of seismic waves generated by an earthquake
   travel through Earth's interior at *different* speeds.                7 _____

8. The instrument geologists use to detect earthquakes is the *seismograph*.                                                               8 _____

9. The point on Earth's surface under which an earthquake originates
   is called the *epicenter*.                                            9 _____

10. Most earthquakes occur within *50 km* of Earth's surface.            10 _____

11. The presence of *sedimentary* rocks on mountaintops is evidence that
    these rocks were probably formed beneath the sea.                    11 _____

12. According to the theory of *continental drift*, all the continents were
    once part of the same huge landmass.                                 12 _____

13. Rock layers that are folded upward are called *synclines*.           13 _____

14. The largest type of volcanic mountain is called a *composite* volcano. 14 _____

15. Both Mount Kilimanjaro and Mount Vesuvius are types of *volcanic*
    mountains.                                                           15 _____

16. A broad, flat region that rises abruptly above the landscape is a *dome*. 16 _____

17. Denser oceanic plates slide under the lighter continental plates in a
    process called *subduction*.                                         17 _____

# Testing Your Knowledge

1. Explain how a fossil sea animal can be found in the rocks on top of a plateau that rises 2 km above sea level. _____

_____

2. Explain how most strong earthquakes occur. _____

_____

3. Describe landforms that are usually found in an earthquake zone. _____

_____

4. Suppose you were at the seashore and noticed that the water suddenly moved away from the shore much more rapidly than usual. Why would it be wise for you to go inland and remain there for several hours? _____

_____

5. What is the difference between fold mountains and fault-block mountains? _____

_____

6. Why is it possible that the size of the Hawaiian islands may increase in the future?

_____

_____

7. By studying seismic wave patterns produced by earthquakes, geologists have learned about the inner structure of Earth. These seismic waves move faster through dense rocks and slower through rocks that are less dense. In addition, as seismic waves travel through Earth, they gradually become weaker. Geologists also know that deeper rocks are denser than rocks closer to the surface. Using this information and Fig. 8-7 on page 105, answer the questions that follow.

   *a.* The densest part of Earth is at its _____ .

   *b.* The strongest seismic waves should be located near the _____ .

   *c.* If seismic waves penetrated all parts of Earth, the waves would travel most slowly through the _____ .

8. Explain why the theories of continental drift and seafloor spreading have been combined into the theory of plate tectonics. _____

_____

# Earthquakes and Eruptions: Can We Predict Them?

The earth beneath your feet is solid and unmoving. Right? Wrong! Huge chunks of Earth's crust are constantly in motion. And, depending on where you live, pockets and rivers of hot, liquid rock may be relatively close to your feet.

Movements within Earth's crust can trigger earthquakes great and small. The liquid, or molten, rock can set loose the most spectacular of natural fireworks—volcanoes.

Unfortunately, earthquakes and volcanoes can cause devastating damage to property and have been known to injure and kill thousands of people. Equally unfortunate is the fact that no one has come up with a way to stop these natural disasters from occurring. But the next best thing to preventing a volcanic eruption or a major earthquake may be predicting one.

At first glance, this seems to make a lot of sense, because predicting geological catastrophes—like predicting where hurricanes will strike—would allow people to prepare for the event. In most cases, "preparing" would mean getting out of harm's way, that is, evacuating a threatened area.

But, first of all, can earthquakes and eruptions be predicted and, if so, how accurately? Usually, major earthquakes and volcanic eruptions occur when great stresses build up a few kilometers below the surface of the ground. These stresses often cause changes that can be measured by special instruments, or sometimes even seen—as in the case of a volcano that starts spewing out steam, rocks, ash, or lava.

Instruments can measure such things as new bulges in the land, tiny sideways movements of huge slabs of Earth's crust past each other, and clusters of small earthquakes. Sometimes, even the unusual behavior of animals may give clues to an upcoming earthquake.

Using such clues, Chinese officials in 1975 predicted an earthquake just hours before it shook a large city. The people were warned and so had time to evacuate the city before its buildings came tumbling down. But other earthquakes have struck Chinese cities without warning, causing great damage and loss of life.

Scientists in the United States are also hard at work searching for ways to predict earthquakes and volcanic eruptions. Even as you read these pages, instruments throughout the state of California are being monitored to detect changes in the earth that may precede an earthquake or volcanic eruption. Many of these instruments are concentrated along part of the *San Andreas Fault*, a long north-south crack in the earth, near the town of Parkfield.

Why Parkfield? In 1985, scientists of the United States Geological Survey (USGS), our country's major earth science agency, predicted that a strong earthquake would shake the city sometime between 1988 and 1992. They based their prediction on statistics. In the last century, the area has been rocked by a strong quake about once every 22 years, and the last quake to strike the area was in 1966.

Although an earthquake did not strike Parkfield in the 1990s as predicted, USGS scientists remain convinced that the area is due for a strong quake, so they are continuing to carefully monitor the area. Some scientists, however, believe that earthquakes are by their nature impossible to accurately predict, at least on a regular basis. It seems that the tell-tale events that precede

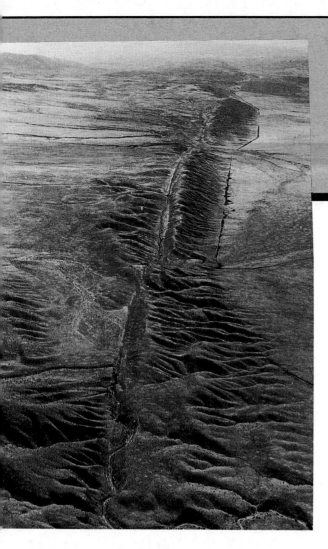

alerted to the danger? Should scientists keep trying to predict geologic disasters? And how certain should they be before issuing a public warning? What do you think?

**1.** *A country in which an earthquake was successfully predicted is*

    a. the United States.    c. Russia.
    b. China.                d. England.

**2.** *The San Andreas Fault is*

    a. an earthquake.
    b. a volcano.
    c. a crack in the earth.
    d. a scientist's mistake.

**3.** *Scientists expected an earthquake in the Parkfield area in the early 1990s because*

    a. the area has frequent large earthquakes.
    b. the area has never had a large quake, so it seemed overdue.
    c. the area has had large quakes at fairly regular intervals in the past.
    d. they observed unusual animal behavior there.

**4.** *What are some possible clues to an upcoming earthquake?*

    _____

    _____

    _____

**5.** *On a separate sheet of paper, express your views on when and under what conditions it is justifiable to issue a warning of a possible earthquake or volcanic eruption. Discuss the various implications of such a warning.*

some quakes are absent before others, or the warning signs appear but the earthquake never happens.

Trying to predict earthquakes and volcanic eruptions has other pitfalls, too. While a successful prediction can save lives and lessen property damage, a prediction that doesn't pan out can have serious unwanted consequences. Many people lose time from jobs, and businesses lose money by closing down. A long-term effect may be a decline in real estate values—who wants to buy a house that's likely to crumble in the near future? Tourists might avoid the area, causing more economic hardship. This, critics say, is what happened in 1982 when USGS scientists warned that a volcanic eruption was a *possibility* at Mammoth Lakes, a major tourist attraction in southern California.

As of this writing, Mammoth Lakes is still in one piece. Should the people have been

# 9

# *Conservation and Protection of the Environment*

## LABORATORY INVESTIGATION

### *THE WATER-HOLDING CAPACITY OF DIFFERENT SOILS*

**A.** Fasten a piece of cheesecloth around the bottom of three funnels. Fill half of the first funnel with dry sand. Fill half of the second funnel with powdered clay. Then, fill half of the third funnel with dry soil containing a lot of humus.

Set up the funnels over separate beakers (see Fig. 9-1).

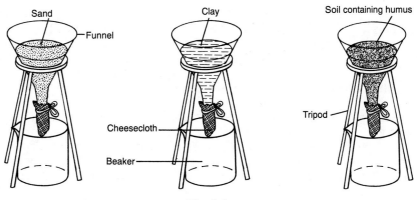

**Fig. 9-1.**

**B.** Pour 250 milliliters of water into each funnel. Observe how the water flows through each funnel. After 5 minutes, remove the funnels from the beakers, and set them aside.

Use a graduated cylinder to measure the amount of water that collected in each beaker. Examine the sample material in each funnel.

**1.** Which soil material absorbed the most water? _____

**2.** Which soil material absorbed the least water? _____

**3.** In a natural setting, which of these soils would probably still be moist a week after a

rainstorm? _____

**4.** On a sloping patch of ground, which of these soils would probably allow rainwater to run off most rapidly? _____

**5.** Which of these soils would probably be best for growing plants? _____

Explain. _____

# ☐ Protecting the Environment ☐

In 1706, the population of the American colonies (not including Native Americans) was about 1 million. By 1800, the population of the newly independent United States had increased to about 5 million. By the early 1990s, there were more than 252 million people living in the United States.

America's earliest European settlers found a country rich in natural resources. There was fertile soil and clean water, plenty of timber and minerals, and forests teeming with wildlife. But as the human population increased, the fertile soils were depleted, and most of the forests were cut down. Scenic rivers became polluted, and huge scars were left on Earth's surface where minerals were extracted from the crust. Herds of buffalo and other wildlife were slaughtered for food, fur, and hides. Because of this carelessness, Americans have used up much of the natural resources that seemed limitless to the colonists 300 years ago.

The U.S. population is still growing. By the year 2000, the population of the United States may reach 300 million. Thus, if present and future generations are to enjoy the remaining natural resources, people must preserve natural areas and use the dwindling natural resources wisely. This practice of preserving natural resources and using them wisely is called **conservation.**

## THE PROBLEM OF SOIL EROSION

About 1900, people began to realize that one of the United States's most important natural resources—the soil—was rapidly being depleted and eroded. Vast areas of once fertile Midwestern farmland were becoming less productive. The topsoil on many farms had been stripped away by rain and wind, because farmers had plowed under the natural grasses that protected the soil from erosion. By the 1930s, much of the soil in the Midwest—the nation's "breadbasket"—had become unproductive or had been carried away by the wind.

Fertile soil takes thousands of years to form. Under natural conditions, soil forms slowly from particles of weathered rock material and the decaying remains of plants and animals. Rapid **soil erosion** is a serious problem because the soil is carried away much faster than it can be formed again.

Soil erosion can be checked by making sure that the ground is completely covered by plants. Plant cover slows down or prevents soil erosion in several ways. Many plants such as grass and clover grow close to the ground surface, forming a natural cover for the soil. These plants prevent the wind from blowing away the soil and slow water down that could otherwise carry away particles of topsoil.

When plants die, their decaying remains form *humus*. Humus-rich soil absorbs and holds more water than any other kind of soil. You observed the water-holding ability of different soils in your laboratory investigation. People often speed up the process of soil erosion by removing the natural plant cover, depleting the soil of materials to make humus, and leaving the topsoil unprotected from harsh weather. Unprotected soil can be rapidly eroded by running water and strong winds.

## PREVENTING SOIL EROSION BY RUNNING WATER

Farmers use several methods to prevent or reduce soil erosion. These methods include *contour plowing, terracing, strip-cropping,* and *damming of gullies.*

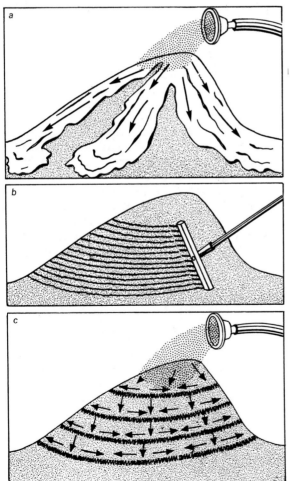

**Fig. 9-2. Demonstrating the need for contour plowing.**

**Contour Plowing.** As you can see in Fig. 9-2a, water sprayed on a mound of soil flows downhill easily. In time, the running water will carve deep grooves into the mound of soil.

In Fig. 9-2b, a rake is used to make a series of small, horizontal grooves on the mound of soil. If you were to spray water on the mound, the water would collect in the grooves before running downhill (see Fig. 9-2c). The grooves act as small dams that slow the downhill flow of water so that the water remains in the grooves long enough for most of it to soak into the soil.

In a similar way, farmers plow deep grooves called *furrows* across the slope of their farmland (see Fig. 9-3). This method is called **contour plowing.** The furrows slow down the water runoff, reducing the amount of soil erosion.

**Fig. 9-3. Contour plowing.**

**Terracing.** Another method of shaping a steep slope to reduce soil erosion is called **terracing.** Terracing involves digging a series of platforms that resemble a giant stairway into a slope (see Fig. 9-4). Each platform has a level surface, giving rainwater more time to soak into the soil.

**Fig. 9-4. Terracing.**

**Strip-Cropping.** When corn or cotton is planted in even rows, with bare soil left in between the rows, the exposed soil can be washed away easily by rain. To reduce soil erosion, farmers plant strips of soybeans, clover, or other crops in the bare soil between rows of corn or cotton (see Fig. 9-5). Because soybeans and clover grow close to the ground and cover the bare soil, these plants are called *cover crops.* **Strip-cropping** greatly reduces soil erosion by giving the soil a better chance to absorb rainwater.

**Fig. 9-5. Strip-cropping.**

**Gully Damming.**    Running water rapidly carves *gullies* into steep slopes in regions of heavy annual rainfall. When a gully begins to form, its growth can be checked easily by making a dam. The **dam** is made by placing boulders, sand, branches, or boards across the gully (see Fig. 9-6). Planting fast-growing grass along the banks of a gully also helps hold the soil in place.

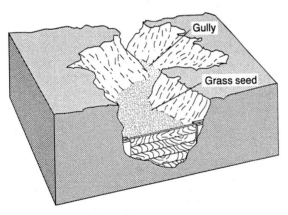

**Fig. 9-6. Gully damming.**

## PREVENTING SOIL EROSION BY THE WIND

The wind can rapidly carry away topsoil from bare fields. For this reason, wind erosion is an especially serious problem in the windy, wide-open spaces of the Plains states. After farmers in this part of the country harvest a crop of wheat, oats, or corn, they immediately plant cover crops to help bind the soil together. This practice reduces soil erosion by the wind. Wind erosion also can be prevented or reduced by planting rows of shrubs or trees next to the fields. These rows of shrubs or trees are called **windbreaks,** or *shelter belts.* Windbreaks and shelter belts reduce the force of the wind before it reaches the fields.

## MINERALS IN THE SOIL

Plants need nitrates and other chemical compounds in the soil for proper growth. *Nitrates* are chemical compounds that contain the elements nitrogen and oxygen.

Most of the nitrates in soil are produced by bacteria called *nitrogen-fixing bacteria.* Some nitrogen-fixing bacteria live in the soil. Others live within the roots of plants in the *legume* family, such as peas, beans, clover, alfalfa, and peanuts. Nitrogen-fixing bacteria take in nitrogen gas from the atmosphere and convert the gas into nitrates. Legumes and other plants can then absorb these nitrates through their roots.

Electrical storms also add large amounts of nitrates to the soil. Lightning causes oxygen and nitrogen in the atmosphere to combine, forming nitrogen compounds. These nitrogen compounds are then brought to the ground by falling rain or snow.

When plants and animals die, their remains are broken down by bacteria and fungi. During this decay, nitrogen gas is released back into the atmosphere. In this way, the nitrogen removed from the air by nitrogen-fixing bacteria is returned to the atmosphere. The process in which nitrogen is taken out of the air and returned to the air by bacteria and fungi is called the *nitrogen cycle.* The nitrogen cycle maintains a relatively constant volume of nitrogen in the atmosphere.

## PREVENTING THE LOSS OF MINERALS FROM SOIL

Growing the same crop in a field year after year uses up the minerals in the soil. If the minerals are not replaced, future crops will not grow properly. In time, most of the minerals in the soil will be exhausted, and the field will be unable to produce crops. The field then will lie exposed to water and wind erosion. To keep the soil productive, farmers restore its minerals by *rotating crops* and by *adding fertilizers* to the soil.

**Crop Rotation.** Growing different kinds of crops in the same field instead of growing the same crop each year is called **crop rotation.** After harvesting one crop, a farmer plants another kind of crop in the same soil. Because different kinds of plants take different minerals from the soil, rotating crops ensures that no single mineral will be rapidly lost from the soil.

Many farmers plant legumes—which help form nitrates—in rotation with crops that remove nitrates from the soil. This helps keep the amount of nitrates in the soil fairly constant over many years.

**Addition of Fertilizers.** Minerals can also be restored to the soil by adding *commercial fertilizers* or *animal manures*. Commercial fertilizers are widely used because they contain the right amounts of minerals (such as nitrates and phosphates) needed for plant growth. Farmers often add animal manures to their fields. Like the decaying remains of plants and animals, the decay of animal manure also adds nitrates to the soil.

## OTHER MINERAL RESOURCES

The mineral nutrients used by plants for proper growth are called **renewable resources.** With proper care of the soil and by using good farming practices, these mineral nutrients can be maintained by natural processes.

In contrast, *coal, oil, natural gas,* and *mineral ores* are **nonrenewable resources.** Once these resources have been removed from the ground, they cannot be replenished for millions of years.

**Fossil Fuels.** Several centuries from now, there may not be much coal, oil, or natural gas left in the ground. These natural resources,

known as **fossil fuels** because they formed from the remains of ancient plants and animals, take millions of years to form and are rapidly being used up.

The fossil fuels that power cars, buses, trucks, boats, trains, and airplanes mostly come from crude oil. Today, supplies of crude oil are being used up faster than new resources are being found. To make fuel supplies last as long as possible, oil companies and the U.S. government are trying to make the best use of the oil that remains.

New methods are being developed to obtain as much oil as possible from the crust. Deeper oil wells are being drilled, and new areas are being explored for oil deposits. Advanced technology allows greater amounts of oil to be recovered from known deposits. Research is also being conducted to find ways to extract oil from *oil shales*. Oil shales are rocks that contain oil trapped inside. However, the extraction of oil from oil shales is currently a very expensive process that involves using large quantities of water.

**Mineral Ores.** Metals such as iron, copper, lead, zinc, and silver are being rapidly consumed by industrialized countries. Each year, huge amounts of these **mineral ores** are imported into the United States because the demand for them is greater than the amount that can be mined in America.

Like fossil fuels, mineral ores are nonrenewable resources. When Earth's supply of these ores has been exhausted, people may have to find substitutes for metals. In fact, plastics have recently been substituted for metals in many products. However, plastics are derived from crude oil, which is also a nonrenewable resource. A better alternative is to reuse, or *recycle,* products made out of metals and plastics.

# Water

## WATER SUPPLY

All living things need water to survive. In nature, water is a renewable resource. However, larger and larger quantities of water are needed each year to irrigate farmland, for use

by industry, and to supply growing cities. Consequently, most of the available fresh water in the United States is being used up.

As shown in Fig. 9-7, industrial use of water and irrigation of farmland consume most of the available water in the United States.

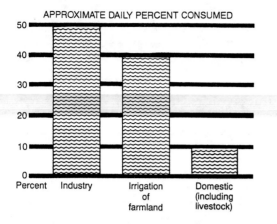

APPROXIMATE DAILY PERCENT CONSUMED

Fig. 9-7. Water consumption in the United States.

The use of water by villages, towns, and cities consumes only a small fraction of the total amount available. Table 9-1 shows the amount of water needed for various domestic purposes.

Water supplies in most parts of the United States are sufficient for current needs. As urban areas grow, however, the water needs for industry, farmland, and domestic uses will also increase. Consequently, the need for water conservation will become increasingly important.

### TABLE 9-1. WATER NEEDED FOR VARIOUS PURPOSES

| Activity | Average Quantity of Water Needed |
|---|---|
| Flushing a toilet | 15 liters |
| Taking a shower | 25 liters |
| Washing dishes | 100 liters |
| Taking a bath | 140 liters |

## WATER POLLUTION

Three hundred years ago, most of the rivers, lakes, and streams in the United States contained clean water. In fact, the country's early settlers could drink safely from almost any river or stream. Then, towns were built along these waterways. As the towns grew, more and more wastes were dumped into the waterways.

At first, the water remained relatively clean because natural processes were sufficient to break down the organic wastes in the sewage. However, as the towns became cities, many rivers, streams, and lakes were turned into open sewers. The increasing volume of sewage dumped into the waterways could not be broken down by natural processes. Today, few waterways in the United States contain water that is safe for people to drink.

Unlike organic wastes, inorganic wastes cannot be broken down easily by natural processes. Many industries have dumped large volumes of inorganic wastes, such as heavy metals and synthetic chemicals, into streams and lakes. Small amounts of some of these substances are deadly to plants, fish, and other animals that live in water. For example, during the 1960s, most of the fish in Lake Erie were killed by chemical pollution; the lake has since been cleaned up a great deal.

In addition, many factories and electric power plants use large volumes of river water as a coolant in their operation. The warmed water is then returned to the river. Unfortunately, such water is too warm for many of the plants and fish living in the river near the factory or power plant. As a result, many organisms die.

Millions of people in the United States get their water from wells drilled into the water table rather than from lakes, rivers, or reservoirs. Like the water in rivers and lakes, groundwater can also be polluted by organic and inorganic wastes.

The pollution of groundwater by sewage is common in rural areas where the water table is near the surface. In areas where detergents seep into the groundwater, a soapy foam may appear in a glass of water drawn from a faucet. Moreover, some detergents (and many fertilizers) contain phosphates. Phosphates are chemical compounds that stimulate the growth of simple water plants such as algae. Because of phosphate pollution, lakes, streams, and rivers can become choked by algae that grow and reproduce at an abnormally high rate.

In coastal regions where groundwater is removed from wells *slowly*, the salt water offshore does not invade the fresh water under the land (see Fig. 9-8*a*). However, when many wells are drilled, large volumes of groundwater are withdrawn *rapidly*. This may cause salt water to seep into wells (see Fig. 9-8*b*). This kind of water pollution has affected many coastal cities such as Los Angeles, California, Miami, Florida, and Atlantic City, New Jersey, as well as some communities on Long Island, New York.

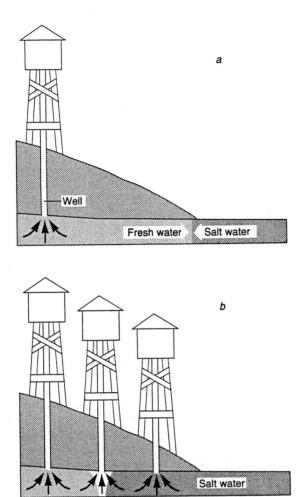

**Fig. 9-8. Saltwater invasion of groundwater in a coastal region.**

## PROTECTING WATER SUPPLIES

To maintain a clean, adequate supply of water in the future, industries and people must practice water conservation. Some important ways to conserve water include *building dams and reservoirs, maintaining forests, preventing water pollution,* and *reducing the waste of water.*

**Building Dams and Reservoirs.** A dam built across a river valley forms an artificial lake, or **reservoir.** Water stored in a reservoir can be used for many purposes, such as irrigating farmland, supplying factories, drinking, washing, recreation, or producing electricity. Dams also help prevent floods by storing excess rainfall rather than allowing the excess rainfall to cause rivers to overflow their banks.

However, a lot of water is lost from reservoirs by evaporation. For example, in arid regions, such as the southwestern United States, evaporation may lower reservoir levels by 1 meter each year. To reduce evaporation, scientists have tried using thin films to cover lakes and reservoirs. The films act as barriers to evaporation, but winds limit their effectiveness, and they can alter the ecology of a water body.

**Maintaining Forests.** Forests should be maintained because they reduce water runoff. You recall that topsoil rich in humus can absorb a large volume of water. This is because the decaying leaves and other organic matter in the humus give topsoil a spongelike texture. Forest soils contain a large amount of humus from the continual supply of decayed leaves and other organic matter. Consequently, forest soils usually hold a lot of water.

In addition, the leaves of trees slow down falling raindrops, while their spreading root systems block the flow of runoff water. Thus, the trees of a forest let rainwater soak into the soil rather than run off rapidly.

Some of the water that soaks into the topsoil seeps downward, becoming part of the groundwater. Later, some of this groundwater may flow into streams, ponds, or lakes. The flow of this groundwater during dry times of the year prevents many streams and lakes from drying up.

**Preventing Water Pollution.** During the past few decades, the federal government, state governments, and private industry have made efforts to prevent water pollution. For example, many towns and cities across the country have built sewage treatment plants. Sewage treatment plants remove solid wastes from the sewage and kill the harmful bacteria in organic wastes.

The solid wastes are disposed of in several ways. Some of the wastes are burned in incinerators. Other solid wastes—containing organic matter—are used as fertilizer to enrich soil on farmland. After the solid wastes are removed, the treated water can be returned safely to the local streams or rivers.

Chemicals such as *polychlorinated biphenyls* (*PCBs*), *pesticides,* and *herbicides,* however, may still cause water pollution. These toxic substances usually pass through sewage treatment plants unchanged. Researchers are looking for ways to remove these pollutants from our waterways.

**Reducing the Waste of Water.** In agricultural areas, many farmers now conserve water by using drip irrigation, in which pipes slowly drip small amounts of water directly into the soil, instead of sprinkler systems. Also, some communities have built large basins to collect rainwater. These artificial basins, called *sumps*, or *recharge basins*, have sandy bottoms that allow water to seep back into the ground, adding to the natural supply of groundwater.

In some communities, factories use large amounts of groundwater for cooling purposes. Instead of allowing these industries to release the water into sewers, local governments have required the factories to return the used cooling water to the ground or to store it in large cooling tanks.

You can also contribute to water conservation by using water carefully in your home. For example, you can store a bottle of water in the refrigerator during hot weather instead of letting water run from the faucet each time you want a cold drink. By attaching a new type of shower head that uses less water and by fixing dripping faucets, you can save many liters of water a day. Toilets can be installed that use less water in each flush. If everyone practiced water conservation, a huge volume of water could be saved each year.

# Air

Like water, air is essential for life. Air is mainly composed of nitrogen, oxygen, carbon dioxide, and water vapor. The percentages of these gases have remained relatively constant for millions of years.

For example, as a result of the nitrogen cycle, nitrogen levels in the air remain at about 78%. Breathing, burning of fossil fuels, and other types of oxidation constantly use up oxygen and release carbon dioxide. Yet, the amount of oxygen in the atmosphere has remained at about 21%, and the amount of carbon dioxide in the atmosphere has remained at between .03% and .04%.

The percentages of oxygen and carbon dioxide are kept relatively constant by the interactions of plants and animals. During photosynthesis, plants use carbon dioxide and water in the presence of sunlight to make food. In this process, plants give off oxygen as a waste product. The oxygen is used by animals and plants during the process of respiration which, in turn, gives off carbon dioxide (see Fig. 9-9). However, the burning of fossil fuels like coal has increased the amount of carbon dioxide in Earth's atmosphere in recent times. This extra carbon dioxide traps more of the sun's heat in the atmosphere, adding to the "greenhouse effect" and increasing Earth's average temperature. Governments are now trying to correct this problem.

## AIR POLLUTION

Although there is an abundant supply of air, its purity must be protected. The number of factories and homes has steadily increased during the past 170 years. Greater quantities of fossil fuel are being burned to run factories and heat homes. As a result, tons of smoke laden with toxic gases (such as sulfur dioxide and carbon monoxide) pour continuously from millions of smokestacks and chimneys and pollute the air. The growing use of cars, buses, and airplanes, which emit exhaust fumes, has added to the problem.

**Acid Rain.** Sulfur dioxide and nitrogen oxides released into the air react with water vapor to form droplets of sulfuric acid and nitric acid. When these droplets fall with rain or snow, the result is acid precipitation, commonly called **acid rain**.

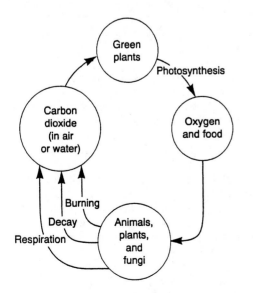

**Fig. 9-9. The carbon dioxide-oxygen cycle.**

**Fig. 9-10. Acid rain damage in a forest.**

## PREVENTING AIR POLLUTION

People showed little concern about air pollution until several disasters happened after World War II. The first known air pollution disaster in the United States occurred in October 1948. Twenty people died and about 6000 people became ill in Donora, Pennsylvania, when gases and dust from steel mills and zinc smelters became trapped in a stagnant air mass over the town. During the 1960s, serious air pollution events in New York City and Los Angeles led to efforts to reduce air pollution throughout the country. The U.S. government passed a strong Clean Air Act in 1990, but it has been weakened due to changes in the law's regulations.

Air is a renewable resource. Natural processes can help eliminate pollutants from the air or reduce the quantity of pollutants to a safe level. However, nature cannot handle the huge quantity of pollutants being produced. To keep air healthful for humans and all other living things, air pollution must be reduced by *controlling exhausts* from smokestacks, chimneys, and motor vehicles and by *assuming personal responsibility* for reducing air pollutants.

Acid rain that falls on the land can cause trees and crops to grow poorly (see Fig. 9-10). When acid rain falls on ponds and lakes, these bodies of water become so acidic that fish and other aquatic organisms cannot survive. In addition, stone buildings and marble monuments weather rapidly in areas where acid rain is common.

**Ozone Depletion.** **Ozone** is a form of oxygen that is found in abundance in a thin layer of the atmosphere between 20 kilometers (km) and 50 km above Earth's surface. This layer of ozone protects living things from the deadly effects of the sun's ultraviolet radiation.

The chemicals called *chlorofluorocarbons* (*CFCs*), used as refrigerants by industries and found in air conditioners and aerosol sprays, deplete the protective ozone layer. Current research has shown that as the ozone layer becomes thinner, more of the harmful ultraviolet rays reach Earth's surface. Ultraviolet rays can cause skin cancer in humans and hinder the growth of food crops and plankton (tiny plants in the sea). The Environmental Protection Agency (EPA) is phasing out the use of CFCs and replacing them with compounds that do not have a harmful effect on the ozone layer.

**Control of Exhausts.** Today, many factory smokestacks and automobile exhaust systems are equipped with devices that remove potentially harmful chemical compounds. Some factories have replaced oil-burning or coal-burning furnaces with electrically heated boilers. Furthermore, electric automobiles and hybrids that alternate between using fuel and using electricity have been introduced. Alternative energy sources, such as solar and wind power, which do not emit any pollutants into the air, are increasingly being used.

**Control of Emissions from Nuclear Power Plants.** During the early 1950s, the federal government backed the development of nuclear power plants to generate electricity. Many people at the time believed that as nuclear energy replaced fossil fuel power plants, the amount of pollutants in the air would decrease. However, nuclear power plants have released dangerous radioactive substances into the environment as a result of accidents in the United States (Three Mile Island, Pennsylvania) and in the former Soviet Union (Chernobyl). To decrease the amount of dangerous radioactivity in the environment, the U.S. government and the governments of other countries must continue to take steps to make nuclear reactors safer.

**Personal Responsibility.** Everyone can help reduce or prevent the pollution of the atmosphere. Some measures that people can take include (1) composting leaves instead of burning leaves outdoors; (2) checking furnaces and incinerators to make sure that they are adjusted properly for complete combustion; (3) insisting that factory smokestacks and cars are built with better pollution-control devices; (4) avoiding use of aerosol sprays and Styrofoam cups, which can cause depletion of the ozone layer in the atmosphere; (5) writing to legislators, asking them to support bills intended to promote clean air; and (6) recycling plastic products that could otherwise give off toxic fumes if incinerated as waste.

# CHAPTER REVIEW

## Science Terms

*The following list contains all of the boldfaced words found in this chapter and the page on which each appears.*

acid rain (p. 128)
conservation (p. 122)
contour plowing (p. 123)
crop rotation (p. 125)
dam (p. 124)
fossil fuels (p. 125)
mineral ores (p. 125)
nonrenewable resources (p. 125)

ozone (p. 129)
renewable resources (p. 125)
reservoir (p. 127)
soil erosion (p. 122)
strip-cropping (p. 124)
terracing (p. 123)
windbreaks (p. 124)

## Matching Questions

*On the blank line, write the letter of the item in column B that is most closely related to the item in column A.*

*Column A*

_____ 1. decaying remains of plants in soil

_____ 2. deep grooves on sloping farmland

_____ 3. rows of shrubs or trees next to open fields

_____ 4. preserving natural resources

_____ 5. planting different crops each year

_____ 6. rows of soybeans between rows of corn

_____ 7. loss of surface soil

_____ 8. artificial lakes created by a dam

_____ 9. has droplets of sulfuric and nitric acids

_____ 10. metals such as iron, copper, lead, and zinc

*Column B*

a. acid rain
b. soil erosion
c. strip-cropping
d. mineral ores
e. contour plowing
f. humus
g. crop rotation
h. conservation
i. reservoirs
j. windbreaks
k. irrigation

# Multiple-Choice Questions

*On the blank line, write the letter preceding the word or expression that best completes the statement.*

1. Soil erosion is slowed down by using a combination of plant cover and
   *a.* clay soil   *b.* sandy soil   *c.* humus soil   *d.* iron-rich soil          1 ____

2. Contour plowing and crop rotation are farming practices that control
   *a.* strip-cropping   *b.* soil erosion   *c.* soil compaction   *d.* terracing          2 ____

3. All of the following plants add nitrogen to the soil *except*
   *a.* alfalfa   *b.* soybeans   *c.* clover   *d.* wheat          3 ____

4. Soil erosion by the wind has occurred in the Plains states primarily because farmers have
   *a.* plowed under the natural grasses   *c.* failed to dig irrigation ditches
   *b.* cut down all the trees   *d.* failed to use organic fertilizers          4 ____

5. The largest consumers of fresh water are
   *a.* private homes and apartments   *c.* commercial fisheries
   *b.* livestock on farms   *d.* industries and irrigation          5 ____

6. The Great Lake that at one time had most of its fish killed off as a result of chemical pollution is
   *a.* Lake Erie   *b.* Lake Huron   *c.* Lake Superior   *d.* Lake Ontario          6 ____

7. A type of pollutant that can affect groundwater supplies of coastal cities is
   *a.* raw sewage   *b.* heated water   *c.* detergents   *d.* salt water          7 ____

8. Sulfur dioxide gas, which is harmful to the lungs and respiratory tract, is largely produced by burning coal and
   *a.* oil   *b.* charcoal   *c.* natural gas   *d.* sewage          8 ____

9. Pollutants that can alter the reproductive cells of animals are pesticides and
   *a.* radioactive substances   *c.* coal fumes
   *b.* carbon monoxides   *d.* sulfuric acid          9 ____

10. An activity that can cause depletion of the ozone layer is
    *a.* burning old leaves   *c.* use of aerosol sprays
    *b.* release of radioactivity   *d.* dumping garbage in landfills          10 ____

# Modified True-False Questions

*In some of the following statements, the italicized term makes the statement incorrect. For each incorrect statement, write the term that must be substituted for the italicized term to make the statement correct. For each correct statement, write the word "true."*

1. Plants such as soybeans and clover are called *cover crops* because they protect soil from erosion.          1 _____

2. *Humus* consists of decayed plant and animal matter.          2 _____

3. In crop rotation, legumes are important because they add *nitrates* to the soil.          3 _____

4. Coal, oil, and gas are *renewable* resources.

4 _____

5. Most of the water supplies are used for *agriculture* and industrial purposes.

5 _____

6. Organic matter in water often is destroyed by natural processes of *decay.*

6 _____

7. The seepage of salt water into wells can pollute the water supplies of *inland* communities.

7 _____

8. The process by which green plants produce food is important in maintaining the *oxygen* content of the atmosphere.

8 _____

# Testing Your Knowledge

1. Why would someone living in New York City be concerned about soil erosion on a farm or in a forest in another state? _____

   _____

2. How do each of the following farming practices reduce soil erosion?

   *a.* contour plowing _____

   *b.* terracing _____

   *c.* strip-cropping _____

   *d.* damming gullies _____

3. Suppose you have a small vegetable garden. Why would you grow different vegetables in different parts of the garden each year? _____

   _____

4. What is the difference between a renewable resource and a nonrenewable resource? Give two examples of each. _____

   _____

5. Some communities obtain their water from underground wells. Why is it wise for these communities to replace a system of individual cesspools with a sewer system that carries sewage away to a central sewage-treatment plant? _____

   _____

6. How does the water level of a river or lake depend upon the conservation of forests?

   _____

7. Name three air pollutants. Describe how each air pollutant can be reduced or eliminated.

   *a.* _____

   *b.* _____

   *c.* _____

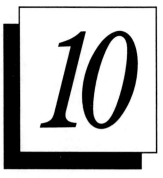

# *The Ocean*

## LABORATORY INVESTIGATION

### *WATER CURRENTS*

**A.** Bottle A contains cold salt water colored with a blue dye. Bottle B contains warm water colored with a purple dye. Measure the temperature of the cold salt solution with one thermometer. Then, measure the temperature of the warm, purple-colored water with a second thermometer.

    **1.** What is the temperature of the salt solution? _____

    **2.** What is the temperature of the warm water? _____

**B.** Fill a large battery jar about three-quarters full with tap water at 20° Celsius. Fill a small bottle with the blue-colored salt solution, and label the bottle A. Cover the bottle loosely with its cap. Hold the covered bottle in the battery jar as shown in Fig. 10-1a. Gently remove the cap from the bottle.

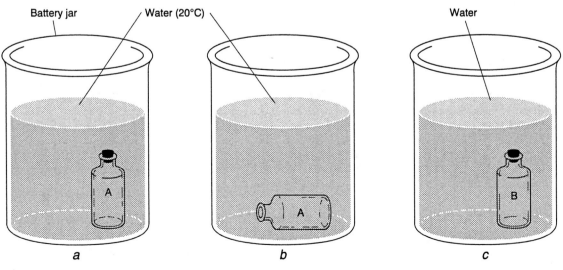

**Fig. 10-1.**

**3.** Describe your observations. _____

_____

**C.** Gently recap the small bottle. Then, carefully lay the small bottle on its side as you see in Fig. 10-1*b*. Now gently remove the cap.

**4.** Describe your observations. _____

_____

**5.** Suppose the salt solution was less dense than the tap water. What do you think would

happen? _____

**6.** Was your prediction correct? _____

Explain the movement of the blue-colored salt solution. _____

_____

**D.** Remove the small bottle from the battery jar. Empty the battery jar. Then, refill the battery jar with cold tap water. Measure the temperature of the water in the battery jar.

**7.** How does the temperature of water in the battery jar compare with the temperature of

water in bottle B? _____

**E.** Fill another small bottle with the warm purple-colored water, and label the bottle B. Cover the bottle loosely with its cap. Lower the bottle into the battery jar as shown in Fig. 10-1*c*. Gently slip the cap off the bottle.

**8.** Describe your observations. _____

_____

**9.** Explain the movement of the purple-colored water. _____

_____

**10.** How does the density of the warm water compare with the density of the cold water?

_____

**F.** Cut out two squares of paper toweling about 10 centimeters on each side. Place several crystals of potassium permanganate on each square. Wrap each piece of paper toweling loosely around the crystals. **CAUTION: Potassium permanganate can easily stain skin and clothing.**

Fill a 1000-milliliter (ml) Pyrex beaker with 750 ml of tap water. Place the beaker on a tripod. Using a pair of tongs, place each of the packages of crystals at opposite sides of the beaker. Heat the beaker on one side as shown in Fig. 10-2.

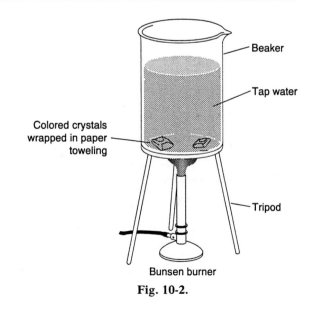

Fig. 10-2.

**11.** Describe your observations. _____

_____

_____

**12.** Add arrows to Fig. 10-2 to show the movement of the colored water. Explain the movement

of the colored water in the beaker. _____

_____

_____

# The World Ocean

The world ocean covers about 71 percent of Earth's surface. Geographers have divided the world ocean into five smaller bodies, which are also called oceans (see Fig. 10-3). The *Pacific Ocean* is the world's largest ocean, covering more of Earth's surface than all of the continents put together. The *Atlantic Ocean* is the second largest ocean, followed by the *Indian Ocean, Antarctic Ocean,* and the *Arctic Ocean.*

On average, the ocean floor lies farther below sea level than the continents rise above sea level. The average depth of the ocean floor is about 4 kilometers (km) below sea level, whereas the average height of the continents is about 0.8 km above sea level. The deepest part of the world ocean is the *Mariana Trench* in the Pacific Ocean. The bottom of the Mariana Trench is almost 11 km below sea level. In comparison, the top of the highest moun-

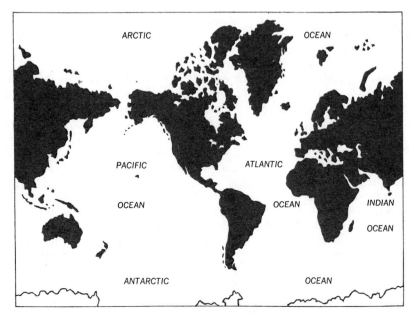

**Fig. 10-3. The oceans of the world.**

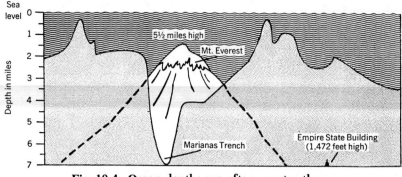

**Fig. 10-4. Ocean depths are often greater than mountain heights.**

tain on land (Mount Everest) is about 8.8 km above sea level (see Fig. 10-4).

Scientists who study the oceans, called *oceanographers*, have slowly accumulated information about different features of the ocean. By using research instruments and vessels, oceanographers have compiled data on the composition and temperature of ocean water, the shape and composition of the ocean floor, the movement of ocean water, and the abundance of useful materials in seawater and under the ocean floor.

## COMPOSITION OF OCEAN WATER

Chemical analysis of ocean water indicates the presence of both dissolved solids and gases.

**Dissolved Solids in Ocean Water.** If you have ever bathed or swum in the ocean, you know that seawater has a salty taste. However, seawater has not always been salty. Oceanographers think that hundreds of millions of years ago, the ocean contained fresh water. In time, the ocean gradually became saltier as rivers carried billions of tons of dissolved mineral salts into it.

About 75% of the salt in seawater is table salt, or sodium chloride. The remainder consists of several other salts (such as magnesium chloride and magnesium sulfate) (see Table 10-1). The weight of the salts in a given volume of water is a measure called the **salinity** of the water. The average salinity of seawater is about 3.5%. In other words, one kilogram of seawater contains about 35 grams of salts.

Gold, silver, uranium, and iodine also are found in seawater, but in much smaller amounts. More than half of all known natu-

rally occurring elements have been found in seawater.

The degree of salinity varies throughout the world ocean. For example, in places where large volumes of fresh water regularly enter the ocean, the salinity is less than 3.5%. Such places include the mouths of rivers, regions of heavy rainfall, and areas where large masses of glacial ice melt into the ocean.

### TABLE 10-1. MOST ABUNDANT SALTS IN SEAWATER

| Salts | Approximate Weight by Percent |
|---|---|
| Sodium chloride, $NaCl$ | 2.70 |
| Magnesium chloride, $MgCl_2$ | .38 |
| Magnesium sulfate, $MgSO_4$ | .17 |
| Calcium sulfate, $CaSO_4$ | .13 |
| Potassium sulfate, $K_2SO_4$ | .08 |
| Calcium carbonate, $CaCO_3$ | .01 |
| Magnesium bromide, $MgBr_2$ | .008 |

**Dissolved Gases in Ocean Water.** The gases dissolved in seawater include nitrogen, oxygen, and carbon dioxide. Dissolved oxygen and dissolved carbon dioxide in seawater are important for the survival of ocean organisms. For example, fish, lobsters, clams, worms, and all other marine organisms require oxygen for their respiration. Ocean plants such as seaweed, other algae, and tiny floating plants called phytoplankton depend on the dissolved carbon dioxide to carry out photosynthesis.

Many marine animals eat plants. Consequently, these animals indirectly depend on dissolved carbon dioxide for their survival. Moreover, the oxygen released by marine plants plays a vital role in maintaining the oxygen content of Earth's atmosphere.

# TEMPERATURE OF OCEAN WATER

The temperature of ocean water varies with its depth and *latitude,* or distance from the equator. Ocean water temperature is fairly uniform to a depth of about 90 meters (m). This results from surface winds churning the warm, upper layers of water. The churning action mixes the warm surface water with the cooler water below.

From a depth of about 90 m to about 900 m, at the equator, the temperature of ocean water falls rapidly. Below 900 m, the temperature continues to fall, but more slowly. The water temperature stops falling when it reaches a few degrees above seawater's freezing point.

Keep in mind that the freezing point of seawater is lower than the freezing point of pure water. Pure water freezes at 0°C. Because of the dissolved salts, seawater freezes at about −2°C. (The freezing point of seawater may vary depending on its salinity in a particular location.)

In general, the surface temperature of ocean water becomes gradually colder traveling from the equator to either the North Pole or the South Pole. Fig. 10-5 shows the temperature differences with increasing depth between waters in high latitudes (south of Greenland) and waters in the middle latitudes (near the United States). Latitude will be discussed more fully in Chapter 11.

## TABLE 10-2. OCEAN TEMPERATURE AT THE EQUATOR

| Zone | Depth (meters) | Temperature (Celsius) |
|---|---|---|
| Surface layer | 0 | 25° |
| Zone of rapid change in temperature | 90<br>300<br>600<br>900 | 24°<br>15°<br>10°<br>5° |
| Zone of gradual change in temperature | 1200<br>1500<br>1800 | 4° |
| Zone of deepest water | Below 1800 | Just above freezing point of seawater |

The temperature of the surface water at the equator may be as high as 33°C. The surface water at the North Pole or near the South Pole may be as low as −2°C, close to the freezing point of seawater. In Table 10-2, you can see how the temperature of seawater at the equator changes with increasing depth. The figures in Table 10-2 vary depending on such factors as time of day, season, and specific location.

The temperature of surface water changes throughout the year. Toward the end of summer, the temperature of ocean water is highest because it has been subjected to months of relatively high air temperatures. Toward the end of winter, after months of cold weather, the temperature of surface water is at its lowest point. Below a depth of 1.8 km, the temperature of ocean water remains uniform, regardless of the season. At these depths, the water temperature remains at −2°C.

# EFFECT OF THE OCEAN ON CLIMATE

The average weather, or *climate,* in the midwestern United States often is much different from the climate along the northeast coast of the United States. That is because large bodies of water greatly influence the average temperature of nearby coastal regions. Water absorbs heat slowly, holds heat well, and gives off heat slowly.

Consequently, in winter the ocean acts as a heat radiator. As the ocean water slowly gives up heat, it warms adjacent coastal regions. In

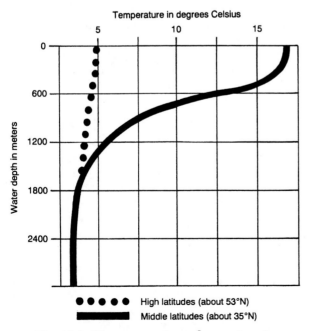

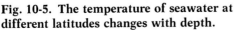

● ● ● ● ● High latitudes (about 53°N)
▬▬▬ Middle latitudes (about 35°N)

**Fig. 10-5. The temperature of seawater at different latitudes changes with depth.**

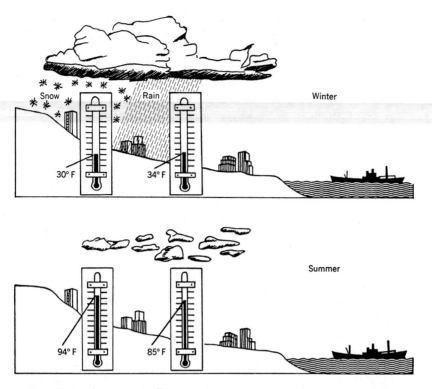

Snow
Rain
Winter
30° F
34° F

Summer
94° F
85° F

**Fig. 10-6. The ocean influences the temperature of a coastal region.**

summer, the ocean acts as an air conditioner. As the ocean water slowly absorbs heat from the atmosphere, it cools adjacent coastal regions. For these reasons, a coastal city such as New York City may be several degrees warmer in winter and cooler in summer than a city located between 5 km and 10 km inland (see Fig. 10-6).

## THE OCEAN FLOOR

Excluding beaches and other coastal areas exposed during low tide, most of the ocean floor is hidden from view. In fact, oceanographers must use special instruments and equipment to explore the deep ocean floor.

**Exploring the Ocean Floor.** In the past, sailors *sounded*, or determined ocean depth, by dropping a measured line with a weight tied to the bottom end over the side of a ship. This method was slow and inaccurate, especially in very deep water.

*Echo Sounding.* Today, oceanographers can measure the depth of the ocean quickly and with great accuracy using an instrument called an *echo sounder.* An echo sounder sends out sound waves through the water. When the

sound waves reach the ocean floor, they are reflected back to the ship (see Fig. 10-7a). The echo sounder detects the echo and records the elapsed time from the moment the sound was transmitted to the moment the echo returns to the ship.

Sound travels about 1460 meters (m) per second in water at 0°C. If you know the speed of sound in water and the length of time the sound has traveled, you can calculate the distance to the ocean floor (speed multiplied by time equals distance). A sound wave that reaches the ocean floor and returns to the ship in 4 seconds travels:

1460 m/second × 4 seconds = 5840 m

This distance of 5840 m includes how long it took the sound wave to reach the ocean floor and return to the ship. To find the distance the sound wave traveled to the ocean floor, you divide the total distance by 2:

$$\frac{5840 \text{ m}}{2} = 2920 \text{ m}$$

Oceanographers can use echo sounding to measure the depth of a section of the ocean floor (see Fig. 10-7b). Depth measurements, recorded as a tracing on a chart, show the actual shape of the ocean floor. As a result of these measurements, narrow strips of the

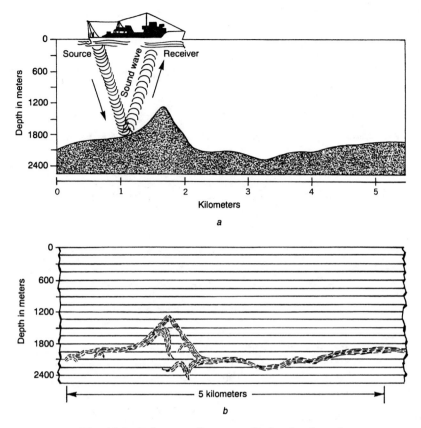

**Fig. 10-7. Echo sounding reveals the depth and shape of the ocean floor.**

ocean floor have been accurately charted. Fig. 10-8 shows a section of the ocean floor between the United States and Africa.

***Bathyscaphs.*** Oceanographers also use specially built submarines, called **bathyscaphs,** to explore the ocean floor. In 1960, the U.S. Navy bathyscaph *Trieste* descended about 11 km to the bottom of the Mariana Trench, the deepest part of the Pacific Ocean.

As a result of underwater research conducted with echo sounders, bathyscaphs, and other special equipment, oceanographers now know there are many kinds of living things that can survive on the ocean floor. In fact,

some of these organisms are similar to forms found near the coastal areas of continents. Moreover, oceanographic research has revealed features such as underwater canyons, plains, volcanoes, plateaus, and mountain ridges.

**Ocean Basins.** The depressions in Earth's crust that contain the oceans are called **ocean basins.** In comparison with the size of Earth, ocean basins are only slight depressions in the crust.

Fig. 10-8 shows a profile of a typical ocean basin. Because this profile is greatly exaggerated, the ocean basin appears deeper than it

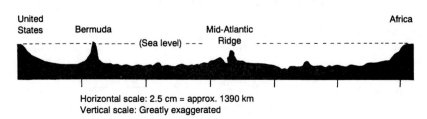

**Fig. 10-8. The shape of the Atlantic Ocean floor.**

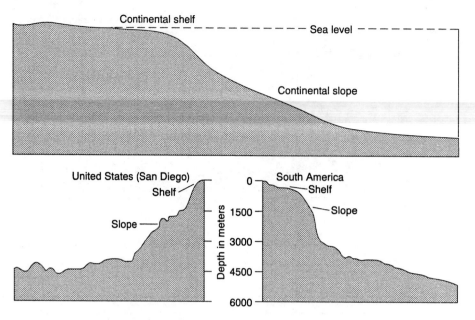

Fig. 10-9. Continental shelves and slopes.

actually is. The ocean basins bordering the continents have very gentle slopes that are nearly flat. Farther offshore, however, the slope becomes much steeper. In very deep water, the slopes level off into broad, flat plains. Mid-ocean mountain ridges, caused by seafloor spreading (discussed in Chapter 8), rise from these deep ocean plains.

**Continental Shelves and Slopes.** The shallow, submerged edges of a continent are called **continental shelves** (see Fig. 10-9). On average, the width of a continental shelf is about 70 km. In some places, such as off San Diego, California, and off the west coast of South America, the continental shelf is only about 5 km wide. In other places, such as off the east coast of the United States, the continental shelf is hundreds of kilometers wide.

From the shoreline, a continental shelf slopes gently toward deeper water. At their farthest distance from the shore, continental shelves are between 100 m and 300 m deep. From the edge of the continental shelf, the ocean floor slopes downward more steeply for several hundred meters. This steeper sloping region is called the **continental slope.** The continental slope levels off at an average depth of about 3.5 km. In some parts of the ocean, these slopes may extend to about 9 km below the ocean surface.

**Underwater Canyons.** Extending outward from the coasts into deeper water are elon-gated depressions in the ocean floor, called **underwater canyons.** The underwater canyons are deep cuts in the continental shelves and slopes. Off the coast of Monterey, California, a 1.5-km deep underwater canyon extends into the Pacific Ocean for about 80 km. The *Hudson River Canyon*, located off the coast of New York City, extends about 320 km into the Atlantic Ocean. Fig. 10-10 shows the general ap-

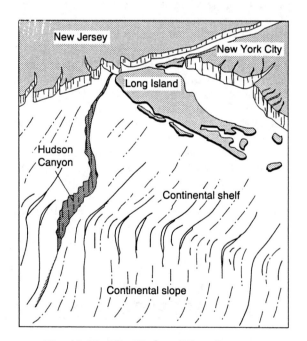

Fig. 10-10. The Hudson River Canyon.

pearance of the Hudson River Canyon if all of the ocean water were drained from the continental shelf off New York and New Jersey.

**Mid-Ocean Ridges.** Long, underwater mountain ranges wind around the world. These mountain ranges, located on the ocean floor, are called **mid-ocean ridges.** As you recall from Chapter 8, these mid-ocean ridges result from seafloor spreading. The separation of crustal plates at the mid-ocean ridges allows molten rock to rise, forming new ocean crust and creating tall mountain ridges under the ocean.

The longest mid-ocean ridge is the Mid-Atlantic Ridge, which divides the Atlantic Ocean into two huge basins (see Fig. 10-11). The Mid-Atlantic Ridge extends for about 16,000 km from Iceland to the tip of South America. The tallest peaks of the Mid-Atlantic Ridge rise above sea level, forming islands such as Iceland and the Azores. Other mid-ocean ridges are found in the Pacific and Indian oceans.

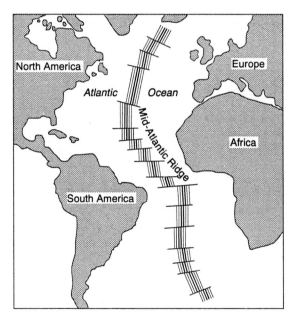

Fig. 10-11. The Mid-Atlantic Ridge.

**Seamounts and Volcanic Islands.** Also found on the ocean floor are underwater mountains called **seamounts.** Most seamounts are formed by underwater volcanoes. When the top of an underwater volcano rises above sea level, a **volcanic island** is formed. Seamounts and volcanic islands are often found in long chains. These volcanic chains are formed by the movement of an oceanic plate riding over a stationary *hot spot* located deep in the mantle. Magma rising from the hot spot pierces the

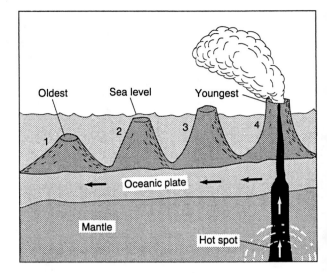

Fig. 10-12. Formation of volcanic islands and seamounts over a hot spot.

crustal plate above and forms a volcano on the seafloor. As the ocean plate moves the volcano away from the hot spot, another volcano is produced behind it (see Fig. 10-12). The Hawaiian islands are a classic example of the formation of volcanic island chains.

**Island Arcs.** Other volcanic islands may form when two oceanic plates collide. As one plate plunges beneath the other, magma is generated. The magma rises to produce a curving chain of volcanic islands called an **island arc** (see Fig. 10-13). The islands of the Lesser Antilles in the Caribbean make up an island arc, formed by subduction of the Atlantic seafloor beneath the ocean crust of the Caribbean Plate.

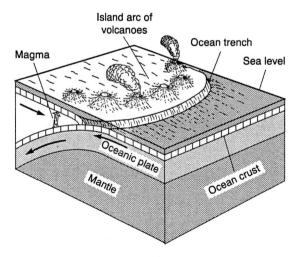

Fig. 10-13. Formation of an island arc of volcanoes.

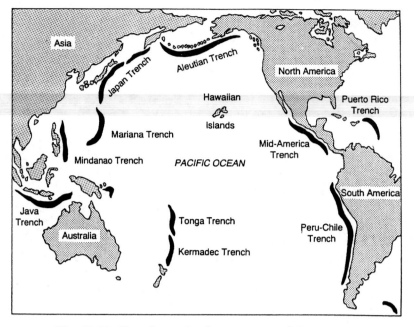

Fig. 10-14. Trenches—the deepest parts of the ocean.

**Coral Islands.** On the summits of underwater volcanoes, coral animals often form **coral islands.** The hard, outer parts of the coral animals build up lime deposits on the submerged peaks of the volcanoes. In time, crustal movements may cause these volcanic peaks to rise above sea level, forming coral islands.

**Trenches.** The deepest parts of the ocean are found at the bottom of long depressions in the ocean floor. These elongated depressions are called **trenches** (see Fig. 10-14). Trenches form where oceanic plates plunge beneath continental plates or other oceanic plates. Trenches are found off mountainous coastlines and island arcs, where subduction is taking place (see the section on Plate Collisions in Chapter 8).

The Pacific Ocean has the deepest trenches on Earth; these include the Mariana Trench, the Mindanao Trench, and the Tonga Trench. They range from about 10 to 11 km deep. The deepest part of the Atlantic Ocean is the *Puerto Rico Trench,* which reaches about 9.5 km below sea level. Other trenches are found off the west coast of South America and off the east coast of Japan.

**Sediments on the Ocean Floor.** Rivers carry huge volumes of dissolved minerals and sediments into the oceans. As you can see in Fig. 10-15, many tons of tiny meteorites, volcanic

dust, cinders, and other particles fall from the atmosphere into the ocean. Most of these particles eventually accumulate on the ocean floor.

Much of the sediment on the seafloor consists of very fine particles called *ooze.* Oozes are made up of the hard parts of dead, microscopic-sized plants (phytoplankton) and animals (zooplankton) that slowly drift down to the bottom of the ocean.

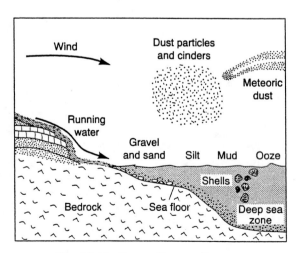

Fig. 10-15. Ocean-floor sediments.

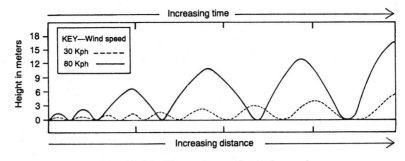

**Fig. 10-16. Wave size and wind speed.**

**Nodules.** Oceanographers have discovered round lumps of metallic ores on the ocean floor. These masses of metallic ores are called **nodules.** Nodules contain large amounts of metals such as manganese, iron, nickel, cobalt, and copper. Some oceanographers think that nodules form when dissolved minerals in seawater are deposited on solid objects such as sand grains, fish bones, and other hard remains of marine organisms. Most nodules are about the size of a potato. However, a nodule brought to the surface from the floor of the Pacific Ocean weighed nearly a ton.

## MOVEMENTS OF OCEAN WATER

Ocean water is constantly moving. The movement is caused by the wind, the sun's heat, Earth's rotation, the moon's gravitational pull on Earth, and underwater earthquakes. The major motions of ocean water are called *waves, currents,* and *tides.*

**The Formation of Waves.** The wind produces most ocean **waves.** As wind blows over the ocean surface, the wind drags the top layers of ocean water along with it. This friction between the air and water causes the ocean surface to ripple and then form small waves.

As the force of the wind increases, the wave sizes also increase (see Fig. 10-16). Strong storm winds that blow for many hours over great distances produce the largest waves. Most storm waves are less than 15 m high. However, there have been reports of storm waves of more than 30 m in height.

**Features of Waves.** As you see in Fig. 10-17, two of the main parts of a wave are the *crest* and the *trough.* The crest is the highest part of a wave, and the trough is the lowest part of a wave. The vertical distance, from the crest to the trough, is the wave's height.

Waves vary in height and in length. The *length* of a wave is either the horizontal distance from the top of one crest to the top of the next crest or the horizontal distance from the bottom of one trough to the bottom of the next trough.

Waves on the ocean surface do not reach the ocean floor, except in shallow areas. In fact, a

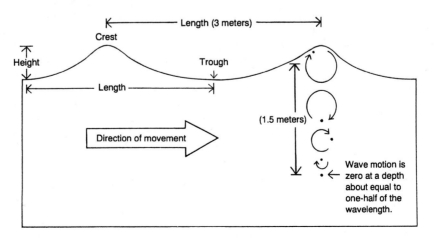

**Fig. 10-17. Wave characteristics.**

wave does not move water very far below the surface. Oceanographers have found that a wave influences the water below to a depth equal to one half the length of the wave. In other words, a wave that is about 3 m from crest to crest will influence water movement only 1.5 m below the surface (see Fig. 10-17).

*Swells.* Ocean waves caused by winds that travel away from the region in which they are formed are called *swells*. A swell is recognized easily because it appears as a series of smooth, rounded waves that rise and fall with a regular motion.

*Breakers.* When an ocean wave or swell moves into shallow water, the lower part of the wave drags along the ocean bottom and slows down. The wave crest, however, continues to move forward at its original speed. As you see in Fig. 10-18, the front of the wave becomes much steeper than the rear of the wave.

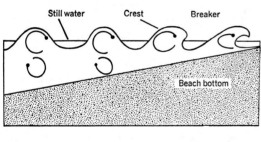

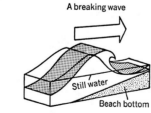

**Fig. 10-18. Breakers.**

Suddenly, the wave becomes unbalanced. The faster-moving crest topples forward, forming a *breaker* that washes up on the beach. Then, the water flows back down the slope of the beach to the waterline, producing an underwater current called an *undertow*. Beaches with long, gentle slopes, such as Waikiki Beach in Hawaii produce huge breakers that

form far from the shore. At times, surfers can get long rides back to shore on these breakers.

**Ocean Currents.** A mass of seawater that flows continuously through a large region of the ocean is called an ocean **current.** Ocean currents that flow along the ocean surface are called *surface currents.* Ocean currents that flow deep below the surface are called *subsurface currents.* Both of these currents are influenced by factors such as the location of landmasses in the current's path and Earth's rotation.

*Surface Currents.* Surface currents are mainly caused by the wind. In fact, ocean water on the surface usually follows a path similar to the prevailing wind pattern in the atmosphere (discussed in Chapter 16).

*Subsurface Currents.* The main cause of subsurface currents is the difference in water density in different areas of the ocean. A mass of ocean water is usually more dense when it becomes either colder or saltier than the surrounding water. In contrast, a mass of water is usually less dense when it becomes warmer or less salty than the surrounding water.

In your laboratory investigation, you studied how differences in density due to temperature or salinity cause water currents. In the ocean, subsurface currents form for similar reasons. As dense water sinks, or less dense water rises, masses of water of a different density move in to replace them. The subsurface currents formed by these water movements may be either horizontal or vertical.

**Major Ocean Currents.** There are two main surface currents that flow along the coastlines of the United States. The *Gulf Stream* flows along the east coast and the *California Current* flows along the west coast. Fig. 10-19 shows the paths of these and other surface currents. You see that ocean currents in the Northern Hemisphere move in a clockwise direction, whereas ocean currents in the Southern Hemisphere move in a counterclockwise direction.

Near the center of the Atlantic Ocean, between the Caribbean Sea and the west coast of Africa, the clockwise circulation of water slows down. This calm region of the Atlantic Ocean is called the *Sargasso Sea*. In the Sargasso Sea, thick masses of seaweed float on the surface. In the past, sailors believed incorrectly that their ships could be trapped by the floating seaweed.

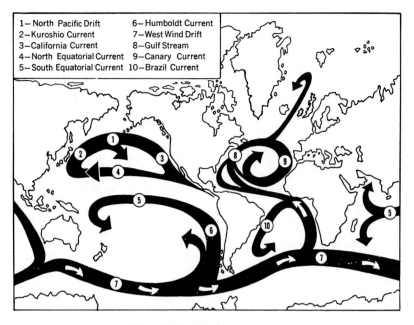

1— North Pacific Drift    6— Humboldt Current
2— Kuroshio Current      7— West Wind Drift
3— California Current     8— Gulf Stream
4— North Equatorial Current  9— Canary Current
5— South Equatorial Current  10— Brazil Current

**Fig. 10-19. Ocean currents.**

***The Gulf Stream.*** The Gulf Stream is about 160 km wide and 0.6 km deep. More water flows in the Gulf Stream than in all of the major rivers on the continents combined. The Gulf Stream flows slowly, usually between 2 km and 3 km per hour. However, off the Florida coast, the Gulf Stream flows about 8 km per hour. Between Florida and Cuba, the Gulf Stream is commonly called the *Florida Current.*

From Florida, the Gulf Stream flows northeast, eventually crossing the Atlantic Ocean. As the Gulf Stream flows northward, a sub-surface current a kilometer below the Gulf Stream flows southwest at a rate of about 13 km per day.

***Polar Currents.*** Some important underwater currents originate near the North and South poles. Ocean water near the poles is dense because (1) it is very cold and (2) it has a high salt content. Polar ocean water is very salty because the salts dissolved in the water do not become part of the ice when the surface layers freeze. Instead, the salts sink to the layer of water below the ice. Consequently, as freezing continues, water under the surface ice becomes saltier and denser.

The downward movement of very cold, dense polar water produces a vertical under-water current. Near the ocean floor, this vertical current begins to move horizontally toward the equator. Because this type of underwater current moves very slowly, oceanographers refer to it as *polar creep.* Measurements have shown that polar water takes several years to reach the equator.

***The Gibraltar Current.*** Other underwater currents are produced when two bodies of water having different salt contents meet. For example, one of these underwater currents forms where the Mediterranean Sea meets the Atlantic Ocean at the Strait of Gibraltar.

The ocean water in the Mediterranean Sea is saltier than the water in the Atlantic Ocean. That is because the warmer surface water of the enclosed Mediterranean Sea evaporates more rapidly than the colder surface water of the Atlantic Ocean. As the water in the Mediterranean Sea evaporates, the dissolved salts remain behind. This makes the water saltier and denser.

The denser water of the Mediterranean Sea sinks and flows under the less dense water of the Atlantic Ocean at the Strait of Gibraltar. As a result, an underwater current flows out of the Mediterranean Sea and into the Atlantic Ocean. During World War II, submarines used this underwater current to drift undetected through the Strait of Gibraltar (see Fig. 10-20).

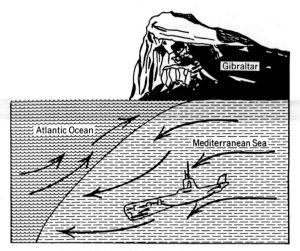

**Fig. 10-20. The Gibraltar Current.**

**Drifts.** A mass of ocean water that is wider, shallower, and slower than an ocean current is called a **drift**. The largest drift is called the West Wind Drift. The West Wind Drift circles Earth in the open ocean south of the continents of Australia, South America, and Africa (refer back to Fig. 10-17).

**Tides.** The regular rise and fall of ocean water along coastlines each day is called the **tide**. Tides are caused mainly by two forces acting on Earth in different ways. One of these forces is the gravitational attraction of the moon. The other force is the tendency of an orbiting object to leave its orbit (*centrifugal* force).

Water on the side of Earth near the moon is more strongly affected by the gravitational attraction of the moon. Water on the side of Earth away from the moon is more strongly affected by Earth's orbit. In general, two bulges, or high tides, occur on Earth at any time with two depressions, or low tides, occurring between them. Tides are discussed in more detail in Chapter 21.

## EFFECT OF THE OCEAN ON THE SHORELINE

The shoreline is constantly changing. Changes in the shoreline are produced by erosion of beaches and coastal rock formations, discharge of sediments into the ocean by rivers, and the constant action of waves and currents. Some shoreline features produced by these agents include *deltas, bars, spits, hooks, lagoons,* and *sea cliffs.*

**Deltas.** As you learned in Chapter 6, a *delta* is a relatively level, fan-shaped deposit of sand and clay particles that forms at the mouth of a river that empties into a large, quiet body of water. In time, the deposit appears above the surface of the water, building up the shoreline into the larger body of water.

A large delta is found at the mouth of the Mississippi River, where the river enters the Gulf of Mexico. Another large delta is located at the mouth of the Colorado River, where the river enters the Gulf of California. The Mississippi River Delta is about 320 km long and over 90 km wide in places. Growth of the delta has largely ceased because dams built upriver have reduced the supply of sediments.

**Bars, Spits, and Hooks.** Streams and rivers deposit sediments at the shore. Then, the force of ocean waves carries the small sand grains and rock fragments into deeper water, where they accumulate. In time, these mounds of ocean-carried sediments are built up above sea level into strips of land called **sandbars**. If the sandbar is not connected to the mainland, it is called an *offshore bar* (see Fig. 10-21). The

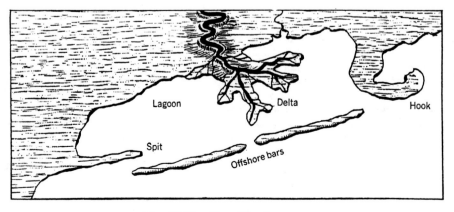

**Fig. 10-21. Shoreline features.**

body of water between an offshore bar and the mainland is a **lagoon.**

Many offshore bars have been built up along the Atlantic and Gulf coasts of the United States. Fire Island and Jones Beach are well-known offshore bars located along the south shore of Long Island, New York. Between Fire Island and Jones Beach is a lagoon called Great South Bay. Several cities, such as Miami Beach, Florida; Galveston, Texas; and Atlantic City, New Jersey, are built upon offshore bars.

A current that flows along the coast, or *offshore current*, produces a sandbar across the opening of a bay. This type of sandbar is called a *spit* (see Fig. 10-21). Sometimes, a current flowing in an opposite direction meets an off-shore current at the end of a spit. Then, the tip of the spit bends inward toward the bay (see Fig. 10-21). These sandbars are called *hooks.* Cape Cod in Massachusetts is an outstanding example of a hook. Fig. 10-22 shows some prominent shoreline features in the New York City area, including two spits (Rockaway and Sandy Hook) and a large lagoon (Jamaica Bay).

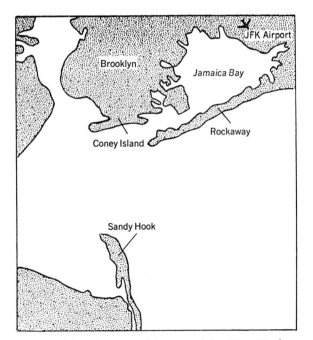

**Fig. 10-22. Shoreline features of the New York City area.**

**Sea Cliffs.** Cliffs can be formed by ocean waves eroding rocky coastlines or coasts that slope steeply beneath sea level. The continual pounding of waves against the shore breaks up rock into fragments that fall into the ocean. The highlands that remain are called **sea**

**cliffs.** In the United States, sea cliffs are found along the coasts of Maine, Oregon, and California (see Fig. 10-23).

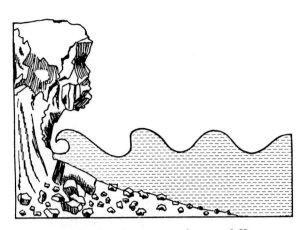

**Fig. 10-23. Formation of a sea cliff.**

**Estuaries.** The submergence of a coast or rise in sea level can produce changes in the shoreline. For example, valleys located along the coast may be submerged, or drowned. Submerged river valleys are called **estuaries.** The lower portion of the Hudson River is an estuary that flows through a valley that is drowned as far as 240 km upstream from its mouth at New York City.

Many drowned valleys are found along the coast of Alaska, Greenland, and Norway. These drowned valleys, however, were originally produced by melting glaciers instead of by rivers. A drowned glacial valley is called a **fiord.**

## FUTURE OF THE OCEAN

Natural resources such as minerals, fossil fuel, food, and fresh water are becoming increasingly scarce. As the world's population continues to grow, oceanographers are developing methods of extracting some natural resources from the ocean. The use of tides to produce electricity is another way that the ocean can be used as a natural resource.

**Minerals and Fossil Fuels.** The continental shelves already have been tapped as a source of oil and natural gas. For example, large amounts of oil are being removed from the bottom of the Gulf of Mexico with special equipment (see Fig. 10-24, page 148).

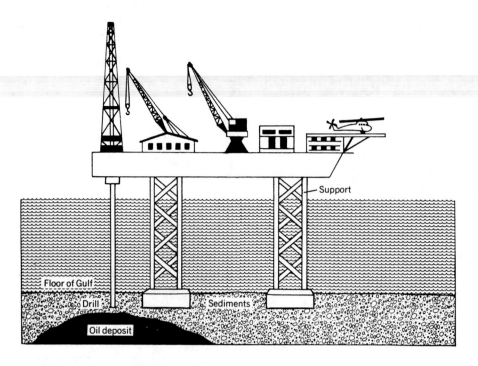

**Fig. 10-24. Removing oil from beneath the ocean floor.**

As you saw in Table 10-1, the ocean also is a huge reservoir of mineral salts, containing the elements bromine, iodine, and magnesium. The ocean contains millions of tons of dissolved precious metals such as gold, silver, and platinum. However, the process of separating these precious metals from seawater currently costs more than the metals are worth.

In addition to the precious metals in the ocean, geologists estimate that about 200 billion tons of minerals in the form of nodules lie on the ocean floor. When inexpensive deep-sea mining methods are developed, nodules may become an important source of metallic ores.

**Food.** Fish and shellfish from the ocean are an important part of the world's food supply. However, some kinds of fish and shellfish are becoming scarce or unfit to eat because of overfishing and pollution of coastal areas. To ensure an adequate supply of fish and shellfish, coastal nations will have to practice conservation. These nations also must avoid destroying breeding grounds and polluting coastal areas where many fish are born and grow to maturity.

**Fresh Water.** For most nations, it is difficult to increase the supply of fresh water. The removal of salt from seawater, or **desalination**, can overcome this problem for coastal nations. Methods of desalinating seawater are already being used in some Middle Eastern countries, including Israel and Saudi Arabia. Desalination involves freezing, evaporating, or passing seawater through special membranes to remove the salts. However, desalination is a costly process that most countries cannot afford to implement.

**Tidal Power.** Where large differences exist between the height of high tide and low tide, the tides have been used to produce electricity. For example, along the coast of Maine, the tides have been harnessed to produce electric power. The incoming tides are allowed to flow behind a large dam. Then, the gates of the dam are closed. When the tide goes out, the water is trapped behind the dam. At low tide, the trapped water is released. As the water passes through the dam, the water turns turbines that generate electricity. Unlike burning coal or oil, this method of producing electric power does not cause either water or air pollution.

# CHAPTER REVIEW

## *Science Terms*

*The following list contains all of the boldfaced words found in this chapter and the page on which each appears.*

bathyscaphs (p. 139)
continental shelves (p. 140)
continental slope (p. 140)
coral islands (p. 142)
current (p. 144)
desalination (p. 148)
drift (p. 146)
estuaries (p. 147)
fiord (p. 147)
island arc (p. 141)
lagoon (p. 147)
mid-ocean ridges (p. 141)

nodules (p. 143)
ocean basins (p. 139)
salinity (p. 136)
sandbars (p. 146)
sea cliffs (p. 147)
seamounts (p. 141)
tide (p. 146)
trenches (p. 142)
underwater canyons (p. 140)
volcanic island (p. 141)
waves (p. 143)

## *Matching Questions*

*On the blank line, write the letter of the item in column B that is most closely related to the item in column A.*

*Column A*

_____ 1. salinity

_____ 2. echo sounder

_____ 3. bathyscaph

_____ 4. continental shelf

_____ 5. trench

_____ 6. ooze

_____ 7. nodules

_____ 8. trough

_____ 9. ridge

_____ 10. drift

_____ 11. delta

_____ 12. lagoon

*Column B*

*a.* ship used to descend to the ocean floor
*b.* deepest parts of the ocean
*c.* mineral lumps found on the ocean floor
*d.* depression in a wave
*e.* ocean wave produced by an underwater earthquake
*f.* wide, shallow, slow-moving masses of ocean water
*g.* measure of salt content in a body of water
*h.* area of water between a sandbar and the mainland
*i.* portion of the ocean floor adjoining the land
*j.* measures ocean depths by means of sound waves
*k.* parts of the ocean floor having a maximum depth of 7 kilometers
*l.* feature formed at the mouth of a river
*m.* very fine sediments found on the ocean floor
*n.* underwater mountain range

# Multiple-Choice Questions

*On the blank line, write the letter preceding the word or expression that best completes the statement.*

**1.** The largest ocean is the
   *a.* Atlantic  *b.* Pacific  *c.* Indian  *d.* Arctic
   1 ____

**2.** The average depth of the ocean is about
   *a.* 1 km  *b.* 2 km  *c.* 4 km  *d.* 11 km
   2 ____

**3.** The average salinity of seawater is about
   *a.* 1.5%  *b.* 3.5%  *c.* 5.5%  *d.* 8.5%
   3 ____

**4.** Of the following compounds, the most abundant in seawater is
   *a.* sodium chloride        *c.* magnesium chloride
   *b.* calcium carbonate      *d.* magnesium bromide
   4 ____

**5.** The salinity of seawater depends *least* upon
   *a.* the presence of mountains along the coasts
   *b.* the flow of rivers
   *c.* the melting of glacial ice
   *d.* areas of heavy rainfall
   5 ____

**6.** Deep ocean water, compared to surface ocean water,
   *a.* has more ice in it        *c.* is much warmer
   *b.* has longer currents       *d.* is much colder
   6 ____

**7.** Seawater freezes at a temperature of about
   *a.* −2°C  *b.* 0°C  *c.* 2°C  *d.* 10°C
   7 ____

**8.** Sound travels through the water at a speed of about
   *a.* 30 m per second        *c.* 1500 m per second
   *b.* 300 m per second       *d.* 1800 m per second
   8 ____

**9.** The continental slope levels off at an average depth of 4000 m, but may extend downward to
   *a.* 5000 m  *b.* 6800 m  *c.* 8000 m  *d.* 11,000 m
   9 ____

**10.** The deepest part of the Puerto Rico Trench extends down about
   *a.* 4 km  *b.* 5 km  *c.* 7 km  *d.* 9 km
   10 ____

**11.** The formation of ocean waves is mainly caused by
   *a.* the wind  *b.* the moon  *c.* Earth's rotation  *d.* Earth's revolution
   11 ____

**12.** The major surface current that flows along the east coast of the United States is the
   *a.* Mexican Current  *b.* Peru Current  *c.* Bermuda Current  *d.* Gulf Stream
   12 ____

**13.** The relatively calm area of the Atlantic Ocean in which thick masses of seaweed float is called the
   *a.* Doldrums  *b.* Driftless Area  *c.* Sargasso Sea  *d.* Coral Sea
   13 ____

**14.** The width of the Gulf Stream is about 160 km, and its depth is about
   *a.* 30 m  *b.* 300 m  *c.* 600 m  *d.* 1000 m
   14 ____

**15.** Subsurface currents that originate in polar regions are called
   *a.* countercurrents  *b.* Arctic surge  *c.* polar creep  *d.* ice flows
   15 ____

**16.** One of the factors that decreases the salinity of seawater is
    *a.* a decrease in rainfall       *c.* the melting of ice
    *b.* a drop in temperature     *d.* an increase in evaporation     16 \_\_\_\_

**17.** The length of the Mississippi River Delta is about
    *a.* 9 km   *b.* 32 km   *c.* 90 km   *d.* 320 km     17 \_\_\_\_

**18.** Nodules on the ocean floor may become an important source of
    *a.* seafood   *b.* salt   *c.* electricity   *d.* metals     18 \_\_\_\_

# *Modified True-False Questions*

*In some of the following statements, the italicized term makes the statement incorrect. For each incorrect statement, write the term that must be substituted for the italicized term to make the statement correct. For each correct statement, write the word "true."*

**1.** The deepest place in the ocean is about *4 km* below sea level.     1 _____

**2.** The substance that makes up about 75% of the salt content of the ocean is *table salt*.     2 _____

**3.** The amount of salt in seawater is called *saltation*.     3 _____

**4.** The echo sounder operates on the principle that *electricity* passes easily through salt water.     4 _____

**5.** The average width of the continental shelf is about *10 km*.     5 _____

**6.** Very deep, canyonlike features on the ocean floor are called *trenches*.     6 _____

**7.** The deepest part of the Atlantic Ocean is found off the island of *Bermuda*.     7 _____

**8.** The salinity of the ocean *decreases* as large amounts of fresh water enter the ocean.     8 _____

**9.** A very large, shallow, slow-moving surface current in the ocean is called a *creep*.     9 _____

**10.** A *delta* will usually form at the mouth of a river that flows into a relatively quiet body of water.     10 _____

**11.** The water between an offshore sandbar and the mainland is called a *spit*.     11 _____

**12.** Submerged, or drowned, river valleys along coastlines are called *estuaries*.     12 _____

# *Testing Your Knowledge*

1. Why does the salinity of seawater vary in different parts of the ocean? _____

_____

2. Why is seawater warmest at the end of summer? _____

_____

3. How can echo-sounding equipment locate fish? _____

_____

4. Draw a diagram of the ocean floor. Label and describe the following: trench, continental shelf, continental slope, and ridge.

5. Why is it safer to be on a submarine than on an ocean liner during a hurricane at sea?

_____

_____

6. Draw a diagram that shows how a breaker forms on a long, gently sloping shore.

7. Suppose you have three bottles of fresh water and salt water. One bottle is filled with fresh water at a temperature of 23°C, the second is filled with salt water at a temperature of 23°C, and the third is filled with fresh water at a temperature of 7°C. If you poured the contents of these bottles slowly into a large tank of fresh water at a temperature of 23°C, what would be the general paths followed by the poured water? Draw a simple diagram to illustrate your answer.

8. How do you think that a larger part of the world's food supply and minerals will eventually

come from the ocean? _____

_____

_____

# The Ocean Floor: Is It the World's Best Garbage Dump?

Every year, the people of the United States produce 300 million tons of *sludge*, roughly enough to form a one-inch-deep ribbon that could circle Earth two times. Sludge is sewage, which contains human waste and other substances, that has been processed in a sewage treatment plant and turned into material with the consistency of watery mud. Obviously, sludge has to be disposed of in a sensible manner, or parts of Earth could end up looking like a huge garbage dump.

The usual way to dispose of most of our country's sludge is to burn it up, pile it up on existing land, or create new land, or *landfills*, usually at the edges of bodies of water. These processes, however, often create unsightly dump sites, foul odors and air pollution, and disrupt the balance of living things in the local environment. They are also a very costly way to get rid of treated sewage.

Mainly because of environmental concerns, state and Federal laws were passed, beginning in the 1970s, that severely restricted the use of landfills as dumping grounds for sludge and other garbage. In the 1970s, there were about 20,000 such dumps operating in the country. By 1995, environmental laws had reduced that number to about 3600, and it is still dropping. But what are we to do with the millions of tons of sewage our society produces?

Dr. Charles D. Hollister, of the Woods Hole Oceanographic Institute on Cape Cod, in Massachusetts, suggests that we investigate the ocean bottom as a potential dump site for sludge. As he puts it, "Does it make sense to close off 70 percent of the earth's surface [under the oceans] without at least studying the idea of using it for waste disposal?"

Dr. Elliot A. Norse, of the Center for Marine Conservation in Washington, D.C., objects to such a study on the grounds that organisms that are living on the ocean floor "have no experience with toxic (poisonous) materials in sewage sludge." The populations of such organisms might be severely reduced, or even wiped out. This, he feels, might cause a kind of chain reaction that would endanger other ocean plants and animals.

On the contrary, say Woods Hole and other oceanographers, there is evidence that sludge *promotes* the growth of at least some bottom-dwelling animals. Scientists point to a sludge dumping site off the Atlantic coast of New York and New Jersey that they studied for ten days in 1989. The ocean floor was found to be teeming with shrimp, fish of all kinds, crabs, and other sea life. The scientists suggest that nutrients in the sludge may have prompted the growth of such animals.

Critics, however, asked whether the organisms might have taken in substances that could be poisonous to people who later ate these animals as seafood. Rodney Fugita, of the Environmental Defense Fund, was especially concerned that poisonous heavy metals, such as cadmium, chromium, lead, mercury, and zinc, may be dumped into sewage by industries and contaminate ocean life.

Recognizing that such problems exist—certainly in relatively shallow waters—some scientists have proposed delivering sludge to very deep spots in the ocean. The sludge would be piped down thousands of feet through hoses or dropped down in chains of leak-proof drums. According to one scientist, Dr. John Edmond of the Massachusetts Institute of Technology (MIT), substances in and on the deep ocean bottom tend to stay put and, thus, would not be a threat to marine organisms that live near the surface and are eaten by people. Edmond makes the point that "this ocean is mixed so slowly that one 'stir' takes over 1000 years."

In any event, special undersea vessels would be designed and built to monitor the reactions of living things to sludge on the ocean floor. If these vessels detected bad effects, the dumping process could be discontinued. But would uncontrollable damage have already been done? As Clifton Curtis, an advisor to the environmental organization Greenpeace International, says: "The ocean, we must all recognize, doesn't have fences. It is one linked ecosystem." Those links could lead to surface waters and to our kitchens.

The Woods Hole researchers are quick to counter with the statement that they are only suggesting that deep ocean sludge dumping be studied, not necessarily adopted at this time. To which Sarah L. Clark of the Environmental Defense Fund replies in no uncertain terms: "The ocean should not be the garbage can for dumping human wastes!"

Which environment should we be more concerned with, the land or the sea? Must we choose? Are there other alternatives? What do you think?

1. *Every year, the people of the U.S. produce*

 a. 3 million tons of sludge.
 b. 30 million tons of sludge.
 c. 300 million tons of sludge.
 d. 3 billion tons of sludge.

2. *The percentage of Earth's surface that is under the ocean is about*

 a. 10.     c. 70.
 b. 50.     d. 90.

3. *Dr. John Edmond said that*

 a. we shouldn't dump sludge in the ocean.
 b. ocean water mixes very slowly.
 c. ocean water mixes very rapidly.
 d. ocean water doesn't mix at all.

4. *What methods have been suggested for delivering sludge to deep spots in the ocean?*

_____

_____

5. *On a separate sheet of paper, explain why you are—or are not—in favor of using the ocean floor as a dumping ground for sludge.*

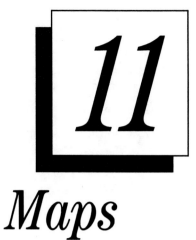

# *Maps*

## LABORATORY INVESTIGATION

### *SHOWING THE FORM OF A ROUND OBJECT ON A FLAT SURFACE*

**A.** Examine the bowl on your desk. Describe its shape. _____

_____

**B.** Find the diameter and height of the bowl using a metric ruler. The diameter of the bowl is

_____ millimeters (mm). The height of the bowl is _____ mm.

**C.** Place the bowl upside down over a sheet of white paper. With a grease pencil, draw a circle around the rim of the bowl where it meets the sheet of paper (see Fig. 11-1). Label this line 0 (zero) mm on the bowl.

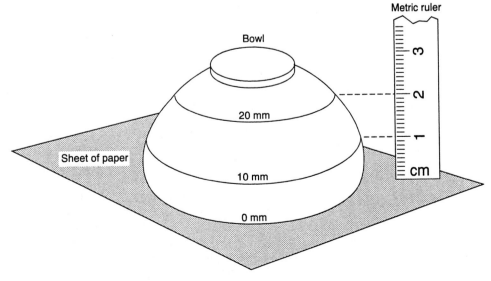

**Fig. 11-1.**

**D.** Draw another line around the bowl at a height of 10 mm above the rim. This line should be parallel to the rim. Label this line 10 mm.

**E.** Draw another line around the bowl at a height of 20 mm above the rim. Continue drawing lines around the bowl at 10-mm intervals until you can go no higher. Label the height above the rim for each of the lines.

**F.** From above, look straight down on the inverted bowl.

   **1.** How do the lines appear? _____

_____

_____

   **2.** Are the lines evenly spaced when observed from above? Explain. _____

_____

_____

   **3.** Do the lines ever cross each other? Explain. _____

_____

_____

**G.** Remove the bowl from the paper. On the paper there should be a circle with the same diameter as the maximum diameter of the bowl. Measure the diameter of this circle. The

diameter of the circle is _____ mm.

**H.** Using a compass and the inverted bowl as a guide, copy the circles on the bowl by drawing circles of the same diameter on the sheet of paper.

   **4.** Measure the distances between the circles on the sheet of paper. What do your measurements show? _____

_____

   **5.** Explain why the circles on the sheet of paper are not 10 mm apart? _____

_____

_____

_____

# Kinds of Maps

A sheet of paper on which a portion of Earth is drawn is called a *map*. Because it is impossible to represent Earth's curvature on a sheet of paper without some distortion, a map of Earth's surface cannot be perfectly accurate.

In spite of the distortion, maps are very useful. For example, a road map indicates the correct highway and direction you should travel to reach a particular destination. A road map can also help you determine the distance between cities and towns. Geologists and geographers use topographic maps. Besides distance and direction, topographic maps can accurately represent the size, shape, and location of natural features on Earth's surface.

## DIRECTIONS AND DISTANCES ON EARTH

As you know, Earth is a huge sphere that has a circumference of about 40,225 kilometers (km) at the equator. Because Earth is so huge, mapmakers have devised a system of

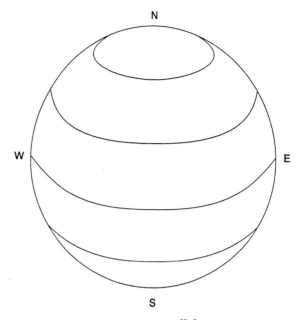

Fig. 11-3. Parallels.

accompanied by a letter—either E or W—indicating whether the meridian is east or west of Greenwich, England.

Greenwich, England, is called the **prime meridian** because this English town has been assigned the value of 0 degrees longitude. Meridians extend east or west of the prime meridian to a maximum of 180 degrees—or exactly halfway around Earth. On a map, meridians and parallels are straight lines that cross each other at right angles.

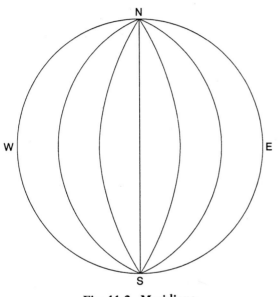

Fig. 11-2. Meridians.

*reference lines* to locate quickly and easily any place on Earth's surface.

First, Earth's surface is divided by a number of equally spaced lines that extend north and south. These equally spaced lines that converge at the North Pole and the South Pole are called *meridians* (see Fig. 11-2).

Second, Earth's surface is divided by a number of equally spaced lines that encircle Earth in an east-west direction and are parallel to the equator. These equally spaced lines are called *parallels* (see Fig. 11-3).

As you see in Fig. 11-4, each meridian and parallel has been assigned a particular number. Thus, individual meridians and parallels can be located quickly on a globe. Besides its number, each parallel is accompanied by a letter—either N or S—indicating whether the parallel is in the Northern Hemisphere or in the Southern Hemisphere. Each meridian is

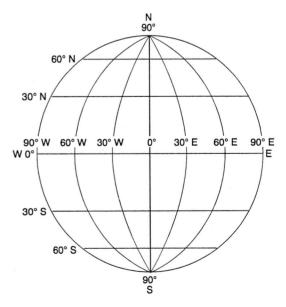

Fig. 11-4. Meridians and parallels on a globe.

# MERIDIANS OF LONGITUDE

Suppose you divide the distance around the equator into 360 equal **meridians of longitude.** Each of the meridians will measure 1/360 of Earth's circumference. The distance equal to 1/360 of Earth's circumference is called a **degree of longitude.** The system of dividing a circle into 360 equal parts also is used to mark off the scale on a protractor. If you measure the distance in kilometers from one meridian of longitude to the next meridian at the equator, the distance will be about 113 km.

Each degree of longitude is subdivided into 60 equal parts called *minutes of longitude,* and each minute is subdivided into 60 smaller equal parts called *seconds of longitude.*

**Length of a Degree of Longitude.** At the equator, a degree of longitude is about 113 km. But, as you travel toward the North Pole or South Pole, the meridians are located closer together. As a result, the distance from one meridian of longitude to the next meridian decreases (see Fig. 11-5).

The farther you travel north or south of the equator, the smaller this distance becomes.

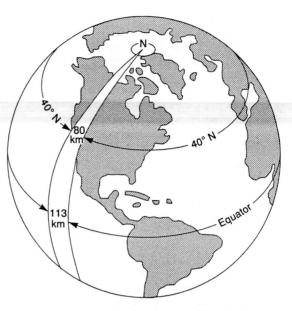

**Fig. 11-5. Length of a degree of longitude.**

For example, in the northern United States, one degree of longitude is equal to about 80 km. At the North Pole and South Pole, all of the meridians of longitude converge. Conse-

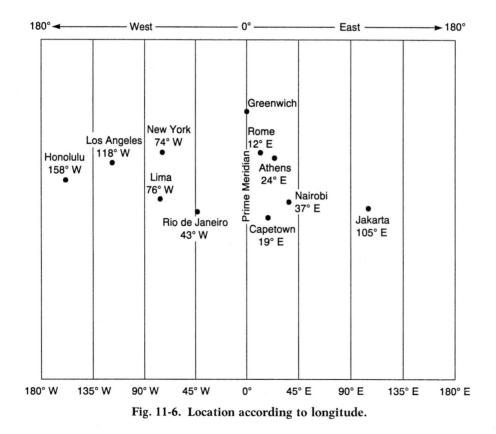

**Fig. 11-6. Location according to longitude.**

quently, the distance between two meridians of longitude at the poles is zero.

**Measuring Distances in Longitude.** Meridians of longitude are measured in *degrees east* or *degrees west* of the prime meridian located in Greenwich, England. Suppose two people travel halfway around Earth in opposite directions. One person travels east from Greenwich, England, and the other person travels west from Greenwich, England. The greatest distance in degrees either of them can be from Greenwich, England, is halfway around Earth, or 180 degrees of a circle. The east and west meridians of longitude meet at the 180 degree meridian, or the *International Date Line.*

A person traveling east would be located east of the prime meridian until he or she reached the 180th meridian of longitude. Thus, all locations east of the prime meridian (up to the 180th meridian of longitude) have *east longitude.* For example, Rome, Italy, is located about 12 degrees E longitude, or 12 degrees east of the prime meridian. Traveling farther east, Athens, Greece, is located at about 24 degrees E longitude, and Jakarta, Indonesia, is located at 105 degrees E longitude (see Fig. 11-6).

Places on Earth located west of the prime meridian (up to the 180th meridian of longitude) have *west longitude.* For example, New York City is located at about 74 degrees W longitude, and Los Angeles, California, is located at 118 degrees W longitude. Honolulu, Hawaii, is located even farther west of the prime meridian, at 158 degrees W longitude.

## PARALLELS OF LATITUDE

Mapmakers divide the distance from the equator to both the North Pole and South Pole into 90 equal parts and draw lines around Earth's circumference. These lines are called **parallels of latitude** (see Fig. 11-7). Each parallel of latitude is numbered in degrees north and south of the equator. The equator is numbered 0 degrees latitude. Unlike the meridians of longitude, distances between parallels of latitude remain equal as you travel from the equator to either the North Pole or the South Pole.

The numbers assigned to parallels of latitude increase from the equator to the poles. For example, Honolulu is located at about 21 degrees N latitude, Los Angeles is located at about 34 degrees N latitude, and New York

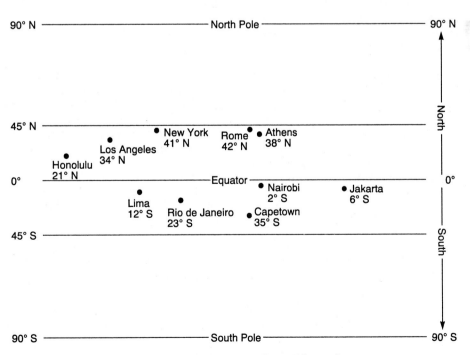

**Fig. 11-7. Location according to latitude.**

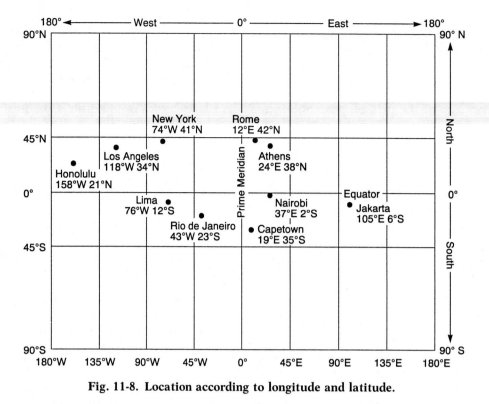

**Fig. 11-8. Location according to longitude and latitude.**

City is located at about 41 degrees N latitude. In Europe, Athens is located at 38 degrees N latitude, and Rome is located at 42 degrees N latitude. In the Southern Hemisphere, Nairobi, Kenya, is located at 2 degrees S latitude, Rio de Janeiro is located at 23 degrees S latitude, and Capetown, South Africa, is located at 35 degrees S latitude.

To locate a specific place on a map, you must know both its longitude and latitude. Fig. 11-8 shows a Mercator projection (discussed later in this chapter) on which the previously mentioned cities are listed with their longitude and latitude. At the North Pole, the latitude is 90 degrees N and at the South Pole the latitude is 90 degrees S.

**Measuring Distances in Latitude.** The distance between parallels of latitude always remain nearly the same. Anywhere on Earth's surface, the distance between one degree of latitude and the next degree of latitude is about 113 km. Like degrees of longitude, degrees of latitude also are divided into minutes and seconds. Each degree of latitude is subdivided into 60 minutes, and each minute is subdivided into 60 seconds. One minute of latitude is about 2 km long, and one second of latitude is about 30 meters (m) long.

## MAKING MAPS

Mapmaking is a complicated process. This is because a three-dimensional shape, such as a sphere (Earth) or a mountain on the sphere, cannot be drawn on a sheet of paper without stretching the sphere or mountain out of shape. You can clearly see why this is a problem by trying to wrap a sheet of cellophane around a rubber ball. To make the cellophane sheet exactly follow the curve of the ball, you must either fold and wrinkle the cellophane or stretch the cellophane over the ball.

If you wrinkle the cellophane sheet, small sections of the cellophane will smoothly fit over the ball. In a similar way, small sections of Earth's surface can be accurately represented on a sheet of paper. Maps of very large areas, however, tend to be distorted on a sheet of paper. In general, the larger the area represented by a map, the more distorted the map becomes.

# MAP PROJECTIONS

Mapmakers transfer the features of Earth's surface onto a sheet of paper by a process called **projection.** Two common types of map projection are the *Mercator projection* and *conic projection.*

**Mercator Projection.** Fig. 11-9 shows the process of making a **Mercator projection.** In this process, a cylinder of paper is wrapped around a transparent model of Earth. A lamp shining inside the model projects Earth's parallels and meridians onto the cylinder.

You can also see a small section of the cylinder in Fig. 11-9. Observe how the parallels of latitude farthest from the equator become distorted, or spread out. Consequently, a world map drawn according to the Mercator projection makes Greenland appear almost as large as North America (see Fig. 11-10). Actually, the surface area of North America is about 12 times larger than that of Greenland.

In spite of this distortion, a map drawn according to the Mercator projection is useful to navigators. That is because any straight line drawn on the map shows the true direction of the line. On a Mercator map, all of the parallels of latitude are aligned due east and west, and all meridians of longitude are aligned due north and south.

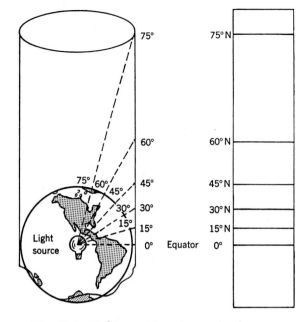

**Fig. 11-9. Making a Mercator projection.**

**Conic Projection.** Fig. 11-11 shows the process of constructing a **conic projection.** In this process, paper cones are placed over a transparent model of Earth. A lamp inside the model projects Earth's meridians and parallels onto the cone. A small strip of the cone (at the point where the cone touches the model)

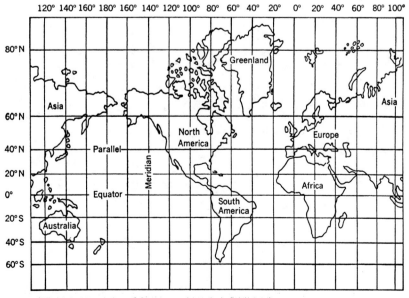

**Fig. 11-10. A world map according to the Mercator projection.**

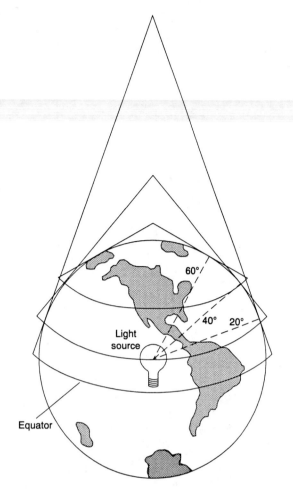

**Fig. 11-11. Making a conic projection.**

accurately represents Earth's shape. Because the narrow strip of a cone matches the curvature of Earth much better than a cylinder, maps drawn according to the conic projection are more accurate than maps drawn according to the Mercator projection.

Fig. 11-12 shows a section of Earth drawn according to a conic projection. You can see that Greenland is much smaller on a conic projection than on a Mercator projection.

There is no projection that produces a completely accurate world map. However, a fairly accurate map can be made of a small area of Earth's surface. A kind of map that shows only a small area is called a *topographic map.*

## TOPOGRAPHIC MAPS

A **topographic map** shows the *physical features* and *elevation* of a small section of Earth's surface. The physical features on a topographic map include valleys, hills, mountains, rivers, lakes, and swamps. The differences in elevation, or the height of the land above sea level, is called the **relief.** The information shown on a topographic map is obtained by *surveyors* on the ground and from *aerial photographs.*

Topographic maps also show artificial features such as highways, bridges, railroads, cemeteries, and different kinds of buildings. Different symbols are used to represent different types of features. Geologists, naturalists, engineers, aviators, hikers, and persons

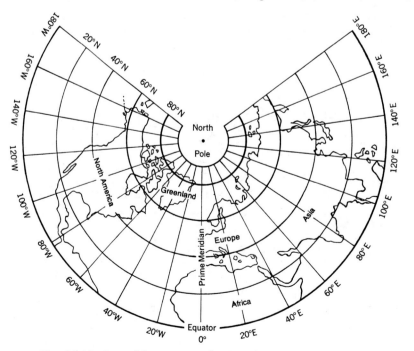

**Fig. 11-12. A world map according to the conic projection.**

who need information about the shape of the land in a particular region use topographic maps. A topographic map is easy to understand once you learn the meanings of the different lines and symbols.

**Area Shown on a Map.** Topographic maps usually show an area that is less than one degree of latitude (N-S) and one degree of longitude (E-W). For example, topographic maps drawn by the U.S. Geologic Survey show an area that is either 7.5 minutes, 15 minutes, or 30 minutes in latitude and longitude.

The surface area shown on a 15-minute map is about four times greater than the surface area shown on a 7.5-minute map. Likewise, the surface area on a 30-minute map is about four times greater than on a 15-minute map. At the equator, a 7.5-minute map shows an area of about 125 square kilometers (sq. km), a 15-minute map shows an area of about 490 sq. km, and a 30-minute map shows an area of about 1970 sq. km.

The features of an area can be shown in greater detail on a map that shows less surface area. For example, a 15-minute map shows much more detail than a 30-minute map, because a 15-minute map covers only one quarter the area of a 30-minute map. However, the 15-minute map does not show as much surface area as the 30-minute map (see Fig. 11-13).

**Map Scales.** Most topographic maps include a **map scale** that shows how the size of the map compares to the actual distance on Earth. Most maps have a scale stated both *verbally* and *graphically*. For example, the scale may state: "1 inch equals 1 mile," or "1 inch equals 10 miles." This phrase is called a *verbal scale*. In many countries the metric system is used. On maps that use the metric system, the scale may state: "1 centimeter equals 1 kilometer," or "1 centimeter equals 10 kilometers." Near the border of a topographic map is a line divided into equal parts that marks off units of distance, such as miles or kilometers. This line is called a *graphic scale*.

Maps also may have scales expressed *numerically*. For example, the scale may read: "1:62,500" or "1:250,000" or "1:1,000,000." A scale of 1:62,500 means that 1 unit on the map represents 62,500 of the same units on Earth. In other words, the map shows Earth 62,500 times smaller than it actually is. On maps drawn to a scale of 1:62,500, 1 inch (in) equals about 1 mile on Earth (1 mile equals 63,360 in).

**Contour Lines and the Shape of the Land.** In your laboratory investigation, you were asked to draw a series of lines around a bowl. These lines are called **contour lines**. Contour lines represent the shape and size of the bowl. Now, imagine that the bowl disappears, leaving the

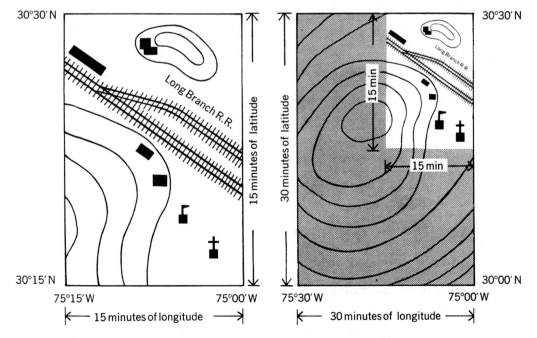

**Fig. 11-13. A 15-minute map and a 30-minute map. (Artificial features not drawn to scale.)**

contour lines suspended in space. Although the bowl is gone, you can still describe its appearance because of the contour lines. The contour lines show you the shape and size of the bowl.

Then, imagine that the contour lines drop straight down onto a sheet of paper. The lines form concentric circles on the sheet of paper that still represent the size and shape of the actual bowl. By using your imagination as you look at the concentric circles, you can easily visualize the actual dimensions of the bowl.

On a topographic map, landforms are shown by contour lines in the same way that the shape of the bowl is shown by concentric circles. Because the contour lines on topographic maps represent the shape, or contour, of the land, topographic maps are sometimes referred to as **contour maps.** In Fig. 11-14, note how the shape of a hill is represented by contour lines.

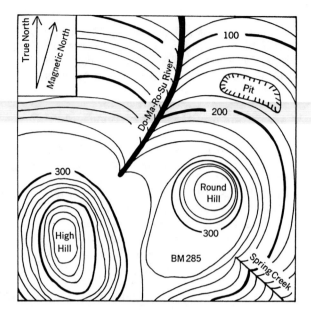

Fig. 11-15. Contour map.

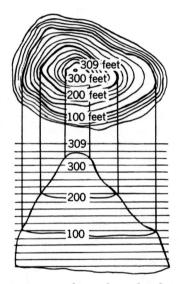

**Fig. 11-14. Contour lines show the shape of a hill.**

**Elevation of the Land.** A surveyor measures the elevation above sea level of many points in an area to construct a contour map. Each of the measurements is stamped on a round metal plate called a **bench mark.** Bench marks are then fastened to the ground. On a topographic map, a bench mark is indicated by the letters BM followed by numbers that indicate the elevation of the bench mark.

After measuring many horizontal and vertical distances, a surveyor plots the data on an outline map. All of the points that have the same elevation are connected by a contour line. Consequently, contour lines indicate both elevation and the shape of Earth's surface in a particular area.

**Contour Intervals.** The difference in elevation between two contour lines is called the **contour interval.** In the upper right section of Fig. 11-15, you will notice that there are five spaces between the 100-foot (ft) line and the 200-ft line. An elevation of 100 ft divided by five spaces indicates that the vertical distance between any two adjacent contour lines is 20 ft. Thus, the contour interval is 20 ft.

The mapmaker must decide whether the contour interval will represent a change in elevation of 1, 5, 10, 20, 50, or 100 ft. The interval used depends on the slope of the land. For gently sloping or nearly level land, a contour interval of 1, 5, or 10 ft may be used. However, for mountainous land with very steep slopes, a contour interval of 50 or 100 ft is commonly used.

As you can see in Fig. 11-15, the spacing of the contour lines on the map helps you visualize the slope of the land. The steepest part of Round Hill is on its northeast side, where the contour lines are spaced closely together. The contour lines are farther apart on the other parts of Round Hill, indicating that the slope is less steep there.

Contour maps help you determine the approximate elevation of a spot that lacks a bench mark. You can easily determine the approximate elevation by using the contour interval. For example, in Fig. 11-15 the contour interval is 20 ft. Thus, to find the height of High Hill, count the contour lines, starting from the one labeled "300." You will find that High Hill is at least 460 ft high, and that it cannot be higher than 479 ft. If High Hill were 480 ft high, the map would have to show another contour line. Similarly, you can determine that the top of Round Hill is between 360 ft and 379 ft high.

To make it easier to interpret the features on a topographic map, every fifth contour line is printed darker than the other contour lines. The elevation is often indicated on the dark contour lines. In addition, the value of the contour interval is usually stated on the bottom of the map.

**Depressions.** You may find *depressions,* or holes, in your path when climbing a hill. A contour map that showed this hill would indicate these depressions with **hachures.** Hachures are tiny comblike markings that point inward from the contour line toward the bottom of the depression. A contour line that has hachures is called a **depression contour.**

In the upper right section of Fig. 11-15, the depression contour is represented by a deep hole, or pit. The rim of this depression, indicated by the closed contour line with hachures, has the same elevation as the contour line just below, or 160 ft.

**Stream and River Valleys.** Contour lines that cross a stream or a river valley indicate the shape of that valley. In Fig. 11-15, you can see that contour lines bend in an uphill direction when crossing a stream or river, such as Do-Ma-Ro-Su River or Spring Creek. The amount of the bend indicates the shape of the stream or river valley. Contour lines that cross a narrow river valley bend sharply, in the shape of a V. Contour lines that cross a wide river valley bend more gently, in the shape of a U.

To help you interpret topographic maps, some of the common symbols used on topographic maps are shown in Fig. 11-16.

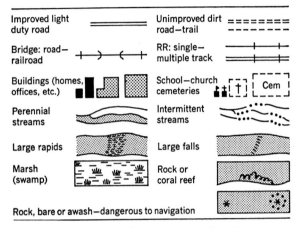

**Fig. 11-16. Symbols on topographic maps.**

# CHAPTER REVIEW

## *Science Terms*

*The following list contains all of the boldfaced words found in this chapter and the page on which each appears.*

# Matching Questions

*On the blank line, write the letter of the item in column B that is most closely related to the item in column A.*

| | Column A | | Column B |
|---|---|---|---|
| ____ | 1. equally spaced lines that converge at the poles | | *a.* map projection |
| ____ | 2. equally spaced lines that run east to west | | *b.* depressions |
| ____ | 3. equal to 1/360th of Earth's circumference | | *c.* contour lines |
| ____ | 4. transferring Earth's features onto paper | | *d.* degree of longitude |
| ____ | 5. shows physical features and land elevations | | *e.* contour interval |
| ____ | 6. represent shape of land on a map | | *f.* meridians of longitude |
| ____ | 7. elevation measurements fastened to the ground | | *g.* hachures |
| ____ | 8. difference in elevation between two areas on map | | *h.* parallels of latitude |
| ____ | 9. another term for holes in the landscape | | *i.* bench marks |
| ____ | 10. tiny markings on a map that indicate holes in the ground | | *j.* topographic map |
| | | | *k.* prime meridian |

# Multiple-Choice Questions

*On the blank line, write the letter preceding the word or expression that best completes the statement.*

**1.** The most accurate representation of Earth's surface can be made on a
   *a.* sheet of paper   *b.* cylinder   *c.* cone   *d.* sphere                     1 ____

**2.** The distance between two meridians is measured in the degrees of
   *a.* latitude   *b.* longitude   *c.* magnitude   *d.* altitude                     2 ____

**3.** In degrees of longitude, the maximum distance that one can travel east from Greenwich, England, is
   *a.* 60   *b.* 90   *c.* 180   *d.* 360                     3 ____

**4.** One degree of longitude equals a distance of zero kilometers at
   *a.* Greenwich, England          *c.* the International Date Line
   *b.* the equator          *d.* the North Pole                     4 ____

**5.** One degree of latitude equals about
   *a.* 30 m   *b.* 2 km   *c.* 48 km   *d.* 113 km                     5 ____

**6.** The scale on a map tells the reader
   *a.* the direction of true north
   *b.* the height of the land
   *c.* how the map compares with the actual size of the area represented
   *d.* the shape of the land                     6 ____

**7.** For which of the following ratios on a map does 1 inch equal about 1 mile?
   *a.* 1:5280   *b.* 1:31,500   *c.* 1:62,500   *d.* 1:250,000                     7 ____

8. Contour lines that are spaced very close together indicate that the land is
   *a.* steep  *b.* swampy  *c.* wide  *d.* flat                                         8 ____

9. A known elevation on a map is indicated by the letters
   *a.* K  *b.* H  *c.* MT  *d.* BM                                                       9 ____

10. A contour line that has tiny comblike lines along the inner edge indicates a
    *a.* mountain  *b.* cliff  *c.* valley  *d.* depression                              10 ____

# Modified True-False Questions

*In some of the following statements, the italicized term makes the statement incorrect. For each incorrect statement, write the term that must be substituted for the italicized term to make the statement correct. For each correct statement, write the word "true."*

1. On maps drawn according to the Mercator projection, Greenland appears *much larger* than North America.                                       1 _____

2. Imaginary lines that encircle Earth and pass through both the North and South poles are called *meridians*.                                      2 _____

3. In degrees, the distance from a parallel to the equator is called the *latitude* of the parallel.                                               3 _____

4. Greenwich, England, is located on the *180 degrees* meridian of longitude.                                                                     4 _____

5. At the equator, one degree of longitude equals *270 km*.                    5 _____

6. A 15-minute map represents a *smaller* area of Earth's surface than a 30-minute map of equal size.                                              6 _____

7. "One inch to 10 miles" or "one inch to 5 miles" are *graphic* statements of map scales.                                                         7 _____

8. Points on a topographic map that are the same elevation are connected by the same *contour* line.                                               8 _____

9. The letters BM on a map indicate the presence of a *bench mark*.           9 _____

10. Contour lines bend *upstream* when they cross a river valley.            10 _____

# Testing Your Knowledge

1. Why does the distance between any two parallels always remain nearly the same while the distance between any two meridians can vary greatly? _____

2. What is the main difference between a 7½-minute topographic map and a 15-minute topographic map? _____

_____

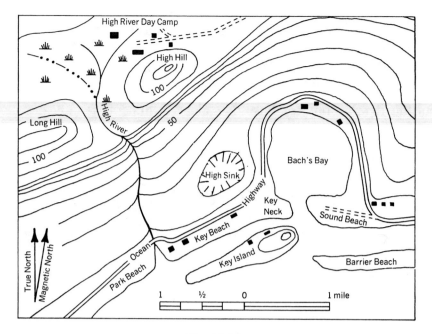

**Fig. 11-17.**

**3.** Study the map in Fig. 11-17. Then answer the following questions.

a. The contour interval on the map is _____ .

b. The elevation of all the contour lines that outline the shape of the coast is _____ .

c. The shortest distance between Key Island and Sound Beach is about _____ .

d. The greatest elevation on Barrier Beach is less than _____ .

e. High River probably flows most rapidly at or near the _____-ft contour line.

f. A depression is located at _____ .

g. In which direction must you move to travel from Park Beach to Key Beach along the

Ocean Highway? _____ .

h. To the nearest foot, what is the highest possible altitude of High Hill? _____ .

# *The History of Earth*

## LABORATORY INVESTIGATION

### *THE REMAINS OF LIVING THINGS*

**A.** Press the outer surface of a small shell into a piece of modeling clay to a depth of about 1 centimeter (cm) (see Fig. 12-1). Coat the impression in the modeling clay with petroleum jelly using a small, soft brush. Brush the impression gently to avoid destroying its fine details.

**B.** Mix three tablespoons of plaster of paris with two tablespoons of water, and add a pinch of table salt. Add enough water to make the mixture have the texture of heavy cream.

**C.** Pour the mixture into the impression in the modeling clay, called a *mold*. Tap the sides of the clay with your finger to dislodge air bubbles that may be trapped in the plaster of paris.

**D.** When the plaster of paris hardens, remove it from the modeling clay. Examine the object you have made. The object you made is called a *cast*. A cast is a replica of an object made in a mold filled with material that hardens.

    **1.** Describe the appearance of the cast you made. _____

_____

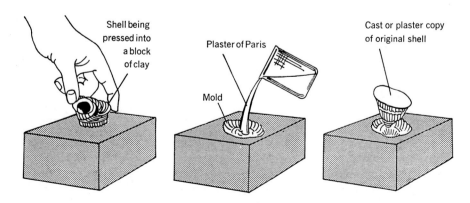

**Fig. 12-1.**

**2.** Why does the cast accurately resemble the outer surface of the shell? _____

_____

**E.** Your teacher will give you a piece of modeling clay containing an impression of an extinct animal. Make a plaster cast of the impression.

**3.** After the plaster hardens, carefully remove the cast from the clay. Describe what you see.

_____

**4.** Do you think that the cast reveals the true appearance of the extinct animal? Explain.

_____

_____

# THE AGE OF EARTH

Early in the nineteenth century, most people believed that Earth was about 6000 years old. Any person who suggested that Earth might be much older was ignored or considered to be antireligious. At that time, however, a few careful observers began to change people's beliefs about Earth's age.

One of these people was *Abraham Werner* (1750–1817), a German geologist. Werner noticed that rocks in coal mines were arranged in layers, or **strata** (see Fig. 12-2). Based on his observations, Werner proposed a theory to account for the origin of the strata.

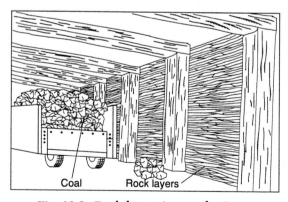

**Fig. 12-2. Rock layers in a coal mine.**

Werner suggested that the entire Earth had initially been covered by a muddy ocean. Over a long period of time, the mud settled in layers on the ocean floor, hardening into rock. The heaviest rock particles in the mud, especially particles of granite, settled to the ocean floor first. Thus, granite became the "oldest" bottom layer in a series of rock strata.

The "younger" rock strata formed on top of the granite layer as lighter particles of sediment accumulated on the ocean floor. Eventually, the entire Earth was covered by thick series of rock strata. Werner concluded that all the rocks in Earth's crust were formed by this slow process.

Another theory about Earth's age was proposed by *James Hutton* (1726–1797), a Scottish geologist. Hutton disagreed with Werner's theory about the origin of granite. Hutton stated that granite was somehow related to the lava that flows out of volcanoes and hardens in fissures in Earth's crust.

In support of his theory, Hutton pointed out that granite is commonly found filling in cracks in limestone, shale, and other layered rocks. He also noticed that limestone strata in contact with granite often had a "baked" appearance, as if the limestone had been heated. These observations led Hutton to conclude that magma had forced its way into the limestone, cooled, and then hardened into granite (see Fig. 12-3).

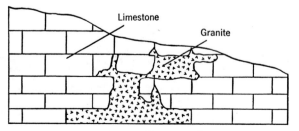

**Fig. 12-3. Granite is often found in cracks in sedimentary rocks, such as limestone.**

People who accepted Hutton's theory also thought that rock particles were formed by the erosion of preexisting rocks. When rock particles settle on the ocean floor, they eventually form layers of sedimentary rock. According to Hutton's theory, the formation of rock strata from sediments had taken many millions of years.

# Determining Earth's Age

About 1850, many people doubted that rocks were millions of years old. That is because some of the methods used by geologists to estimate Earth's age were crude. They included (1) estimating the rate at which sediments are deposited, (2) estimating the rate at which rocks erode, and (3) estimating the time required for the oceans to become as salty as they are today.

Sediments are continuously eroded from rock formations and deposited in the ocean. Consequently, nineteenth-century geologists used the rate of erosion and the rate of sedimentation to determine the amount of time needed to erode rocks and deposit sediments. The scientists estimated that it takes between 12,000 and 30,000 years to form about 1 meter (m) of sedimentary rock. Based on this figure, a bed of sedimentary rock about 4 kilometers (km) thick would take about 50 million years to form.

Geologists have determined that the Colorado River erodes the Grand Canyon at the rate of about 1 m every 3000 years. Now the Grand Canyon is about 2000 m deep. Thus, the Colorado River took about 6 million years to carve the Grand Canyon.

By comparing the total amount of salts in the ocean today with the amount being carried into the ocean yearly, scientists estimate that it took about 500 million years for the oceans to reach their current salinity of 3.5%. Although all of these early methods for estimating Earth's age are not very accurate, they suggest that Earth is much older than 6000 years. For example, because the ocean took about 500 million years to reach its current salinity, Earth must be much older than 500 million years.

## ABSOLUTE DATING

More accurate estimates of Earth's age have been made possible by the discovery of *radio-activity* and the invention of instruments that can measure the amount of radioactivity in rocks. The use of radioactivity to make accurate determinations of Earth's age, called **absolute dating**, depends upon comparing the amount of radioactive material in a rock with the amount that has decayed into another element.

**Radioactive Elements.** To date, the most accurate method of measuring Earth's age lies in studying the radiations given off by atoms of **radioactive elements.** The atoms of radioactive elements are unstable. Unstable atoms are continuously breaking down, or undergoing **radioactive decay.** In radioactive decay, different kinds of rays and particles, called *radiations*, are emitted from atoms.

As a radioactive element decays, it changes into another element. The radioactive element that decays is called the *parent element*. The new element that results from the radioactive decay of the parent element is called the *daughter element*.

For example, the radioactive decay of *uranium-238* results in the formation of *lead-206*. The rate at which a parent element decays into a daughter element is constant and is not affected by heat, pressure, or ordinary chemical or physical reactions.

**Radioactive Decay.** The time required for one half of a given amount of a radioactive element to decay is called the **half-life** of that element. For example, the half-life of uranium-238 is about 4.5 billion years. Thus, after 4.5 billion years only half of an original mass of uranium-238 in a rock would remain. The decayed half would have changed into several other radioactive elements in sequence before it finally became lead. It would take another 4.5 billion years for half of the remaining half-block of uranium-238 to also change into lead, and so on (see Fig. 12-4).

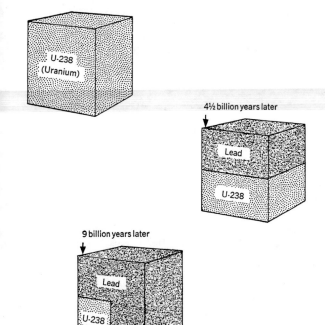

**Fig. 12-4. Half-life of uranium-238.**

The half-lives of radioactive elements range from milliseconds to billions of years. Uranium-238 with a half-life of 4.5 billion years is commonly used to calculate the age of very old rocks. *Radioactive carbon* (carbon-14) is another radioactive element. Carbon-14, with a half-life of 5730 years is used to date organic materials that are up to about 75,000 years old.

**Uranium Dating.** Because of its very long half-life, uranium-238 is useful in dating rocks that are billions of years old. By comparing the amount of lead-206 in a rock to the amount

of uranium-238 present, a geologist can calculate the rock's age.

Using the **uranium dating** method, geologists have calculated that rocks found in the Black Hills of North Dakota are about 1.5 billion years old and that some rocks in northern Canada are almost 4 billion years old. Earth must have existed before the formation of the most ancient rocks in the crust. Therefore, Earth must be older than 4 billion years. Geologists think that Earth is about 4.6 billion years old.

**Carbon Dating.** In the cells of all plants and animals, both carbon-12 (ordinary carbon) and *carbon-14* occur in fixed proportions. When a plant or animal dies, the carbon-14 in its cells begins to decay into the element *nitrogen*. Consequently, the amount of carbon-14 in the cells of the dead plant or animal decreases at a constant rate.

The amount of carbon-12 does not change in plant or animal cells. Thus, by measuring the proportion of carbon-14 to carbon-12 in an organic substance, a geologist can use **carbon dating** to calculate the age of the substance. For example, suppose a geologist determined that half of the carbon-14 in an ancient bone had changed into nitrogen. The geologist could then infer that the bone must be about 5730 years old.

Carbon-14, or radioactive carbon, is used to date organic materials that are up to about 75,000 years old. Using carbon dating, scientists were able to determine that the Dead Sea Scrolls were 2000 years old. The Dead Sea Scrolls were written on parchment, an ancient writing material made out of animal skins.

# Geologic Time and the Evolution of Life

Geologists think that Earth formed from clouds of contracting gas and dust particles in space. The original surface of Earth was covered by partially molten or completely molten material, with volcanic eruptions occurring

regularly. These volcanic eruptions emitted gases that formed Earth's first atmosphere, consisting of carbon dioxide, ammonia, methane, and water vapor.

As Earth's surface cooled, an original crust

formed, with some elevated areas and some very large basins. When the water vapor in the atmosphere cooled, it condensed into raindrops. Rain fell continuously for millions of years, gradually filling in the large basins to form the world ocean.

## BEGINNINGS OF LIFE

Living things probably evolved in the world ocean that formed more than 3 billion years ago. At that time, it is thought, all living things were microscopic, single-celled organisms. Over hundreds of millions of years, larger, more complex organisms, such as *worms* and *jellyfish* evolved in the ocean. The bodies of these early life forms were soft. Animals with hard parts, such as shells, appeared much later. The shells of these early animals are often found today preserved as fossils in sedimentary rocks.

The next major group of animals to appear on Earth were the *fish*. Fish have gills, which enable them to breathe underwater. Over time, some fish evolved simple lungs, which allowed them to breathe air from the atmosphere.

In time, some of the fish with primitive lungs and lobe fins evolved into *amphibians* (see Fig. 12-5). Frogs and newts are examples of amphibians. Amphibians have gills at birth and spend the early part of their life underwater. As they mature, amphibians develop lungs and can live on land. Amphibians, however, must lay their eggs in water.

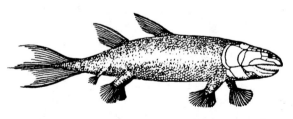

**Fig. 12-5. Primitive lobe-finned fish were the ancestors of amphibians.**

Some amphibians evolved into *reptiles*. Some reptiles later evolved into *birds* and *mammals*. And, over time, plants evolved from simple, one-celled aquatic organisms, such as algae, into enormous land plants such as the giant redwood tree.

## FOSSILS

The study of extinct plants and animals indicates that living things have evolved from life forms that previously existed. Evidence of the extinct plants and animals that previously existed have been found in sedimentary rocks as *fossils*. A **fossil** is the remains or trace of an ancient organism that has been preserved naturally in Earth's crust. Fossils have helped geologists understand the evolution of life on Earth.

**Fossils in Rocks.** Few fossils are found in metamorphic rocks, and virtually none are found in igneous rocks. The magma from which igneous rocks are formed is so hot that any organism trapped in the magma usually is destroyed, leaving no remains.

Fossils are usually found in sedimentary rocks. That is because clamlike animals living in the mud, and other marine organisms with hard body parts, were covered relatively quickly by the constant deposition of sediment on the ocean floor. As layer upon layer of sediment accumulated, the fossils became embedded in the slowly forming sedimentary rock strata. The oldest fossils known are the traces of 3.5-billion-year-old bacteria found in sedimentary rocks.

A fossil indicates the form of an ancient organism. A fossil also can provide some information about the behavior of an ancient organism as well as how it became a fossil. In some cases, the entire body of an extinct animal has been preserved. In other cases, only the footprints of the extinct animal have been preserved. From the appearance of the footprints, scientists try to reconstruct some of the animal.

**Fossils in Ice.** The hard parts, especially the teeth and bones of an animal, are more likely to be preserved as fossils. The soft parts decay rapidly. In some cases, however, even the soft parts of an animal may be preserved for thousands of years.

For example, a group of Russian explorers discovered the carcass of an elephantlike animal called a *woolly mammoth*, embedded in the ice of a Siberian glacier. Woolly mammoths lived during the last ice age, about 15,000 years ago (see Fig. 12-6).

**Fossils in Tar.** Some of the best-preserved animal remains have been discovered in natural tar pits. When an animal accidently fell into the tar, it became trapped. The trapped animal sank deep into the tar, where bacteria,

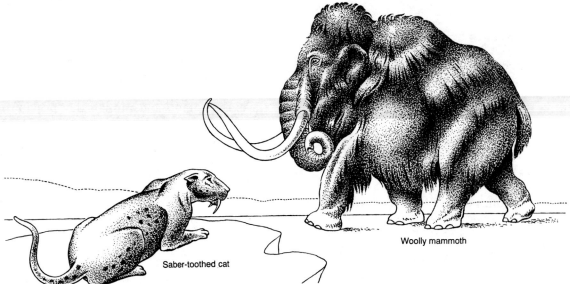

**Fig. 12-6. Saber-toothed cat and woolly mammoth.**

animals, and the weather could not destroy all of its remains.

When bones are removed from a tar pit and cleaned, they give a good idea of the animal's original appearance. Bones of the extinct *saber-toothed cat* have been preserved as fossils in tar pits. Many of these remains have been removed from the *La Brea Tar Pits*, located in Los Angeles, California (see Fig. 12-6).

**Fossils in Amber.** Pine trees produce a sticky substance called *resin*. Resin produced by some extinct species of pine trees hardened and turned into **amber,** or *fossil resin*. Millions of years ago, insects sometimes became trapped in the hardening resin. Today, you can observe the form and color of these prehistoric insects clearly preserved in such amber.

**Fossil Molds and Casts.** Sometimes, the bones and shells embedded in sedimentary rocks were dissolved completely by natural processes. However, the spaces previously occupied by individual bones and shells retain their original shapes. This kind of hollow space in a rock is called a **fossil mold.**

Sculptors often use clay to make a hollowed mold. They fill the mold with plaster of paris to get a solid impression, or *cast* of the mold. Many statues are casts made in this way. In the laboratory investigation, you made a cast of a shell with plaster of paris. Similarly, if a hollow fossil mold left in sedimentary rock by a dissolved shell or bone becomes filled with sediments or minerals that later harden, then a **fossil cast** is formed.

**Fossil Tracks.** When birds or other animals walk over a layer of mud, they leave tracks, or footprints, in the mud. If the mud hardens into stone, the imprints of their feet remain as **fossil tracks.** There are many famous fossil tracks of dinosaur feet. In a similar way, the trails of earthworms and many other impressions of extinct organisms have been left in sediments that subsequently hardened into rock.

**Replaced Fossil Remains.** Fossil remains that have a different composition from the living plant or animal are called *replaced remains*. For example, many *petrified trees* found in Arizona no longer consist of wood, but are composed of the mineral quartz. The original wood has been "changed to stone," or *petrified*, as described in Chapter 5 (see Fig. 5-13). Petrified dinosaur bones also have been found in Arizona and other parts of the southwestern United States.

## FOSSIL RECORD

As you learned in Chapter 6, running water carries rock particles such as sand and clay into the ocean. In time, these sediments accumulate and harden into sandstone and shale, as described in Chapter 3. At the same time, the shells and hard parts of small marine animals sink to the ocean floor and become embedded in slowly forming sedimentary rocks such as sandstone, shale, and limestone.

In some places, sedimentary rock strata are more than 12 km thick. One of the most magnificent displays of sedimentary rock strata is the *Grand Canyon* of the Colorado River. The

Grand Canyon is located where the Colorado River has carved a channel through more than 2 km of rock, exposing many different sedimentary rock strata.

Sedimentary rock strata can be compared to a pile of newspapers. Every day another newspaper is placed on top of the pile. Assuming that the pile is not disturbed, the oldest newspaper is always on the bottom of the pile, and the most recent newspaper is on top. Like a pile of newspapers, when sedimentary rock strata are not disturbed by natural forces, the oldest strata are on the bottom, and the more recent strata are on top. Geologists call this phenomenon the *principle of superposition* (see Fig. 12-7). Using the principle of superposition, geologists can determine the relative age of undisturbed sedimentary rock strata and the fossils inside them.

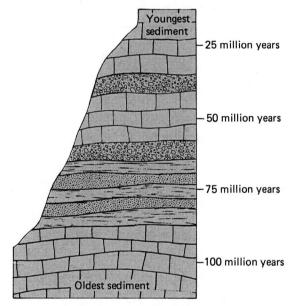

Fig. 12-7. Undisturbed sedimentary strata.

The appearance of igneous rock masses called batholiths, dikes, sills, and lava flows, discussed in Chapter 3, indicate the relative age of sedimentary rock strata and fossils. For example, a dike that cuts through sedimentary rock strata is younger than the strata it cuts through. A sill is younger than any sedimentary rock beneath it (see Fig. 12-8).

**Fossils as Timekeepers.** Sedimentary rock strata are not always left piled neatly on top of each other like a stack of newspapers. Instead, the crustal forces that build mountains and cause earthquakes have often twisted, folded, or caused the strata to break apart. In addition, weathering and erosion may have re-

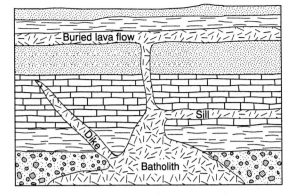

Fig. 12-8. Relative ages of rock strata.

moved entire sequences of sedimentary rock strata. Then, newer sediments may have been deposited on top of these disturbed strata, changing and confusing their original sequence.

When rock strata are disturbed, how can the strata be accurately dated? In the late 1700s, an English civil engineer named *William Smith* (1769–1839) noticed that certain rock strata always contained characteristic kinds of fossils, no matter what the position of the strata. Smith confirmed his observation of fossils in widely scattered parts of England. Although the locations of the strata were different, the similarity of the fossils in similar types of rock strata led Smith to conclude that the rocks were of the same age.

William Smith proposed the idea that certain easily identified and widely scattered fossils could be used to date periods of Earth's history. Any fossils used to date strata in this way are called **index fossils.** Index fossils are often used to *correlate*, or match, rock strata that are separated from each other by many kilometers (see Fig. 12-9). By studying index fossils in disturbed and undisturbed strata, and by radioactive dating, geologists have

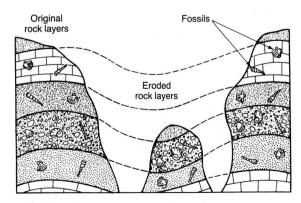

Fig. 12-9. Correlation of rock strata by index fossils.

been able to figure out the order in which fossils and rock strata were originally deposited.

Using index fossils, geologists have determined the probable order in which plants and animals appeared on Earth. The *fossil record* indicates that the simpler forms of life appeared first, followed by the more advanced and complex life-forms. Over hundreds of millions of years, more complex life-forms evolved from simple organisms.

## GEOLOGIC TIME

Earth's history extends over more than four billion years, and is reckoned in terms of a **geologic time scale.** *Paleontologists,* or scientists who study the relationship between fossils and Earth's history, have divided this huge period of time into four large time units called *eons.* Eons are divided into smaller units of time called *eras.* An era refers to a time interval in which particular plants and animals were dominant, or present in great abundance. The end of an era is most often characterized by (1) a general uplifting of the crust, (2) the extinction of the dominant plants or animals, and (3) the appearance of new life-forms.

Each era is divided into several smaller divisions of time called *periods.* Some periods are divided into smaller time units called *epochs.* Eras, periods, and epochs represent distinctive time units of Earth's history. However, you should note that an eon, era, period, and epoch are not exact units of measurement, such as a day, hour, or minute. In fact, no two geologic units represent the same amount of time.

In this chapter, only the five major eras are described. These major eras are listed in Table 12-1.

### TABLE 12-1. MAJOR ERAS OF EARTH HISTORY

| Era | Years Ago (began) |
|---|---|
| Cenozoic (recent life) | 65 million |
| Mesozoic (middle life) | 245 million |
| Paleozoic (ancient life) | 570 million |
| Proterozoic (earlier life) | 2.5 billion |
| Archean (primitive life) | 4 billion |
| Hadean (before life appears) Formation of Earth | 4.5 billion |

## PRECAMBRIAN TIME

**Precambrian** time makes up most of Earth's history. This unit of time, consisting of about four billion years, is divided into three eons. They are the *Proterozoic Eon,* the *Archean Eon,* and the *Hadean Eon.*

Little is known about Precambrian time. That is because heat, pressure, and chemical reactions have greatly changed the rocks formed during the Proterozoic, Archean, and Hadean eons. Fossils have not been discovered in rocks of the Hadean Eon and are extremely rare in rocks of the Archean and Proterozoic eons. In the Archean Eon, most plants and animals consisted entirely of soft parts, which decayed rapidly after death.

Carbon remains have been found in Archean rocks. This has convinced many paleontologists of the existence of simple, one-celled organisms during early Precambrian time. During the Proterozoic and Archean eons, living things probably were either plantlike marine organisms, such as bacteria and algae, or soft-bodied marine animals, such as protists, jellyfish, and worms.

## PALEOZOIC ERA

The **Paleozoic Era** is divided into three ages. Each of the ages is characterized by marked differences in the dominant life-forms. These ages are called the *Age of Invertebrates,* the *Age of Fishes,* and the *Age of Amphibians.*

**Age of Invertebrates.** Animals without backbones are called *invertebrates.* Sometimes, the body of an invertebrate is enclosed in a shell. Worms, clams, lobsters, and insects are familiar examples of invertebrate animals.

The oldest sedimentary rocks of the Paleozoic Era contain the fossils of numerous invertebrates that had shells or other hard parts. Two outstanding examples of shelled invertebrates that flourished during the Paleozoic Era are *trilobites* and *brachiopods* (see Fig. 12-10). Many types of large seaweed also flourished in the ocean during the Paleozoic Era.

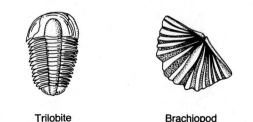

Trilobite          Brachiopod

**Fig. 12-10. Common fossils from the Paleozoic Era.**

**Age of Fishes.**   Primitive types of fish appeared early in the Paleozoic Era. Fish were the first animals with backbones, or *vertebrae.* Such animals are called *vertebrates,* and also include the amphibians, reptiles, birds, and mammals. Fish became the dominant life-form by the middle of the Paleozoic Era.

At the same time that fish appeared in the ocean, more invertebrates, such as spiders and scorpions, evolved on land. Plants, such as mosses and ferns, also first appeared during the Paleozoic Era.

**Age of Amphibians.**   All of the earliest life-forms lived in the ocean. In time, however, some marine vertebrates developed the ability to survive out of water for relatively long periods of time. Amphibians evolved from these air-breathing, aquatic animals. Frogs, toads, and salamanders are examples of familiar amphibians. Amphibians evolved into the first reptiles.

Insects also developed toward the end of the Paleozoic Era. Large swarms of giant flying insects, some with wingspans of almost 1 m, flourished in the swampy fern forests during the late Paleozoic Era. These swampy forests later changed into the fossilized remains that form the coal layers that are mined today.

The Paleozoic Era ended with the extinction of some dominant life-forms and drastic changes in the physical appearance of Earth. The trilobites, which had lived in the ocean for millions of years, became extinct. In addition, landforms changed. Large sections of North America rose above sea level. For example, at the end of the Paleozoic Era, the Appalachian Mountains rose in the eastern United States.

## MESOZOIC ERA

The **Mesozoic Era** is commonly known as the *Age of Reptiles.* That is because *dinosaurs* were the dominant life form during most of the Mesozoic Era. Dinosaurs ranged in length from about 10 cm to about 30 m. The best-known dinosaurs include the *brontosaurs, tyrannosaurs,* and *triceratops.* Most paleontologists think that snakes, turtles, and lizards probably evolved from very ancient lizardlike reptiles that later also gave rise to the ancestors of the various dinosaurs.

Unlike amphibians, all reptiles lay their eggs on land. Inside the leathery shell of a reptile egg is food for the developing *embryo.* When it emerges from its protective shell, the young reptile is able to live on its own.

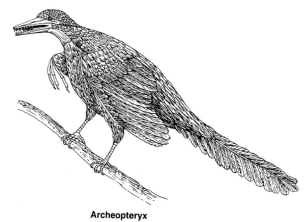

**Archeopteryx**

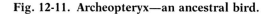

**Fig. 12-11. Archeopteryx—an ancestral bird.**

Birds first appeared during the Age of Reptiles. These birds evolved from certain types of reptiles. Some paleontologists think that birds evolved directly from dinosaurs. One of the first birdlike animals was the strange-looking one that scientists call *archeopteryx.* Archeopteryx had physical characteristics of both birds and reptiles, including teeth, claws on the front of its wings, and a feathered tail (see Fig. 12-11).

The Mesozoic Era ended with the rising of the Rocky Mountains and the extinction of many life-forms, including the dinosaurs. Some scientists now think that the dinosaurs and other dominant life-forms that disappeared at the end of the Mesozoic Era were victims of a large asteroid that struck Earth. The theory is that the impact of the asteroid caused huge amounts of dust to be thrown upward into the atmosphere. The dust particles blocked incoming sunlight, which caused the Earth's climate to become much cooler. The climate change proved fatal to the dinosaurs and many other dominant life-forms.

## CENOZOIC ERA

The most recent era of geologic time, including the present, is called the **Cenozoic Era.** The Cenozoic Era also is called the *Age of Mammals.* Mammals made their first appearance during the Mesozoic Era. However, only at the end of the Mesozoic Era, about 65 million years ago, did mammals begin to increase in variety and numbers. Today, mammals are a dominant form of animal life.

Nearly all mammals have body hair, or fur. Mammals feed their young milk, which is produced by special glands in the mother. Rats, dogs, cats, kangaroos, whales, horses, and human beings are examples of mammals.

The ancestors of horses, mice, squirrels, dogs, monkeys, elephants, and many other mammals first appeared during the Cenozoic Era. Saber-toothed cats and woolly mammoths also were among the mammals that appeared during this era. However, the saber-toothed cat, woolly mammoth, and many other extremely large mammals became extinct about 12,000 years ago, toward the end of the last ice age (see Chapter 7).

## HUMAN BEGINNINGS

Archaeologists have found the fossilized bones and teeth of humanlike primates, or *hominids*, that date back millions of years. The most primitive apelike humans, called *australopithecines*, lived from about one to four million years ago and, evidently, walked upright on two feet. Their bones have been found in eastern and southern Africa. *Homo habilis*, dated at over one-and-a-half million years, was a tool user and had a bigger brain than that of the australopithecines.

The discovery of humanlike bones in caves in Europe, Asia, and Africa indicates that a more advanced hominid called *Homo erectus*

lived from about one million to 500,000 years ago. The first true humans, *Homo sapiens*, lived in Africa, Asia, and Europe more than 200,000 years ago. These humans are called *Neanderthals*. Neanderthals lived in caves and other shelters, used stone tools and fire, and hunted animals for food and clothing.

About 35,000 years ago, Neanderthals were replaced by *Cro-Magnons*, a more modern *Homo sapiens*. The Cro-Magnons looked much as people look today. Cro-Magnons were good hunters and fishers, and fashioned a great variety of tools. They also had remarkable artistic ability, producing beautiful paintings of animals on cave walls in southern France and Spain.

## THE FUTURE OF HUMANS

Throughout geologic time, new life-forms have continually evolved from existing organisms. Many different kinds of plants and animals have become extinct after having lived for millions of years.

Compared to the vast amount of time that has passed since life first appeared on Earth, it seems that the *Age of Humans* has just begun. The human brain probably surpasses the brain of any other animal in intelligence. This high level of intelligence has enabled peo-

## TABLE 12-2. HIGHLIGHTS OF EARTH HISTORY

| Era | Years Ago (began) | Dominant Animal Life | Dominant Plant Life | Major Geologic Changes |
|-----|-------------------|----------------------|---------------------|------------------------|
| Cenozoic | 65,000,000 | Mammals, Birds | Flowering plants | Ice ages; Alps and Himalayas rise; Appalachians rise again |
| Mesozoic | 245,000,000 | Reptiles (Dinosaurs) | Cone-bearing plants, Ferns | Rocky Mountains and Sierra Nevada rise; the Palisades form; the Atlantic Ocean opens up |
| Paleozoic | 570,000,000 | Amphibians, Fishes, Invertebrates | Mosses, Ferns | Swamps; Taconic, Green, and Appalachian mountains rise |
| Proterozoic | 2,500,000,000 | Worms, Sponges (Invertebrates) | Algae | Great volcanic and earthquake activity |
| Archean | 4,000,000,000 | Unknown simple forms of life | Bacteria | Formation of Earth continues |
| Hadean | 4,500,000,000 | No life; only chemical elements and compounds | | Formation of Earth, oceans, and atmosphere |

ple to control their environment and sometimes even to change the environment to suit human needs. As a result, the needs of other organisms are often forgotten. Humans now have to learn to use their intelligence to try to protect Earth's environment and the millions of other inhabitants that also share a long and complex evolutionary history.

## HIGHLIGHTS OF EARTH HISTORY

Table 12-2 summarizes the most important geologic changes that have occurred during the history of Earth. The table also shows the evolution and dominance of some groups of living things during the past four billion years.

# CHAPTER REVIEW

## Science Terms

*The following list contains all of the boldfaced words found in this chapter and the page on which each appears.*

absolute dating (p. 171)
amber (p. 174)
carbon dating (p. 172)
Cenozoic Era (p. 177)
fossil (p. 173)
fossil cast (p. 174)
fossil mold (p. 174)
fossil tracks (p. 174)
geologic time scale (p. 176)

half-life (p. 171)
index fossils (p. 175)
Mesozoic Era (p. 177)
Paleozoic Era (p. 176)
Precambrian (p. 176)
radioactive decay (p. 171)
radioactive elements (p. 171)
strata (p. 170)
uranium dating (p. 172)

## Matching Questions

*On the blank line, write the letter of the item in column B that is most closely related to the item in column A.*

### Column A

_e_ 1. earliest forms of life on Earth

_f_ 2. hollow space in the shape of an organism

_g_ 3. layers of rocks that are many years old

_i_ 4. minerals hardened into shape of an organism

_b_ 5. atoms that emit rays and particles

_d_ 6. wood that has changed to stone

_j_ 7. Ages of Amphibians and of Fishes

_a_ 8. fossil resin that preserves insects

_h_ 9. Age of Mammals

_c_ 10. Age of Reptiles

### Column B

a. amber
b. radioactive elements
c. Mesozoic Era
d. petrified
e. single-celled organisms
f. fossil mold
g. strata
h. Cenozoic Era
i. fossil cast
j. Paleozoic Era

# Multiple-Choice Questions

*On the blank line, write the letter preceding the word or expression that best completes the statement.*

1. At the beginning of the nineteenth century, most people believed that Earth was about
   a. 6000 years old                  c. 100 million years old
   b. 6 million years old             d. 5 billion years old                    1 *a*

2. The salts in seawater were added to the oceans mainly by
   a. landslides  b. salt mines  c. rivers  d. evaporation                      2 *b*

3. When radioactive uranium decays, it changes into
   a. lead  b. salt  c. tin  d. copper                                          3 *a*

4. The rate of radioactivity of uranium-238 can be speeded up by
   a. heat  b. pressure  c. chemicals  d. none of these                         4 *d*

5. The age of a sample of wood can be calculated by carbon dating up to about
   a. 100 years                       c. 4.5 million years
   b. 75,000 years                    d. 500 million years                      5 *c*

6. The first living things probably were
   a. dinosaurs  b. jellyfish  c. single-celled organisms  d. bony fish         6 *c*

7. Amphibians probably evolved from
   a. insects  b. dinosaurs  c. reptiles  d. fish                               7 *d*

8. The rock type in which most fossils are found is
   a. volcanic  b. igneous  c. metamorphic  d. sedimentary                      8 *d*

9. A hollowed-out space in a rock that retains the form of an animal that was buried in the rock is called a
   a. cast  b. mold  c. print  d. petrified form                                9 *b*

10. A piece of wood that has been changed to stone is referred to as
    a. cast  b. molded  c. crystallized  d. petrified                          10 *d*

11. The fossils that are most useful in dating strata are called
    a. datum fossils  b. index fossils  c. easy fossils  d. true fossils       11 ____

12. The eras that comprise most of Earth's history (about three billion years) are the
    a. Archean and Proterozoic         c. Paleozoic and Cenozoic
    b. Cenozoic and Mesozoic           d. epochs                               12 *a*

13. The oldest evidence of life on Earth is found in rocks containing
    a. decayed logs                    c. carbon deposits
    b. salt crystals                   d. coal deposits                        13 *c*

14. In the Age of the Amphibians, the dominant life-form was probably an early "cousin" of the
    a. snake  b. spider  c. crab  d. frog                                      14 *d*

15. Which of the following animals is most closely related to the dinosaur?
    a. frog  b. fish  c. bat  d. bird                                          15 *d*

16. The first true humans that lived in Europe, Asia, and Africa over 100,000 years ago were the
    a. Neanderthals  b. Cro-Magnons  c. Australopithecines  d. *Homo erectus*  16 *a*

# Modified True-False Questions

*In some of the following statements, the italicized term makes the statement incorrect. For each incorrect statement, write the term that must be substituted for the italicized term to make the statement correct. For each correct statement, write the word "true."*

1. According to estimates based on how rapidly rock erodes, it has taken the Colorado River about *six million* years to cut the Grand Canyon.

   1 _____

2. Elements that give off rays when they decay are called *radioactive*.

   2 _____

3. The half-life of a substance is the time it takes for *all* of a given quantity of that substance to change into another substance.

   3 _____

4. When uranium-238 decays, it changes into *gold*.

   4 _____

5. Earth is between four and five *million* years old.

   5 _____

6. The age of a prehistoric dwelling can be determined by measuring the radioactive *uranium* contents of its remains.

   6 _____

7. The first life-forms probably appeared on Earth about *500,000* years ago.

   7 _____

8. The parts of prehistoric animals most likely to be preserved as fossils are *muscles*.

   8 _____

9. When sediments are undisturbed, the oldest layers are found at the *top*.

   9 _____

10. Trilobites are invertebrates that lived about 550 *thousand* years ago.

    10 _____

11. The first vertebrates were primitive *reptiles*.

    11 _____

12. Present-day cousins of the dinosaurs are *elephants*.

    12 _____

13. The La Brea site in California is noted for the many remains of extinct animals preserved in its *salt* deposits.

    13 _____

14. Early humans called *Cro-Magnons* were capable hunters, fishers, and artists.

    14 _____

# Testing Your Knowledge

1. Describe one method of relative dating and one method of absolute dating. _____

   _____

   _____

2. How would scientists calculate the age of an animal skin found in a cave? _____

   _____

**3.** If fossils of worms have never been found in sediment, how can scientists conclude that these invertebrates existed hundreds of millions of years ago? _____

_____

**4.** Can trees of the Petrified Forest in Arizona be used as firewood? Explain. _____

_____

**5.** The energy of sunlight from millions of years ago is released when coal is burned. Explain.

_____

_____

**6.** Why have fossils not been found in the oldest rocks on Earth? _____

_____

**7.** Number the following life-forms in the order of their appearance in Earth's history. (The oldest form should be number 1, and the most recent form should be number 6.)

reptiles _____ amphibians _____

fish _____ mammals _____

worms _____ one-celled organisms _____

# *The Atmosphere*

# LABORATORY INVESTIGATION

## *GASES IN THE ATMOSPHERE*

You will receive three closed bottles labeled oxygen and three closed bottles labeled carbon dioxide. All of the bottles are closed with rubber stoppers. Oxygen and carbon dioxide are two gases that are present in air.

**A.** Place a sheet of white paper behind one bottle of carbon dioxide.

   **1.** What is the color of carbon dioxide? _____

**B.** Fill a medicine dropper with limewater. Remove the rubber stopper from one bottle of carbon dixoide, and quickly add 10 drops of limewater to the bottle. Replace the stopper, and shake the bottle.

   **2.** Does any change take place in the bottle? Explain. _____

_____

**C.** Loosen the stopper of another bottle of carbon dioxide. Light a wooden splint, remove the stopper, and insert the burning splint into the bottle of carbon dioxide.

**CAUTION: Wear safety goggles. Do not allow the flame to touch the sides of the bottle.**

   **3.** What happens to the burning splint? _____

_____

**D.** Add 10 drops of limewater to an empty bottle. Stopper the bottle, and shake.

   **4.** What happens to the limewater? _____

_____

   **5.** What does this indicate about the composition of the air around us? _____

_____

Remove the stopper from the same bottle of air, and insert a burning splint for a few seconds. Remove the splint. Stopper the bottle, and shake.

**6.** What happens to the limewater? _____

_____

**7.** What can you conclude from your observations? _____

_____

_____

**E.** Place a sheet of white paper behind the bottle of oxygen.

**8.** What is the color of oxygen? _____

_____

**F.** Loosen the stopper on the bottle of oxygen. Light a wooden splint, and let it burn for a few seconds. Then blow out the flame.
While the splint is still glowing, quickly remove the stopper from the bottle, and carefully thrust the glowing splint into the bottle. Repeat this procedure three times.

**9.** What happens to the glowing splint? _____

_____

Insert a glowing splint into a bottle of air.

**10.** What happens to the glowing splint? _____

_____

_____

**11.** What does this experiment indicate about the composition of air? _____

_____

_____

**12.** Compare and contrast the effects of carbon dioxide and oxygen on a burning splint. ___

_____

_____

**G.** Wet the inside of a test tube with water. Insert a wad of wet, untreated steel wool inside the test tube. Holding your thumb over the mouth of the test tube, shake the test tube. Place the test tube in a jar or beaker of water as shown in Fig. 13-1.

**13.** What is the length of the test tube? _____ centimers (cm).

**14.** What does the test tube contain besides the wad of steel wool? _____

**15.** What is the height of the water inside the test tube? _____ cm.

Mark the level of the water with a grease pencil.

**H.** Set the test tube aside. After 24 hours have passed, examine the contents of the test tube.

**16.** Describe the contents of the test tube. _____

_____

**17.** Mark the level of the water inside the test tube with a grease pencil. What is the height of the water inside the test tube?

_____ cm.

**18.** Approximately how much of a change has taken place in the water level inside the

test tube? _____ cm.

**19.** What might have caused the change in

the water levels? _____

_____

_____

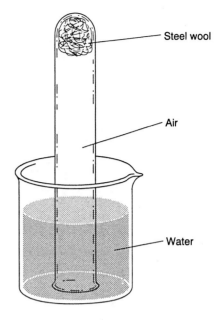

Steel wool

Air

Water

**Fig. 13-1.**

**20.** What percentage of the test tube's length does this change in the water level represent? (Divide the value obtained in question 18 by the value obtained in question 13, and multiply

the result by 100.) _____

**21.** Does the test tube contain the same quantity of air that it contained originally? Explain.

_____

# ☐ Earth's Original Atmosphere ☐

The **atmosphere** is a mixture of gases and solids that completely covers the lithosphere and hydrosphere. Careful observation and study by weather scientists, or _meteorologists,_ have explained many of the processes that occur in the atmosphere. Your life is greatly influenced by atmospheric events. For example, violent storms in the lower part of the atmosphere occasionally cause extensive property damage and even death for humans.

When Earth first formed, the atmosphere was much different from the current atmosphere. Scientists think that the elements hydrogen and helium were abundant in Earth's original atmosphere. Hydrogen and helium, however, are the lightest elements. Thus, most of these gases escaped rapidly into outer space.

As Earth slowly cooled and hardened into solid matter, the lava from volcanic eruptions

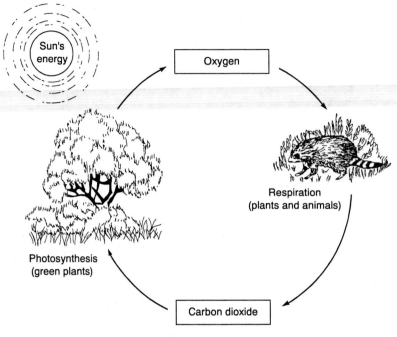

**Fig. 13-2. Plants use carbon dioxide during photosynthesis.**

released additional gases into the atmosphere. These gases consisted mainly of the compounds *ammonia* ($NH_3$), *methane* ($CH_4$), *water vapor* ($H_2O$), and *carbon dioxide* ($CO_2$). In time, sunlight decomposed, or broke down, some of these gaseous compounds into single elements, such as *nitrogen, oxygen,* and *hydrogen*.

At the same time that most of the lightweight hydrogen was escaping into outer space, nitrogen was accumulating near Earth's surface. That is because nitrogen is too heavy to escape Earth's gravity. Thus, large amounts of nitrogen gradually accumulated in the atmosphere.

**Oxygen in the Atmosphere.** Earth's early atmosphere probably contained a lot of carbon dioxide and not much oxygen. However, after algae and other aquatic plants evolved (about 3.5 billion years ago), oxygen increasingly became a major ingredient of the atmosphere. By about 400 million years ago, the atmosphere probably contained as much oxygen as it does today.

During *photosynthesis*, plants give off oxygen as a waste product (see Fig. 13-2). In the presence of sunlight, plants take in carbon dioxide from the air and use the carbon dioxide to make food through photosynthesis. Consequently, as plants became abundant, a greater amount of oxygen was added to the atmosphere, and more and more carbon dioxide was removed from the atmosphere.

**Loss of Carbon Dioxide from the Atmosphere.** Some of the carbon dioxide present in Earth's early atmosphere dissolved in the ocean. The carbon dioxide combined with calcium compounds in seawater to form *carbonate* compounds (see Chapter 2). Many marine animals and some plants used these carbonate compounds to build their shells and other hard parts. Thus, the amount of carbon dioxide in the atmosphere decreased gradually because (1) carbon dioxide was increasingly used by plants in photosynthesis and (2) large amounts of carbon dioxide dissolved in the ocean and were used by marine organisms.

# Present Composition of the Atmosphere

Paleontologists think that the proportion of gases in the atmosphere has remained relatively constant for about the past 400,000 years. This figure corresponds roughly with the appearance of abundant plants on Earth. (In Chapter 9, you learned how the nitrogen cycle maintains a constant level of nitrogen in the atmosphere—about 78% by volume.)

As stated above (and shown in Fig. 13-2), during photosynthesis, oxygen is given off by plants, and carbon dioxide is removed. During respiration, animals give off carbon dioxide and remove oxygen. This exchange of oxygen and carbon dioxide is called the *carbon dioxide-oxygen cycle* (refer back to Fig. 9-9).

The composition of the atmosphere will probably remain relatively constant for the next several hundred years. Humans, however, may cause changes as they continue to add pollutants to the atmosphere. For example, the enormous amount of coal, oil, and other fuels burned since 1900 has led to a 15% increase in the amount of carbon dioxide in the atmosphere. The $CO_2$ in the air in recent decades is as follows: 1970, 0.0321%; 1980, 0.0330%; 1990, 0.0340%; and 2000, 0.0369%.

Many scientists think that the emission of carbon dioxide from the burning of fossil fuels may be the main cause of global warming. Like the glass on a greenhouse, carbon dioxide allows sunlight to enter the atmosphere. As sunlight strikes Earth's surface, the light energy is changed into heat energy. Carbon dioxide prevents heat from leaving the atmosphere in a similar way that the glass on a greenhouse prevents heat from escaping a greenhouse. Consequently, an increase in atmospheric carbon dioxide will cause a warming of Earth's climate. This warming phenomenon is called the *greenhouse effect*.

**An Experiment with Steel Wool.** In your laboratory investigation, you wedged a piece of untreated steel wool into the bottom of a test tube. After 1 day, you observed the water level inside the test tube. You discovered that when oxygen is removed from the air, the volume of gas remaining in the test tube is about 4/5 of its original volume. Thus, oxygen must occupy about 1/5 of the gases in the atmosphere.

The gas that remained in the test tube was mainly nitrogen. Nitrogen makes up about 4/5 of the gases in the atmosphere. The composition of a sample of pure, dry air is shown in Table 13-1 below.

**TABLE 13-1. COMPOSITION OF PURE, DRY AIR**

| Gas | Approximate Percentage (by volume) |
|---|---|
| Nitrogen ($N_2$) | 78.08 |
| Oxygen ($O_2$) | 20.95 |
| Argon (Ar) | 0.94 |
| Carbon dioxide ($CO_2$) | 0.03–0.04 |
| Neon (Ne) Helium (He) Krypton (Kr) Xenon (Xe) Hydrogen ($H_2$) Nitrous oxide ($N_2O$) Methane ($CH_4$) Other gases | 0.003 total |

**Water Vapor in the Atmosphere.** Water as a gas is called **water vapor.** Water vapor is always present in the atmosphere. However, the amount of water vapor in the atmosphere varies in different parts of the world. The percentage of water vapor in the atmosphere varies from a low figure of about 0.5% to as much as 4% of the total volume of air.

The amount of water vapor in the atmosphere is called the **humidity.** On very humid days, the water vapor content of the air may approach 4%. On very dry days, the water vapor content of the air may be less than 1%.

Water vapor displaces other gases when it enters the atmosphere. Thus, any increase in the amount of atmospheric water vapor will reduce the percentages of the other atmospheric gases. For example, on a very humid day, the oxygen content of the atmosphere may be about 18% instead of about 21% by volume.

**Dust in the Atmosphere.** In addition to gases, the atmosphere also contains dust particles. **Dust particles** include sands, clays, soils, sea salts, and ashes from erupting volcanoes and forest fires.

Dust particles in the atmosphere are related to weather phenomena such as rain and snow. That is because water vapor collects on the surface of dust particles. Under the proper conditions, the water vapor condenses into water droplets. The water droplets form clouds, which may bring precipitation as rain or snow.

Dust also has an effect on the sky's blue color. The sky looks blue during the day because of *light scattering*. Light scattering occurs when air molecules separate the *blue* wavelengths of light from the other wavelengths, or colors, of sunlight (see Fig. 13-3). As the amount of dust increases, the blue color tends to decrease.

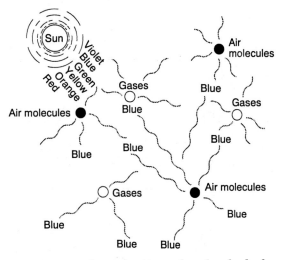

**Fig. 13-3. Light scattering makes the sky look blue.**

When you look at the sky during the middle of the day, your eyes receive scattered blue wavelengths. Thus, you see the color blue. However, at sunrise or sunset, sunlight travels to the observer at a lower angle, passing through much more dust than during the middle of the day. In this case, the blue wavelengths are removed, and long wavelengths come through more readily. Consequently, at sunrise and sunset the sky often looks yellow, orange, or red.

The sky appears black to an astronaut who is traveling in outer space. That is because beyond the atmosphere there are no gas or dust particles to scatter and reflect sunlight.

The relationship of dust particles to fog and smog is discussed in Chapter 17.

# PROPERTIES OF GASES IN THE ATMOSPHERE

The properties of many atmospheric gases are important to people and all other organisms. For example, gaseous oxygen supports respiration. *Combustion*, or burning, also requires oxygen. Thus, if there was no oxygen in the atmosphere, it would be impossible for people to burn fuels, such as wood and coal. Even more important, most organisms on Earth could not breathe without oxygen.

Another important form of oxygen is *liquid oxygen*. Liquid oxygen is oxygen gas cooled to the point at which it condenses into a liquid. Liquid oxygen (called LOX) is often used in rocket engines because LOX lets fuels burn with great energy. This gives rocket engines enough force to escape Earth's atmosphere.

Unlike oxygen, carbon dioxide does not support combustion. You verified this fact during your laboratory investigation. In fact, carbon dioxide often is used to extinguish fires, especially gasoline fires. Carbon dioxide extinguishes fire by cutting off the supply of oxygen needed for the combustion of gasoline (see Fig. 13-4).

Suppose a piece of steel wool is heated over a flame until the steel wool glows red. What would happen if the glowing piece of steel wool were dropped into a bottle of gaseous oxygen? From the results of your laboratory investigation, you would predict that the piece of steel wool would begin to burn. If you perform this experiment, you will find that the steel wool burns with a brilliant white light until it is burned completely.

This experiment shows that oxygen supports combustion. Because the atmosphere contains oxygen, the logs in a campfire con-

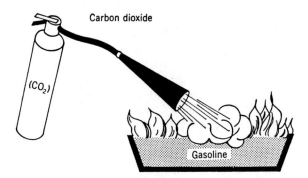

**Fig. 13-4. A carbon-dioxide fire extinguisher.**

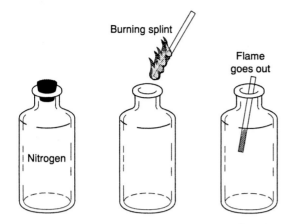

**Fig. 13-5. Nitrogen does not support combustion.**

tinue burning for a long time after the logs are ignited.

If you tried the same experiment using gaseous nitrogen instead of oxygen, the result would be different. The red-hot piece of steel wool would not burst into flame. As you can see in Fig. 13-5, when you plunge a burning wood splint into a bottle of nitrogen gas, the flame is immediately extinguished.

It's a good thing that nitrogen does not support combustion. If the atmosphere consisted mainly of oxygen gas, fires would burn with great intensity and would be very difficult to extinguish. In fact, because the atmosphere consists mainly of gaseous nitrogen (78%), burning materials usually can be extinguished in a short time.

The main atmospheric gases and some of their uses are described below.

**Nitrogen.** **Nitrogen** does not burn, nor does it support combustion. Nitrogen, however, is an important ingredient in fertilizers and in the manufacture of explosives, such as TNT and nitroglycerine. Nitrogen also is used to produce dyes, chemicals, and plastics.

**Oxygen.** **Oxygen** is an extremely active chemical element. Although oxygen does not burn, it does support combustion. Your body and most animals use oxygen to release the energy in food. Oxygen also is a waste product given off by plants during photosynthesis.

Oxygen makes up almost 50% of the weight of rocks and minerals in Earth's crust. When combined with iron compounds in rocks or manufactured products made of iron and steel, oxygen forms rust. Oxygen is an important ingredient in tap water and in many household products, including hydrogen peroxide and laundry bleaches.

**Carbon Dioxide.** Plants take in **carbon dioxide** from the air to carry out photosynthesis. Most living things (including plants) also give off carbon dioxide as a waste product of respiration. Carbon dioxide is produced by the combustion of fuels, such as wood, coal, oil, and natural gas. Carbon dioxide performs a vital function for all living things by reducing the amount of heat that escapes from the atmosphere into outer space.

Water and carbon dioxide combine to form *carbonic acid*. Carbonic acid is used to make the bubbles of gas, or fizz, in soft drinks. Fire extinguishers also contain carbonic acid, which is an effective agent used to extinguish oil and electrical fires. The solid form of carbon dioxide is called *dry ice*. Because the temperature of dry ice is $-79°$ Celsius (C), it is sometimes used as a refrigerant.

**Argon.** **Argon** is a chemically inert, or inactive, element in the atmosphere. One of the major uses of argon gas is in light bulbs. Light bulbs that are filled with argon gas last longer.

**Water Vapor.** The weight of air decreases as the quantity of water vapor in the air increases. That is because a given quantity of water vapor is lighter than equal quantities of either nitrogen or oxygen. Like carbon dioxide, water vapor also contributes to the greenhouse effect by preventing the escape of heat from the atmosphere into outer space. When water vapor condenses into liquid water droplets, the water vapor forms clouds, rain, snow, and other forms of precipitation.

**Neon.** Like argon, **neon** is an inert chemical element. When an electric current is passed through neon, it glows red. For this reason, neon is often used in electric signs and billboards.

**Helium.** **Helium** is the second lightest chemical element. Like argon and neon, helium is an inert element. Helium does not burn and does not support combustion.

Helium has many uses. Helium is used to give buoyancy, or lift, to balloons and blimps. Scuba divers use helium instead of nitrogen in their breathing equipment because helium does not cause the "bends," a problem divers face when surfacing from great depths. In addition, many old and valuable documents are preserved by being stored in containers filled with helium.

**Hydrogen.** **Hydrogen** is the lightest chemical element. Hydrogen is a very active element, which can burn in air. Blimps were formerly filled with hydrogen to give them buoyancy, or lift. However, mixtures of hydrogen and oxygen are explosive. Consequently, blimps today are filled with helium rather than with hydrogen. Hydrogen is an element that is an important ingredient in water, fossil fuels, animal and vegetable oils, and all acids.

# Structure of the Atmosphere

The atmosphere extends thousands of kilometers above Earth's surface. The gases in the atmosphere are mixed continually by local winds and global air currents close to Earth's surface. Consequently, the lower regions of the atmosphere over the entire Earth have a similar composition.

About 90 percent of the atmospheric gases are found within 30 kilometers (km) of Earth's surface. The remainder of the atmospheric gases extend upward for thousands of kilometers. However, at an altitude of about 150 km, gas molecules are widely separated. For this reason, the highest regions of the atmosphere are said to consist of "thin" air.

The atmosphere has been classified into several regions, or layers. The atmospheric layers are based on differences in composition and temperature at different altitudes.

## REGIONS OF THE ATMOSPHERE

Scientists have divided the atmosphere into regions based mainly on temperature differences. The four main regions, or layers, of the atmosphere are the *troposphere, stratosphere, mesosphere,* and *thermosphere* (see Fig. 13-6). Each of these atmospheric regions has distinctive air temperatures and other physical characteristics.

**Troposphere.** The layer of the atmosphere closest to Earth's surface is called the **troposphere.** The prefix *tropo* comes from a Greek word that means "turning or moving." In fact, the large-scale movements of air masses in the troposphere produce the various weather conditions on Earth. These air movements are described in Chapter 16.

Over the North Pole and the South Pole, the troposphere extends from Earth's surface to an altitude of about 10 km. Over the equator, the troposphere extends to an altitude of about 16 km. Most of the water vapor and dust particles in the atmosphere are found in the troposphere.

Air temperature in the troposphere decreases with increasing altitude. For every

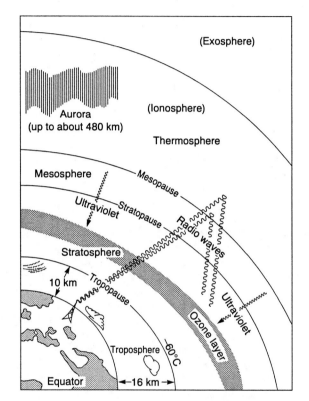

**Fig. 13-6. Regions of the atmosphere.**

1000 meters of increased altitude, the air temperature (of calm air) decreases by about 6.5°C. The troposphere ends at the altitude at which air temperature does not decrease with increasing altitude.

*Tropopause.* The highest region of the troposphere is called the *tropopause.* The tropopause is characterized by very low air temperatures. The air is so cold at the tropopause, that clouds at this altitude consist only of ice crystals.

The lowest air temperature (about −80°C) at the tropopause has been recorded over the equator. That is because the tropopause is highest above Earth's surface in equatorial regions. Over the continental United States, the air temperature averages −60°C at the tropopause.

**Stratosphere.** Above the tropopause is a layer of the atmosphere called the **stratosphere.** From the tropopause, the stratosphere extends about 48 km above Earth's surface.

The atmospheric gases in the stratosphere are similar to those found in the troposphere. However, the stratosphere contains very little water vapor and dust particles. Consequently, clouds are rarely seen in the stratosphere. Only a rare type of cloud that shines at night (*noctilucent cloud*) may be seen occasionally in the stratosphere. Noctilucent clouds consist of a very thin layer of dust particles or ice crystals.

The stratosphere usually contains a lower region of nearly steady temperature and an upper region in which the temperature increases with altitude. The temperature of the lower region is about −55°C. The temperature of the upper region of the stratosphere increases to about −2°C.

Within the upper regions of the stratosphere, there is a region consisting of a type of oxygen called ozone. This region is called the *ozone layer.* The ozone layer protects humans and all other organisms by absorbing most of the harmful ultraviolet radiation in sunlight.

**Stratopause.** The boundary between the stratosphere and the mesosphere is called the *stratopause.* The stratopause is characterized by increasing air temperature because of the ozone layer. Ozone heats the air in this part of the atmosphere by absorbing ultraviolet rays.

**Mesosphere.** From the stratopause to an altitude of about 80 km above Earth's surface is a region of the atmosphere called the **mesosphere.** The air temperature of the mesosphere decreases with increasing altitude. Extremely strong winds blow in the mesosphere. These winds blow from west to east during the winter and from east to west in the summer.

**Mesopause.** The boundary between the mesosphere and the thermosphere is called the *mesopause.* The lowest air temperatures in the atmosphere occur at the mesopause. Over the North Pole and the South Pole, the air temperature in this region approaches −110°C.

**Thermosphere.** The uppermost layer of the atmosphere is called the **thermosphere.** The thermosphere begins at the mesopause and gradually fades into outer space. The chemical composition of the thermosphere differs from the other three layers of the atmosphere. The lower regions of the thermosphere contain individual atoms of oxygen instead of oxygen molecules. The upper regions of the thermosphere consist of individual helium and hydrogen atoms.

The air temperature rises rapidly from the mesopause upward through the thermosphere. At 200 km above Earth's surface, the air temperature rises to about 600°C and then levels off.

**Ionosphere.** When high-energy solar rays strike gas molecules in the thermosphere, the collisions cause the gas particles to lose electrons. When atoms lose electrons, the atoms acquire a positive charge, and are called *ions.* Most of the ions are produced in the lower part of the thermosphere—between 80 km and 200 km above Earth's surface. The region of the atmosphere containing electrically charged, or ionized, gases is called the *ionosphere.*

The ionosphere plays an important role in long-distance radio communication. The electrically charged ions reflect radio signals back to Earth that would otherwise keep traveling into outer space.

During times of increased sunspot activity (see Chapter 20), gas particles in the ionosphere are disturbed. Because increased sunspot activity and disturbances in the ionosphere occur simultaneously, scientists think these two phenomena are related. Not surprisingly, increased sunspot activity interferes with radio communications.

**Auroras.** During periods of increased sunspot activity, another phenomenon produced in the ionosphere is the beautiful *aurora borealis,* or *northern lights.* In the Southern Hemisphere, this phenomenon is called the *aurora australis,* or *southern lights.* Auroras appear as shimmering bands, folds, and streamers of red, green, blue, and violet light.

Auroras are produced when charged particles emitted by the sun are captured by Earth's magnetic field. The solar particles smash into the thermosphere in a ring around each of Earth's magnetic poles, releasing light energy.

**Exosphere.** The upper part of the thermosphere is called the *exosphere.* The exosphere begins at an altitude of about 480 km above Earth's surface and gradually fades into outer space.

The exosphere contains an extremely small amount of atmospheric gases, mostly individual atoms of helium and hydrogen. At the upper part of the exosphere (about 1000 km), rapidly moving atoms of lightweight hydrogen escape Earth's gravity into outer space.

# CHAPTER REVIEW

## Science Terms

*The following list contains all of the boldfaced words found in this chapter and the page on which each appears.*

argon (p. 189)
atmosphere (p. 185)
carbon dioxide (p. 189)
dust particles (p. 188)
helium (p. 189)
humidity (p. 187)
hydrogen (p. 189)

mesophere (p. 191)
neon (p. 189)
nitrogen (p. 189)
oxygen (p. 189)
stratosphere (p. 191)
thermosphere (p. 191)
troposphere (p. 190)
water vapor (p. 187)

## Matching Questions

*On the blank line, write the letter of the item in column B that is most closely related to the item in column A.*

### Column A

_____ 1. water vapor in Earth's atmosphere

_____ 2. sand, clay, soil, and ash in the air

_____ 3. active element that supports combustion

_____ 4. gas produced by burning of fuels

_____ 5. gas that makes up nearly 80% of Earth's atmosphere

_____ 6. layer of Earth's atmosphere closest to Earth

_____ 7. layer of Earth's atmosphere containing ozone

_____ 8. warmest, uppermost layer of Earth's atmosphere

_____ 9. northern lights in the ionosphere

_____ 10. southern lights in the ionosphere

### Column B

*a.* oxygen
*b.* thermosphere
*c.* nitrogen
*d.* troposphere
*e.* aurora australis
*f.* humidity
*g.* aurora borealis
*h.* carbon dioxide
*i.* dust particles
*j.* stratosphere
*k.* mesosphere

## Multiple-Choice Questions

*On the blank line, write the letter preceding the word or expression that best completes the statement.*

**1.** Oxygen became a significant part of Earth's atmosphere after the emergence of
*a.* plants   *b.* dinosaurs   *c.* volcanoes   *d.* fossils

**2.** Because of the increased combustion of fuels, a gas that has increased in quantity in Earth's atmosphere is
*a.* nitrogen  *b.* oxygen  *c.* carbon dioxide  *d.* helium

2 _____

**3.** The volume of nitrogen in Earth's atmosphere at sea level is about
*a.* 4%  *b.* 21%  *c.* 78%  *d.* 96%

3 _____

**4.** The *least* abundant gas in Earth's atmosphere is
*a.* oxygen  *b.* neon  *c.* carbon dioxide  *d.* argon

4 _____

**5.** On a very humid day, the amount of water vapor in Earth's atmosphere, by volume, may be as high as
*a.* 0.5%  *b.* 0.04%  *c.* 4%  *d.* 50%

5 _____

**6.** The gas in Earth's atmosphere that normally makes up about 20% of the total volume of air is
*a.* hydrogen  *b.* carbon dioxide  *c.* nitrogen  *d.* oxygen

6 _____

**7.** The second lightest element is
*a.* neon  *b.* oxygen  *c.* hydrogen  *d.* helium

7 _____

**8.** The part of Earth's atmosphere closest to Earth's surface is called the
*a.* ionosphere  *b.* stratosphere  *c.* thermosphere  *d.* troposphere

8 _____

**9.** The air temperature in the tropopause over the continental United States is about
*a.* −30°C  *b.* −60°C  *c.* −180°C  *d.* −273°C

9 _____

**10.** The part of Earth's atmosphere that shields Earth from harmful ultraviolet radiation is called the
*a.* equatorial bulge             *c.* ozone layer
*b.* ionic layer                   *d.* protective layer

10 _____

**11.** The layer of Earth's atmosphere that reflects radio waves back to Earth is called the
*a.* ionosphere  *b.* radiation zone  *c.* aurora borealis  *d.* ozone layer

11 _____

**12.** Which of the following causes the aurora borealis?
*a.* gases escaping from Earth
*b.* particles reflected off the moon
*c.* particles given off by the sun
*d.* radiation from Earth's magnetic field

12 _____

# Modified True-False Questions

*In some of the following statements, the italicized term makes the statement incorrect. For each incorrect statement, write the term that must be substituted for the italicized term to make the statement correct. For each correct statement, write the word "true."*

**1.** Green plants and seawater remove *carbon dioxide* from Earth's atmosphere.

1 _____

**2.** On very humid days, the amount of water vapor in Earth's atmosphere may be as much as *96 percent* of the total volume of air.

2 _____

**3.** The bubbles in soda water consist of *carbon dioxide* gas.

3 _____

**4.** Neon gas glows *yellow* when electricity passes through it.    4 _____

**5.** Priceless documents are stored in *hydrogen* gas to keep them from deteriorating.    5 _____

**6.** Most of the dust and water vapor in Earth's atmosphere are found in the *stratosphere.*    6 _____

**7.** The ionosphere consists of electrically charged *gases.*    7 _____

**8.** In the *mesosphere,* some gas particles escape into outer space.    8 _____

# Testing Your Knowledge

**1.** How did the emergence of green plants on Earth cause the composition of the atmosphere to change? _____

_____

_____

_____

**2.** List two of the four major regions of Earth's atmosphere, and briefly describe the special characteristics of each region.

*a.* _____

_____

_____

*b.* _____

_____

_____

# Ozone Layer Damaged! Did the Clouds Do It?

The huge red eye didn't blink, nor did its green pupil contract and expand with the brightening and dimming of sunlight. For this eye was not part of a living thing. It was an eye in the sky, a strange cloud tens of kilometers long and a few kilometers thick. It was also the object of the attention of more than 20 scientists aboard a NASA DC 8 flying 10 km above the Antarctic peninsula a few years ago.

What were the scientists doing at the bottom edge of Earth's *stratosphere*, a layer of atmosphere that extends from 10 km to 50 km above Earth's surface? They were searching for the answer to a deadly puzzle. The puzzle? What was punching a hole through the ozone layer in the stratosphere? But why should the scientists, or we, care? And what is ozone anyway?

Ozone is a form of oxygen. But, unlike the oxygen you breathe, which consists of two atoms of oxygen bonded together, a molecule of ozone consists of three atoms of oxygen. This might be no more than an interesting curiosity except that this oxygen triplet possesses a very important property. It absorbs ultraviolet (UV) radiation streaming in from the sun. So what, you might ask? So this: UV radiation can harm or kill organisms. It can also cause skin cancer in human beings.

For a number of years, scientists had known that substances called *chlorofluorocarbons* (CFCs) were at least partly responsible for destroying ozone high in the sky. And this meant that we human beings down on Earth's surface were also responsible. That's because CFCs are chemicals that we use as coolants in refrigerators and air conditioners. Until being banned recently in many countries, CFCs were also used in various kinds of aerosol cans and in the manufacture of products such as insulating foams.

When CFCs escape into the air—and they do—they rise into the stratosphere, where they can remain for 50 to 100 years. During that time, the CFCs break up, releasing free chlorine. The highly reactive chlorine atoms can do two things. They can react with ozone—thus thinning out the ozone layer— to form a new compound called chlorine monoxide ($ClO$), or they can react with nitrogen—leaving the ozone layer undamaged—to form chlorine nitrate ($ClONO_2$).

Obviously, as far as people are concerned, a reaction with nitrogen is preferable to one with ozone. But this doesn't happen to a great enough extent, even though there is plenty of nitrogen in the stratosphere. As a matter of fact, about 80 percent of the stratosphere consists of nitrogen. So why doesn't the nitrogen hook onto the chlorine? That's the question the scientists in the DC 8, and many other scientists, were trying to answer.

Perhaps the red-eyed clouds were to blame, reasoned some scientists. But a study of these clouds revealed that they were made of pure water. They had no effect on the nitrogen in the stratosphere. But there were other clouds in the stratosphere, so thin and so spread out that they were invisible to the unaided eye. What were *they* made of?

The answer was as startling as it was revealing. These invisible clouds contained a mixture of water and . . . nitric acid ($HNO_3$)! The nitrogen that might otherwise have neutralized the chlorine set loose from CFCs was already tied up. This left the chlorine atoms free to break up ozone molecules.

But then again, if we didn't pump CFCs into the atmosphere to begin with, no harm would come to the ozone layer. So the ultimate responsibility rests with us, not the clouds above our heads.

What can we do to save the ozone layer? Is enough being done about it? What do you think?

1. *The stratosphere extends above Earth's surface from*

   a. 0 km to 10 km.
   b. 10 km to 20 km.
   c. 10 km to 50 km.
   d. 50 km to 100 km.

2. *Ozone consists of a molecule of*

   a. 1 oxygen atom.
   b. 2 oxygen atoms.
   c. 3 oxygen atoms.
   d. 2 chlorine atoms.

3. *The ozone layer absorbs*

   a. ultraviolet radiation.
   b. nitrogen.
   c. CFCs.
   d. water molecules.

4. *Describe how nitrogen might protect the ozone layer and why its effect is limited.*

   _____

   _____

   _____

   _____

5. *On a separate sheet of paper, express your views on what people—or society—might do to protect the ozone layer. Include measures already undertaken by various governments, and a discussion of the impact of your proposal.*

# Atmospheric Pressure

## LABORATORY INVESTIGATION

### *WEIGHING AIR*

**A.** In the following investigation, use a 1-liter (l) or 2-l plastic bottle that has a tight-fitting cap.

**1.** Is any substance inside the bottle? Explain. _____

**2.** Weigh the capped bottle on a platform balance (see Fig. 14-1). What is the weight of the capped bottle? _____ grams (g).

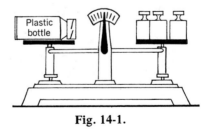

**Fig. 14-1.**

**B.** Remove the bottle from the balance, and unscrew the cap. Place the bottle on a table, and flatten the bottle as much as you can without breaking it. Reseal the bottle with the cap.

**3.** Did you change the contents of the bottle? Explain. _____

_____

**4.** Weigh the flattened, sealed bottle. What is the weight of the bottle now? _____ g.

**5.** Did you observe any difference in weight? Explain. _____

**6.** Why is it difficult to find the exact weight of the air inside the bottle, using your equipment?

_____

_____

**C.** Obtain a large tin can, and remove its cap.

**7.** Weigh the tin can on a platform balance (see Fig. 14-2a). What is the weight of the tin can?

_____ g.

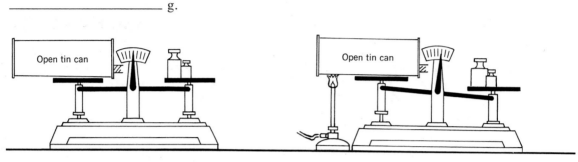

**Fig. 14-2.**

**D.** Heat one end of the tin can with a Bunsen burner for 2 minutes (see Fig. 14-2b). Use a pair of gloves to handle the tin can.

**CAUTION: Do not touch the hot tin can with your bare hands.**

**8.** What is the weight of the tin can now? _____ g.

**9.** Did you observe any difference in weight between the heated and unheated container?

Explain. _____

_____

**E.** Let the tin can cool on the platform balance. As the tin can cools, observe the position of the needle on the balance.

**10.** Does the movement of the needle indicate that the tin can is increasing or decreasing in

weight? Explain. _____

_____

## AIR EXERTS PRESSURE

You learned in Chapter 13 that people actually live at the bottom of an "ocean" of air that extends hundreds of kilometers above Earth. The enormous volume of air that makes up the atmosphere presses down on your body with great force. Your body is not crushed by this force because the air pressure inside your body exerts a force that balances the force exerted by the outside air pressure, or **atmospheric pressure.**

People sometimes feel uncomfortable when the atmospheric pressure changes rapidly. For example, when you ride to the top of a mountain in a car, you sometimes feel an unpleasant sensation inside both ears (see Fig. 14-3). This discomfort results from the atmospheric pressure changing more rapidly than the air pressure inside your ears. The difference in air pressure causes your eardrums to bulge outward slightly, causing this uncomfortable sensation. You can equalize, or balance, the air pressure on both sides of your eardrums by opening your mouth, swallowing, or chewing gum.

Atmospheric pressure varies in different locations and at different times. Most people do not notice these differences because they are too slight to be sensed. There is, however, an

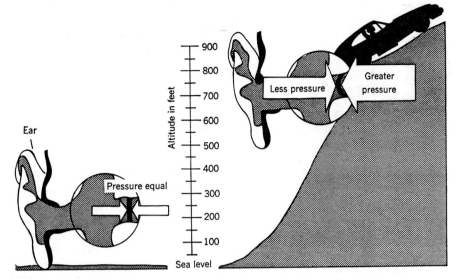

**Fig. 14-3. Changing air pressure affects the ear.**

important connection between these slight changes in atmospheric pressure and changes in the weather. You must understand the factors that cause air pressure and how air pressure is measured to appreciate the connection between atmospheric pressure and weather changes.

## CAUSE OF AIR PRESSURE

People often say that a drinking glass is empty when it does not contain water. In fact, the glass is not empty. It is actually filled with air. In your laboratory investigation, you discovered that air fills an empty container.

Air and water have many properties in common. Air and water occupy space and have weight. Under most conditions, air and water expand when warmed and contract when cooled. Also, both air and water are *fluids*, or substances that can flow.

**Weight of Air.** One cubic foot of air weighs about 1.29 ounces (37 g) at sea level. The air in a classroom 30 feet (9 meters) long, 20 ft (6 m) wide, and 10 ft (3 m) high has a volume of 6000 cubic feet (170 cu m). Thus, the air in this classroom weighs about 484 pounds (lb) (220 kilograms [kg]), figured as 6000 × 1.29/16.

Imagine an experiment in which the air in the classroom is reshaped into a tall column of air, 1 ft (.3 m) square and 6000 ft (1829 m) high. If this column of air could be placed on a platform balance, the balance would indicate that the air weighed about 484 lb (220 kg) (see Fig. 14-4).

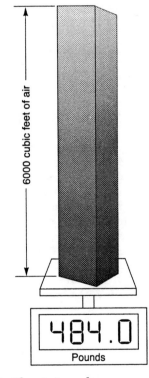

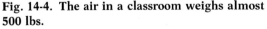

**Fig. 14-4. The air in a classroom weighs almost 500 lbs.**

In this column of air, the bottom layer would have about 484 lb of air pressing on it. Because air can be compressed, more of the air molecules would be squeezed into the lowest layer of air. In contrast, the highest layer of air would contain fewer air molecules than would any layer beneath it.

The imaginary column of air is similar to

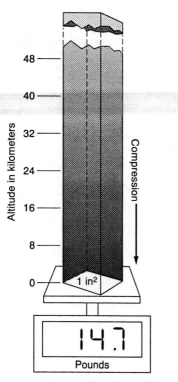

Fig. 14-5. **Air exerts pressure (14.7 lb per square inch).**

Earth's atmosphere. In Earth's atmosphere, a column of air is *densest*, or contains a maximum number of molecules, closest to the ground. In fact, about 50 percent of the weight of Earth's atmosphere is concentrated below an altitude of 6 kilometers (km), and about 99.9 percent of the atmosphere's weight is found within 50 km of Earth's surface, although the atmosphere extends upward for thousands of kilometers.

If it were possible to weigh a vertical column of air 1 square inch (sq in) in cross section that extended from sea level to the top of Earth's atmosphere, the column would weigh about 14.7 lb (6.7 kg). Fig. 14-5 shows that all of the air molecules in the column would press down on the balance with a combined force of about 14.7 lb. Thus, *atmospheric pressure at sea level exerts a force of about 14.7 lb per square inch (1.05 kg per square centimeter).*

**Direction of Air Pressure.** Air presses down on the surface of an 8-in (20-centimeter) by 10-in (25-cm) sheet—an area of 80 sq in (500 sq cm)—of paper with a total force of about 1176 lb (534 kg):

80 sq in × 14.7 lb/sq in = 1176 lb (534 kg)

The weight of Earth's atmosphere exerts a much greater force on your desk than on the sheet of paper. Why doesn't your desk collapse under this enormous force? The reason is that Earth's atmospheric pressure exerts force in *all* directions. Your desk does not collapse because the air pressure pushing upward is balanced by the air pressure pushing downward.

You can demonstrate that Earth's atmospheric pressure exerts force in all directions by performing the simple experiment shown in Fig. 14-6. All you need to perform this experiment is a small drinking glass with a smooth rim and an index card. (Perform this experiment over a sink!)

First, completely fill the drinking glass with water. Press a wet index card over the rim of the glass. Do not hold the card. Then, quickly turn the glass upside down. Surprisingly, the index card does not fall off the rim of the drinking glass.

Why does the index card remain pressed to the drinking glass even when you turn the glass upside down? Remember that the water in the drinking glass does not weigh much—perhaps 8 oz (203 g). When you turn the glass upside down, only the weight of the water is pushing downward on the index card, but the air is pushing upward with a force of about 14.7 lb per sq in (1.05 kg per sq cm) against the other side of the index card. Consequently, the index card remains firmly pressed against the rim of the drinking glass.

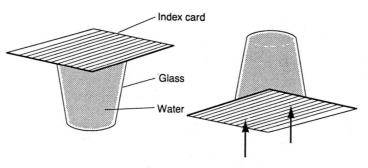

Fig. 14-6. **Air exerts pressure in all directions.**

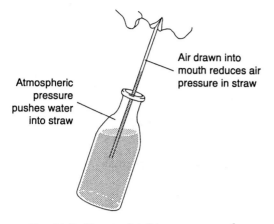

Atmospheric pressure pushes water into straw

Air drawn into mouth reduces air pressure in straw

**Fig. 14-7. How a drinking straw works.**

**Using Air Pressure.** People use air pressure every day. Air pressure is used to operate drinking straws and machines such as vacuum cleaners. A vacuum cleaner works by blowing air from the inside to the outside of a tank. As air is blown out of the tank, the air pressure inside the tank is reduced. Because the air pressure is greater outside the vacuum cleaner, the outside air rushes into the vacuum cleaner through a hose attached to the tank.

Use of a drinking straw is based on the same principle as the operation of a vacuum cleaner. When you suck through one end of a drinking straw, you remove air from inside the straw. This reduces the air pressure inside the straw. Remember that the outside air pressure is constantly pushing against the surface of the liquid. Consequently, the outside air pressure, which is now greater than the air pressure inside the straw, pushes the liquid through the straw and into your mouth (see Fig. 14-7).

## MEASURING AIR PRESSURE

**Barometers.** An instrument that measures atmospheric pressure is called a **barometer.** The first barometer was invented in 1643 by an Italian scientist named *Evangelista Torricelli* (1608–1647). Torricelli filled a long glass tube with mercury. The glass tube was sealed at one end. Torricelli held his thumb over the open end of the glass tube to prevent the mercury from escaping while he turned the glass tube upside down. He then placed the open end of the glass tube into a dish of mercury. When Torricelli removed his thumb, the column of mercury inside the tube moved down slightly, stopping about 30 in (76 cm) above the level of mercury in the dish (Fig. 14-8).

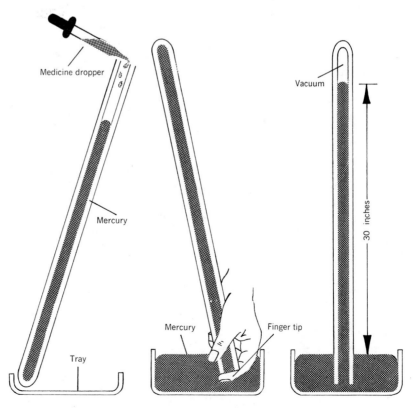

Medicine dropper

Mercury

Tray

Mercury

Finger tip

Vacuum

30 inches

**Fig. 14-8. Torricelli's mercury barometer.**

The space above the column of mercury became a *partial vacuum*. That is because only a few air molecules replaced the mercury that moved downward. Because the weight of the mercury was greater than the weight of the column of air pressing down on the mercury in the dish, the column of mercury moved downward inside the glass tube. The column of mercury stopped moving downward when the pressure it exerted equaled the pressure exerted by the bottom of the column of air pressing against the surface of the mercury in the dish.

**Mercury Barometer.** The device that Torricelli invented is called a **mercury barometer.** In a mercury barometer, the weight of a column of mercury is balanced against the atmospheric pressure. Changes in atmospheric pressure are determined by observing how the level of the mercury column changes. When the atmospheric pressure decreases, the column of mercury drops in the tube. The column of mercury rises in the tube when the atmospheric pressure increases.

**Measuring Air Pressure.** Fig. 14-8 shows that if the column of mercury in the glass tube was measured, the mercury would be about 30 in (76 cm) high. The length of the mercury column is measured from the level in the tube to the level in the dish. Readings on a mercury barometer usually range from a low of about 29 in (74 cm) to a high of about 31 in (79 cm).

The unit used by weather scientists, or *meteorologists*, to measure atmospheric pressure is the **millibar.** The scale of a barometer, however, can be marked off in either inches, centimeters, or millibars. For example, readings of 29.92 in of mercury, 76 cm, and 1013.2 millibars represent the same atmospheric pressure. Thus, a mercury level of 29.92 in exerts a pressure of 1013.2 millibars. A change of 1 in, or 2.5 cm, of mercury and a change of 33.86 millibars represent the same change in air pressure. Fig. 14-9 shows a comparison of the two barometric scales.

Can a barometer filled with water instead of mercury measure atmospheric pressure? You may recall that water is about 14 times lighter, or less dense, than mercury. Consequently, a column of water would have to be about 14 times taller than a column of mercury to measure the same atmospheric pressure at sea level. A barometer filled with water would have to be about 35 ft (12 m) high! (29.92 in × 14 = 418.88 in/12 = 34.9 ft)

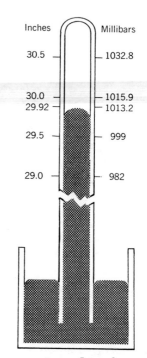

Fig. 14-9. **A comparison of two barometer scales.**

In the seventeenth century, a German scientist named *Otto von Guericke* (1602–1686) made a water barometer that was more than 10 m tall. People noticed that stormy weather usually followed a sharp drop in the water level of von Guericke's barometer. Most of the townspeople believed that von Guericke's water barometer actually caused the bad weather because of some magical property. As a result of this superstition, von Guericke was forced to remove the barometer.

**Aneroid Barometer.** A barometer that does not contain mercury, water, or any other liquid is called an **aneroid barometer.** The main part of an aneroid barometer is a thin, round, sealed metal can from which most of the air has been removed. The sides of the can are flexible so that they can move in or out in response to changes in atmospheric pressure. For example, an increase in air pressure slightly pushes in the top of the can. A spring attached to the top of the can prevents the can from totally collapsing. When the air pressure decreases, the spring pulls out the flexible side of the can.

As the top of the can moves in and out, the movement is transferred to a needle by a series of levers. The levers magnify the slight movements of the can. The needle moves around a dial marked in units of pressure (Fig. 14-10).

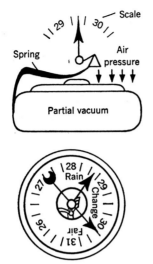

**Fig. 14-10. An aneroid barometer.**

Aneroid barometers are portable and easy to use because they are small, lightweight, and liquidless. However, they are not as accurate as mercury barometers. In scientific research, where great accuracy is needed, an aneroid barometer is usually checked against the readings of a mercury barometer.

**Barograph.** An instrument that keeps a continuous record of changes in atmospheric pressure is called a **barograph**. A barograph is an aneroid barometer attached to a clock-driven drum. A sheet of graph paper is wrapped around the drum. Instead of a needle, a pen is attached to the levers of the aneroid barometer. The pen touches the sheet of graph paper. As the drum turns, the pen draws a line showing changes in the atmospheric pressure for a given time.

## FACTORS THAT CHANGE AIR PRESSURE

Atmospheric pressure is caused by the weight of air. Thus, whenever the weight of the air changes, the air pressure also changes. Air pressure varies with changes in *altitude, temperature,* and *humidity.*

**Altitude and Air Pressure.** The higher you are above sea level, the less air there is pressing down on you. Thus, as altitude increases, the atmospheric pressure decreases. If your school building has three or more stories, you can investigate this relationship between air pressure and altitude.

First, record the air pressure with an aneroid barometer on the top floor or roof of your school. Then, carry the barometer to the ground floor or to the basement, and record the air pressure again. The air pressure at ground level or in the basement will be about 1/25 of an inch (0.1 cm) more than the air pressure on the roof or the top floor.

Atmospheric pressure decreases by about 1 in (2.5 cm) for every 1000-ft (305-m) increase in altitude, because Earth's atmosphere at high altitudes is less dense than at sea level.

If the scale of the barometer is marked off in feet instead of inches or millibars, a barometer also can be used to indicate the **altitude,** or height above sea level. A barometer that indicates altitude is called an **altimeter.**

**Temperature and Air Pressure.** Two cities located at sea level but distant from each other, such as New York City and Miami, Florida, may have different atmospheric pressures. For example, New York City might have an air pressure of 30.50 in (775 millimeters [mm]), while Miami, Florida, at the same moment might have an air pressure of 29.00 in (737 mm). Thus, altitude is not the only factor responsible for differences in atmospheric pressure.

The lower air pressure at Miami indicates that the air over Miami is lighter, or less dense, than the air over New York City. If the air were heavier, or denser, over Miami, the air would push down with greater force against the mercury of the barometer, causing a higher barometric reading.

Differences in atmospheric pressure often are caused by differences in weight between masses of warm air and cold air. Smoke rising from a chimney illustrates that hot air is lighter than cold air. You probably have noticed that the air near the ceiling of a room is warmer than the air near the floor.

If you weighed a large metal container on a scale, you would be weighing both the container and the air molecules inside the container. If you heated the container (as you did in the laboratory investigation), and then weighed the container while it was still hot, the metal container would weigh less. After heating, the container weighs less because there are fewer air molecules inside.

Heat causes the air to expand and makes the air molecules inside the container move faster. Because the air molecules are moving faster, many of them escape through the opening at the top of the container.

When air is heated in the atmosphere, the air also becomes lighter and expands. When the air in the atmosphere cools, it contracts and becomes heavier.

A given volume of warm air weighs less than an equal volume of cold air. Consequently, a mass of warm air exerts less air pressure on Earth's surface than does a mass of cold air.

**Humidity and Air Pressure.** In addition to air temperature, *humidity* causes changes in atmospheric pressure. When water vapor enters the atmosphere, the water vapor displaces some of the heavier gases in the atmosphere. You recall that a given volume of water vapor (gaseous water) weighs less than an equal volume of dry air at the same temperature. Consequently, moist air weighs less than dry air. When the atmosphere is moist, atmospheric pressure is lower than when the atmosphere is dry.

A mass of warm, moist air exerts less pressure than does a mass of cold, dry air. Therefore, on a day when the air in Miami is warm and moist and the air in New York City is cold and dry, Miami will have a lower atmospheric pressure than New York City.

## WORLD AIR PRESSURE BELTS

Although the atmospheric pressure changes continuously worldwide, some places tend to have consistently higher or lower air pressure. These regions form belts of high or low air pressure that encircle Earth. Fig. 14-11 shows the names and locations of these global **air pressure belts**. (World air pressure belts are discussed in more detail in Chapter 16.)

Earth's polar regions always have high atmospheric pressure. That is because the air in the polar regions is extremely cold and dry. In contrast, the air over equatorial regions is very warm and moist. Consequently, a belt of low atmospheric pressure persists in the region surrounding the equator.

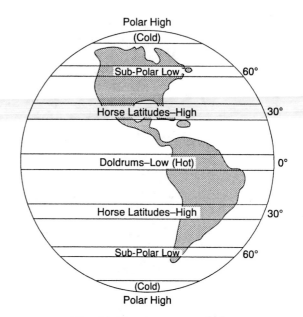

Fig. 14-11. Air pressure belts.

Temperature differences, however, do not always determine the location of a region of high or low atmospheric pressure. Other factors such as the location of land and water areas, Earth's rotation, and the upward and downward movements of large air masses also determine regions of high or low atmospheric pressure. Many other factors affect atmospheric pressure, so that the air pressure belts are not actually as simple as those shown in Fig. 14-11.

Although the location of world air pressure belts may vary, it is important to know their normal positions. That is because world air pressure belts play an important part in determining global wind patterns and the movement of storms. Winds are discussed in Chapter 16; storms are discussed in Chapter 18.

# CHAPTER REVIEW

## Science Terms

*The following list contains all of the boldfaced words found in this chapter and the page on which each appears.*

air pressure belts (p. 204)
altimeter (p. 203)
altitude (p. 203)

aneroid barometer (p. 202)
atmospheric pressure (p. 198)
barograph (p. 203)

barometer (p. 201)
mercury barometer (p. 202)
millibar (p. 202)

# Matching Questions

*On the blank line, write the letter of the item in column B that is most closely related to the item in column A.*

| *Column A* | *Column B* |
|---|---|
| _____ 1. force of atmospheric pressure at sea level | *a.* barograph |
| _____ 2. tool used to measure atmospheric pressure | *b.* altitude |
| | *c.* altimeter |
| _____ 3. keeps a record of changes in pressure | *d.* temperature and humidity |
| _____ 4. average reading on mercury barometer scale | *e.* 14.7 lb per square inch |
| _____ 5. refers to the height above sea level | *f.* millibars |
| | *g.* 30 inches of air pressure |
| _____ 6. also cause changes in air pressure | *h.* barometer |
| _____ 7. units used to measure atmospheric pressure | |

# Multiple-Choice Questions

*On the blank line, write the letter preceding the word or expression that best completes the statement.*

**1.** About 50 percent of the atmosphere's weight is concentrated between sea level and an altitude of about
*a.* 1.5 km   *b.* 6 km   *c.* 48 km   *d.* 160 km          1 _____

**2.** The air pressure that is normally exerted against each square inch of your body at sea level is about
*a.* 1.5 lb   *b.* 5 lb   *c.* 10 lb   *d.* 15 lb          2 _____

**3.** Your body is not crushed by Earth's atmospheric pressure because
*a.* your bones are very strong
*b.* your muscles push back against the air pressure
*c.* your body fluids absorb the pressure
*d.* the pressure inside your body balances the outside air pressure          3 _____

**4.** Devices that depend on air pressure to function include vacuum cleaners and
*a.* fire extinguishers   *b.* steam engines   *c.* drinking straws   *d.* elevators          4 _____

**5.** The space above the column of mercury in the tube of a barometer is
*a.* filled with an inert gas
*b.* a partial vacuum
*c.* connected to the atmosphere through a small opening
*d.* filled with alcohol          5 _____

**6.** The mercury in a barometer tube moves downward until it
*a.* is equal in weight to an equal volume of air
*b.* is equal in weight to a column of air having the same diameter as the inside of the tube and extending to the top of the atmosphere
*c.* reaches the level marked 29.92 inches
*d.* fills the reservoir at the bottom of the column of mercury          6 _____

**7.** The number of millibars equal to 1 inch of mercury is
  *a.* 29.92   *b.* 33.86   *c.* 1013.2   *d.* 2992.0

7 _____

**8.** Water is not used in liquid barometers because water
  *a.* evaporates in hot weather
  *b.* stains the glass
  *c.* is colorless
  *d.* barometers would have to be more than 10 m high

8 _____

**9.** An instrument designed to draw a written record of changing air pressure is called
  *a.* a barograph          *c.* a thermograph
  *b.* a mercury barometer          *d.* an altimeter

9 _____

**10.** In an aneroid barometer, the part that expands and contracts as the air pressure changes is a
  *a.* sealed can   *b.* coiled spring   *c.* steel disk   *d.* sensitive needle

10 _____

**11.** A volume of air weighs the least when it is
  *a.* hot and moist   *b.* hot and dry   *c.* cold and moist   *d.* cold and dry

11 _____

**12.** The *least* important factor in the location of air pressure belts over Earth is the
  *a.* location of areas of land and water
  *b.* upward and downward movement of large air masses
  *c.* rotation of Earth
  *d.* amount of rainfall in a given area

12 _____

# Modified True-False Questions

*In some of the following statements, the italicized term makes the statement incorrect. For each incorrect statement, write the term that must be substituted for the italicized term to make the statement correct. For each correct statement, write the word "true."*

**1.** A cubic foot of air weighs about *37 grams* at sea level.

1 _____

**2.** Nearly 100% of the total weight of Earth's atmosphere is found *above* an altitude of 50 km.

2 _____

**3.** The barometer was invented in 1643 by the Italian scientist *Evangelista Torricelli.*

3 _____

**4.** The column of mercury in a liquid barometer is supported by a column of air in the atmosphere that has a weight of almost *15 lb* per square inch.

4 _____

**5.** A unit used to indicate atmospheric pressure is the *millibar.*

5 _____

**6.** A barometer using water would have to be approximately *1 m* tall to measure air pressure accurately.

6 _____

**7.** A barometer that does not use a liquid to measure air pressure is an *aneroid* barometer.

7 _____

**8.** A type of barometer used to measure altitude is a *barograph.*

8 _____

**9.** A belt of high pressure is usually found at the *equator.*

9 _____

**10.** The different pressure belts in Earth's atmosphere are caused by differences in temperature and *humidity.*

10 _____

# *Testing Your Knowledge*

1. How can a barometer be used to measure changes in altitude? _____

   _____

2. Why is air several thousand feet above sea level lighter in weight than air at sea level?

   _____

   _____

3. How can a barometer located inside a building measure the atmospheric pressure? _____

   _____

4. How can a column of air 2.5 cm in diameter and hundreds of kilometers high support a column of mercury 2.5 cm in diameter and only 76 cm high? _____

   _____

5. What two conditions would cause an air mass at sea level to weigh less? How do these conditions affect the quantity of air? _____

   _____

6. Why do the interiors of continents become centers of high atmospheric pressure in the winter?

   _____

   _____

# Heating the Atmosphere

## LABORATORY INVESTIGATION

### *HEATING AND COOLING: SAND AND WATER*

**A.** Set up two thermometers and two beakers (see Fig. 15-1). Place each thermometer about 3 centimeters (cm) above the bottom of its beaker.

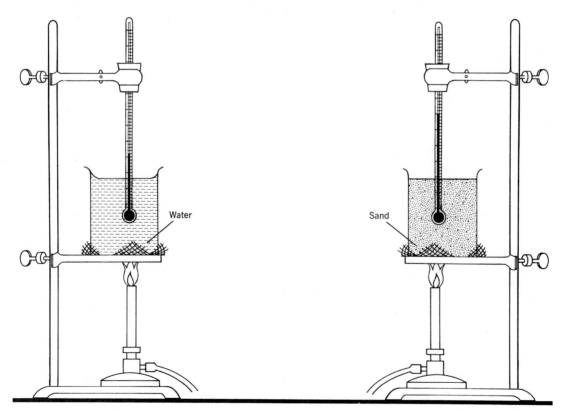

Water

Sand

**Fig. 15-1.**

Fill one beaker with water and the other beaker with sand. Both the water and the sand should be at room temperature, or about 20° Celsius (C).

**1.** What is the temperature of the beaker filled with water? _____

**2.** What is the temperature of the beaker filled with sand? _____

Record the temperatures of both beakers at 0 minutes (min) in Table 15-1.

**B.** Light two Bunsen burners, and adjust the flames so that they are about the same size. Place the burners under the beakers. Using a stopwatch, or a watch with a second hand, read the thermometers at intervals of 1 min, and record the temperature readings in Table 15-1. Remove the burners after 5 min.

**CAUTION: Quickly remove both burners if the sand thermometer records a temperature of 65°C before the 5 min are up. Wear safety goggles.**

Continue recording the sand and water temperatures for 10 min after the burners are removed, also at 1-min intervals. Record the temperatures in Table 15-2.

**TABLE 15-1.**

| *Minutes* | *Temperature* | |
|---|---|---|
| | Sand | Water |
| Begin 0 | | |
| 1 | | |
| 2 | | |
| 3 | | |
| 4 | | |
| 5 | | |

*Heating*

**TABLE 15-2.**

*Cooling (Heat Removed)*

| *Minutes After* | *Temperature* | |
|---|---|---|
| | Sand | Water |
| 1 | | |
| 2 | | |
| 3 | | |
| 4 | | |
| 5 | | |
| 6 | | |
| 7 | | |
| 8 | | |
| 9 | | |
| 10 | | |

**C.** Plot your temperature readings on Graph 15-1, using a solid line for the sand and a dotted line for the water.

**D.** Use the graph to answer the following questions:

**3.** Over the 5-min heating period, which substance heats up faster?

_____

**4.** Over the 10-min cooling period, which substance cools off faster?

_____

**5.** Which substance holds heat longer? _____

How does the graph show this? _____

_____

**6.** Would a sandy beach or a body of water heat up more quickly on a hot summer day?

Explain. _____

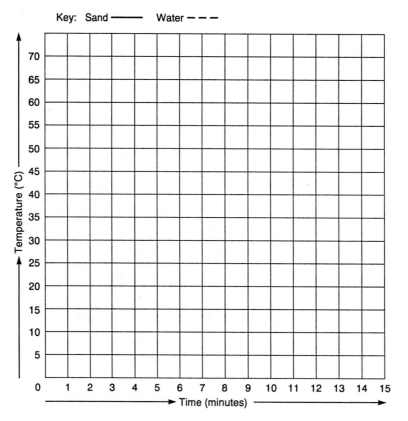

**Graph 15-1.**

**7.** Would a sandy beach or a lake cool off more quickly at the end of the summer? Explain.

## HEAT ENERGY

The temperature of Earth's atmosphere is always changing. Air temperature often changes from hour to hour, from day to day, and from one season to the next. These temperature changes ordinarily depend on the amount of *heat energy* in the atmosphere, in Earth's landmasses, and in bodies of water.

The particles that make up substances are in constant motion. **Heat energy** is a term that describes how fast these particles, or atoms and molecules, are moving. For example, the faster the particles in a substance such as water move, the hotter the water. The water is said to contain a large amount of heat energy.

The slower the motion of particles in water, the colder the water is. The water is said to contain a small amount of heat energy. Clearly then, molecules in boiling water move faster than do molecules in cold water. Thus, a quantity of boiling water contains more heat energy than does an equal quantity of cold water.

Because the gas molecules that make up Earth's atmosphere are in constant motion, the atmosphere contains heat energy. Like the atmosphere, the hydrosphere and the lithosphere also contain heat energy because the molecules in water and in rocks are in motion.

The continuous changes that affect heat energy in Earth's atmosphere, in the ocean, and on the land cause the *weather*. To understand the weather, you must understand the source of the heat energy that causes the weather.

## ENERGY FROM THE SUN HEATS EARTH

Most of Earth's heat energy is supplied by our star, the sun, which is located about 150 million kilometers from Earth. The sun emits heat in the form of **radiant energy.** Radiant energy from the sun travels through outer space as *rays*, or **radiation.**

When radiant energy strikes an object, some of the energy is absorbed by the object. As a result, the molecules in the object move faster. As you just learned, when the molecules in an object move faster, the object becomes hotter, or contains more heat energy. Because the radiant energy of the sun is changed into heat energy, the sun's rays warm Earth.

Fig. 15-2 shows the different forms of radiant energy. *Ultraviolet (UV) radiation* is an invisible form of radiant energy given off by the sun. When UV radiation strikes your skin, some of the radiation is changed into heat energy, and you feel warmth. However, if you stay in the sun too long, the UV radiation may produce painful sunburn. In extreme cases, UV radiation can cause skin cancer. *Radio waves, X rays,* and *infrared (IR) radiation* are

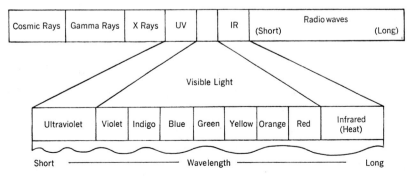

**Fig. 15-2. The different forms of radiant energy.**

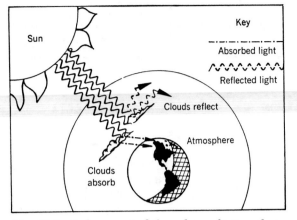

**Fig. 15-3. Only a part of the solar radiation that enters the atmosphere reaches Earth.**

other forms of radiant energy given off by the sun.

*Visible light* is also a form of radiant energy given off by the sun. Sunlight is visible because your eyes are sensitive to this form of radiant energy. Likewise, a radio receiver is sensitive to radio waves, and photographic film is sensitive to X rays. Visible light also is absorbed by many objects, thereby warming them.

The sun's radiant energy reaches Earth almost unchanged after traveling through outer space. However, only about 80% of the sun's radiant energy reaches Earth's surface. The remainder of the radiant energy is absorbed by Earth's atmosphere and by the clouds. Some radiant energy also is reflected back into outer space by clouds, water vapor, and dust particles suspended in the atmosphere (see Fig. 15-3).

The sun's rays that are mainly responsible for heating Earth are visible light and invisible infrared radiation. These rays are changed to heat energy when absorbed by substances in Earth's atmosphere and on Earth.

The amount of solar radiation absorbed by Earth's atmosphere depends mainly on the amount of water vapor, dust, and other substances in the air. However, the atmosphere is heated most by the solar radiation absorbed by Earth's surface. The amount of solar radiation absorbed by Earth and changed into heat is determined mainly by (1) the kind of material present on Earth's surface, (2) the substances in Earth's atmosphere, and (3) the angle at which solar radiation strikes Earth.

**Earth Materials.**  Fig. 15-4 shows that different kinds of surfaces vary in their ability to absorb solar radiation. Dark-colored land surfaces absorb more solar radiation than do light-colored land surfaces, and rough land surfaces absorb more solar radiation than do smooth land surfaces. In general, cropland and woodland (forests) absorb up to 90% of the incoming solar radiation. Sandy regions absorb about 80%. Freshly fallen snow, which is a very good reflector of sunlight, only absorbs about 25% of the solar radiation that strikes it.

The ability of water to absorb solar radiation is greatly influenced by the angle at which the sun's rays strike the water. When the sun is low in the sky, or near the horizon, water absorbs about 50% or less of the solar radiation that strikes it. When the sun is directly overhead, water absorbs about 97% of the incoming solar radiation.

Land and water also differ greatly in their ability to hold heat. Land areas heat up quickly during the day, but do not retain the heat. In the evening, land areas quickly give up heat. In contrast, water heats up slowly but retains heat for a long time. In other words, water gives up heat slowly.

Scientists use the term **specific heat** to describe the heat storage ability of substances.

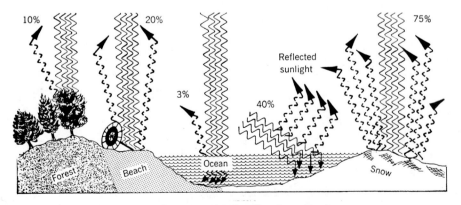

**Fig. 15-4. Absorption and reflection of solar radiation by different land surfaces.**

## TABLE 15-3. SPECIFIC HEATS OF COMMON SUBSTANCES

| (cal/g-°C) | | |
|---|---|---|
| Water | = | 1.0 |
| Ice | = | .5 |
| Water vapor | = | .5 |
| Dry air | = | .24 |
| Basalt | = | .20 |
| Granite | = | .19 |
| Iron | = | .11 |
| Copper | = | .09 |
| Lead | = | .03 |

Water has a specific heat that is about three times greater than that of land areas. Put simply, water can store about three times more heat than can an equal mass of land materials such as sand and rock. (See Table 15-3 for a list of specific heats of some common substances.)

Although land areas heat up more quickly than do bodies of water, only the top layers of soil or sand heat up. For example, on a hot day at the beach, you probably have noticed that the top layer of sand is very hot. However, if you dig just a few centimeters into the sand, it feels much cooler.

You observed this phenomenon when the bulb of your thermometer was buried in sand during the laboratory investigation. After several minutes of heating, the heat reached the bulb, causing a rapid rise in temperature. In contrast, the water temperature increased at a steady rate over the heating period.

Solar radiation can readily travel through water. To what depth it travels depends on the type of energy that the solar radiation contains. For example, the red part of visible light has less energy than either the blue or green parts. The red part of visible light is rapidly absorbed by the top layers of water and is changed into heat.

The blue and green parts of visible light travel deeper into the water before being absorbed. Water usually looks blue-green because part of the blue and green light is reflected back to your eyes. In fact, most of the solar radiation is absorbed before it travels more than 10 meters (m) into the water. Some light, however, has been detected more than 600 m below the ocean's surface.

Because water is partly transparent to solar radiation, bodies of water are heated to a much greater depth than are land areas. Consequently, water can store more heat energy than can land. The amount of heat energy stored in the ocean is increased by currents that mix the warmer surface water with colder deep water.

**Effect of Gas and Dust Particles on Sunlight.** The amount of sunlight that reaches Earth is affected by clouds, dust, water vapor, and other substances in the atmosphere (see Fig. 15-5). Clouds reflect a large percentage of solar radiation. On overcast days, thick clouds reflect large amounts of solar radiation back into space. Then, only about 45 percent or less of the sun's radiant energy reaches Earth's surface.

Water vapor and carbon dioxide in Earth's atmosphere also absorb some of the infrared radiation from sunlight and are heated. Consequently, both of these gases radiate some heat energy back into the atmosphere.

Dust particles in Earth's atmosphere block solar radiation. The radiation that strikes dust particles may be absorbed by them, reflected back into space, or scattered through the atmosphere. As you learned in Chapter 13, the sky's blue color is caused by the scattering of sunlight by dust and gases in the air. The solar radiation absorbed by dust is changed into heat energy and added to the atmosphere.

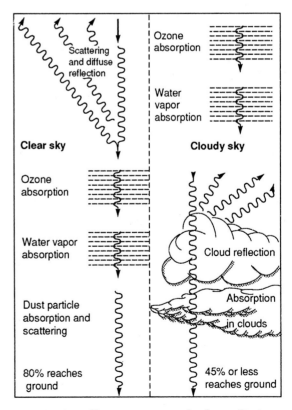

**Fig. 15-5. Different amounts of solar radiation reach Earth.**

Most of the UV radiation from the sun is absorbed by oxygen molecules in the upper part of Earth's stratosphere. As a result, oxygen is changed into ozone. Ozone is a special kind of oxygen molecule that has three atoms instead of two. The absorption of UV radiation by ozone is important because exposure to large amounts of UV radiation is harmful to living things.

**Angle at Which Solar Radiation Strikes Earth.** The angle at which solar radiation strikes Earth's surface determines the amount of heat that any place on Earth receives. At the equator, the sun passes almost directly overhead throughout the year. Thus, solar radiation strikes Earth with greatest intensity along the equator. Consequently, land areas and bodies of water along the equator absorb more of the sun's heat than do areas to the north or south of the equator.

In higher latitudes, such as in the northern United States and Canada, the sun always appears lower in the sky than it does along the equator. In the United States and Canada, Earth's surface curves away from the direct rays of the sun. For this reason, the solar radiation is spread out over a larger part of the surface than it is near the equator (see Fig. 15-6). Because solar radiation is less direct at higher altitudes, the United States and Canada receive much less heat annually than do equal areas along the equator.

At higher latitudes, solar radiation strikes Earth's surface at a slanting angle, or less directly than at the equator. Thus, the sun's rays must travel a greater distance through the atmosphere to reach Earth's surface. Consequently, more of the solar radiation is absorbed, reflected, or scattered by the atmosphere. This further reduces the amount of radiation received at the surface.

Solar radiation strikes Earth at different angles at different times of the year. For example, the Northern Hemisphere and the Southern Hemisphere receive different amounts of solar radiation in winter than they do in summer. This is because Earth's axis is tilted about 23½ degrees in relation to the sun. The tilting of Earth's axis, which is responsible for the seasons, is described in greater detail in Chapter 19.

## TRANSFER OF HEAT ENERGY TO THE ATMOSPHERE

Sunlight absorbed by Earth's surface is the most important factor responsible for heating the troposphere. The heat energy absorbed by Earth is transferred back to the atmosphere by the processes of *conduction*, *convection*, and *radiation*. Solar radiation and the different forms of radiant energy were discussed earlier in this chapter.

**Conduction.** The process by which heat energy is transferred from molecule to molecule in a substance is called **conduction**. For example, if you hold one end of an iron rod in your hand and hold the other end of the rod in a flame, heat will travel through the iron rod to your hand (see Fig. 15-7).

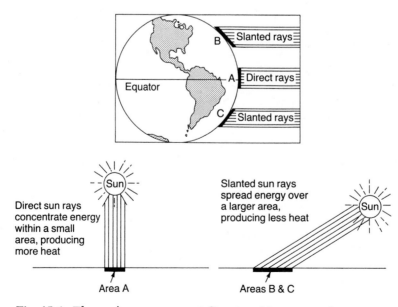

Fig. 15-6. The sun's rays are most direct and intense at the equator.

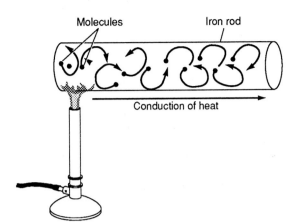

**Fig. 15-7. Transfer of heat by conduction.**

Heat also travels through solid Earth materials by conduction. The molecules in the heated part of a rock, for example, move faster as they absorb heat from sunlight. The faster-moving rock molecules at the surface collide with the slower-moving molecules nearby, increasing their speed as well. Thus, the heat is transferred. Conduction continues as long as the rock is being heated. In this way, some heat is transferred beneath Earth's surface.

During the day, the part of Earth's atmosphere touching the ground is heated by conduction. Heat energy from the warmed ground surface is transferred directly to air molecules. The atmosphere is a poor conductor of heat because air molecules are far apart. Consequently, the part of the atmosphere that does not touch the ground does not receive heat by conduction. The main portion of the atmosphere, which is not in contact with Earth's surface, is heated by convection and radiation.

**Convection and Convection Currents.** The transfer of heat through a substance by currents moving inside the substance is called **convection.** It takes longer to heat a container of water than to heat a metal rod because water is a poor conductor of heat. However, the bottom layer of water in a beaker heated from below does heat readily by being in contact with the beaker.

The heated water becomes lighter; then cooler, denser water sinks and forces it upward. The cooler water is then heated, continuing the cycle. The moving water forms **convection currents** that gradually heat the entire container of water (see Fig. 15-8a). Convection currents occur readily in liquids and gases because currents can move easily through such substances.

When air is heated, it expands, or occupies more space. As gas molecules in the air absorb heat, they move faster, colliding more strongly with each other. These collisions push the gas molecules farther apart. As a result, fewer gas molecules occupy the original space, and the air in this space weighs less than before it was heated.

In contrast, cooler air has less heat energy than does an equal amount of warm air. The molecules in cool air move slower and are closer together. Cool air, therefore, weighs more than does an equal amount of warm air.

A mass of warm air is lighter than an equal mass of cool air. Consequently, cool air sinks and forces warm air to rise into the atmosphere. The upward movement of warm air and the downward movement of cool air form convection currents. Fig. 15-8b shows an example of a convection current in air.

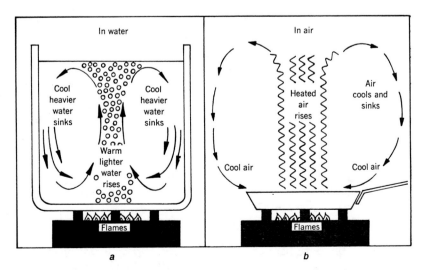

**Fig. 15-8. Convection currents in water and in air.**

Another example of a convection current can be observed in a room heated by a radiator. Air in contact with the radiator is heated by conduction. The heated air expands and gets lighter. Cooler air moves in from below and pushes the lighter warm air upward. The denser air soon becomes hot, and it, too, rises. The convection current continues as long as the air away from the radiator is cooler than the air surrounding the radiator (Fig. 15-9).

Land and water surfaces are similar to a radiator in a room. When a cool air mass meets a body of warm water, the air mass is heated by conduction. The warmed air expands and rises. Cooler, heavier air that has replaced the warmed air is heated, and it also rises. A convection current results from this. The movement of air masses in Earth's atmosphere is similar to the example of a convection current shown in Fig. 15-9.

Large amounts of heat energy are added to Earth's atmosphere by the upward movement of heated air. As the heated air rises, it cools slowly. The higher the heated air rises from Earth's surface, the cooler the air becomes. Consequently, the warmest part of the troposphere is usually in contact with Earth's surface.

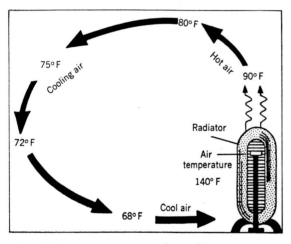

**Fig. 15-9. Air in a room is heated by convection.**

## GREENHOUSE EFFECT

Heat energy given off by Earth's surface is absorbed by water vapor and carbon dioxide in the atmosphere. The absorption of heat energy by water vapor and carbon dioxide causes these gases to become warmer. In this way, large amounts of heat given off by Earth are trapped in the atmosphere. Of the two gases, water vapor absorbs more heat energy from Earth's surface.

The trapping of heat energy by water vapor, carbon dioxide, and other gases in Earth's atmosphere is known as the **greenhouse effect.** This effect is illustrated by greenhouses used by gardeners and farmers.

Most solar radiation passes through the transparent glass roof of a greenhouse. The solar radiation is then absorbed by plants and other objects inside the greenhouse. Thus, the solar radiation is changed into heat energy. The objects in the greenhouse radiate heat energy in all directions. However, the glass roof of the greenhouse prevents the heat from escaping. Instead, the heat energy remains trapped in the greenhouse (see Fig. 15-10).

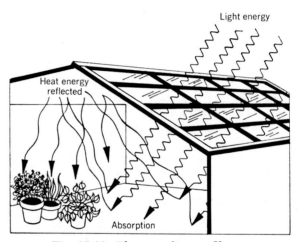

**Fig. 15-10. The greenhouse effect.**

In Earth's atmosphere, carbon dioxide and water vapor are similar to the glass roof of a greenhouse. Together, these gases trap large amounts of heat energy, preventing much of Earth's heat from escaping into outer space. On overcast nights, dense, low clouds influence the greenhouse effect by redirecting heat energy back to Earth (see Fig. 15-11). As a result, cloudy nights are usually warmer than clear nights.

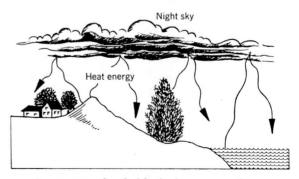

**Fig. 15-11. Clouds block the escape of heat radiated by Earth at night.**

## TEMPERATURE CHANGES CAUSED BY EXPANSION OR COMPRESSION OF AIR

The expansion and compression of large air masses affect the temperature of Earth's atmosphere. When air is compressed, it gets hotter. For example, when you inflate a bicycle tire with a hand-held air pump, the pump chamber becomes hotter as you continue to inflate the tire. The downward movement of the plunger compresses the air molecules entering the pump chamber. As the air molecules are pushed tightly together, heat energy in the pump chamber increases. You can feel this heat with your hand.

In contrast, air becomes cooler when it expands. You can sense this by holding your hand in a stream of air escaping from an inflated bicycle tire. When the air was pumped into the tire, the air was compressed. When the air escapes from the tire, the air molecules rapidly move apart. Thus, the volume of the escaping air increases. As the distance between the air molecules increases, the heat energy in the air decreases. For this reason, the escaping air feels cool on your hand.

On a gigantic scale, a similar heating and cooling effect takes place in Earth's atmosphere. When an air mass rises, it moves into regions where the atmospheric pressure surrounding the air mass has less pressure. The reduction in air pressure allows the air mass to expand. As the air expands, it becomes cooler. A mass of cool, dry air will decrease in temperature at a rate of about 3°C for each 300 meters it rises.

The opposite effect occurs when an air mass sinks toward Earth. As the air mass sinks to a lower altitude, the increasing atmospheric pressure compresses the air. As the air is compressed, it becomes warmer. If the air mass is dry, its temperature will increase at a rate of about 3°C for each 300 m it sinks.

The heating and cooling of sinking and rising air masses affect the weather. This process is described in Chapter 18.

## DAILY TEMPERATURE CHANGES

Earth receives most of its heat during the day. The rise and fall of daily air temperature is shown in Fig. 15-12. At sunrise, the air temperature usually begins to increase. At noon, when the sun is at its highest point in the sky, the air temperature is still increasing. Because Earth gains more heat than it loses during the early afternoon hours, the hottest time of the day usually occurs in the midafternoon.

In the late afternoon, when the sun moves lower in the sky, Earth begins to lose the heat accumulated during the day. The air temperature begins to fall and continues to drop through the night. The temperature reaches its lowest point just before sunrise. Occasionally, weather changes affect this pattern. For example, a weather change might cause air temperatures to fall throughout the day or to rise throughout the night (see the right-hand section of Fig. 15-12).

## SEASONAL AND DAILY TEMPERATURE CHANGES

As you know, winters in the Northern Hemisphere are always colder than the summers.

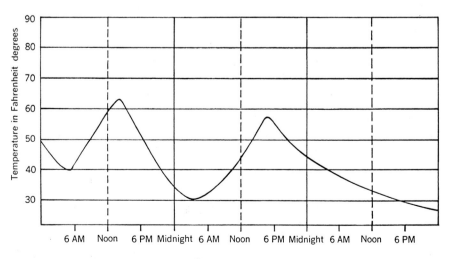

**Fig. 15-12. Daily temperature variations.**

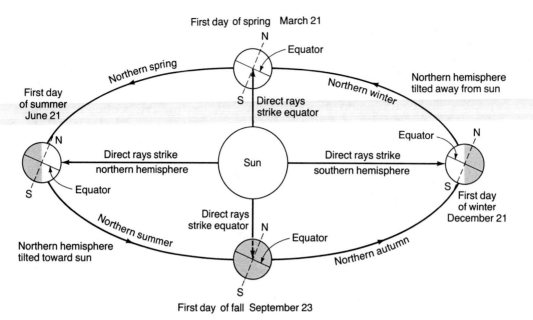

**Fig. 15-13. The tilt of Earth's axis causes the seasons.**

The temperature change from season to season occurs because Earth's axis is not vertical, or perpendicular to the plane of its orbit around the sun. Instead, Earth's axis is tilted at an angle of 23½ degrees from the vertical (see Fig. 15-13).

For example, the North Pole is tilted toward the sun during summer. At this time, the solar radiation is concentrated over the Northern Hemisphere. During winter, the North Pole is tilted away from the sun. Then, the sun's radiation strikes the Northern Hemisphere at a slanting angle and is therefore less direct.

In summer, the Northern Hemisphere is heated more strongly than during winter. In winter, solar radiation is less concentrated in the Northern Hemisphere.

Days and nights also are warmer during summer because the sun shines on the Northern Hemisphere for more hours each day in the summer than during the winter. In the north during winter, the sun shines for fewer hours, and because of the tilt of Earth's axis, less solar radiation reaches Earth's surface in winter.

At the beginning of spring and fall, Earth is not tilted either toward or away from the sun. Then, solar radiation is equally divided between the Northern Hemisphere and the Southern Hemisphere. At the beginning of spring and fall, days and nights are about equal in length. Thus, spring and fall temperatures in the Northern Hemisphere and the Southern Hemisphere are not as extreme as are summer or winter temperatures.

## HOTTEST AND COLDEST DAYS OF THE YEAR

The first days of summer are not the hottest days of the year. Likewise, the first days of winter are not the coldest days of the year. You already learned that the hottest time of the day is midafternoon and that the coldest time of the day is just before sunrise. There is a time lag between the period of maximum concentration of the sun's rays and maximum air temperature. A time lag also exists between minimum concentration of the sun's rays and minimum air temperature.

Likewise, summer temperatures steadily increase from June 21, when the sun is at its highest point in the sky and the day is longest. During summer (when the sun is high in the sky), Earth continues to absorb heat energy. The average monthly air temperature usually is highest during July.

In winter, the lowest average monthly temperatures occur during January. That is because it takes about a month from December 21 (the shortest day) for Earth to lose most of the heat accumulated during the summer and the fall.

## THE OCEAN'S INFLUENCE ON THE TEMPERATURES OF COASTAL AREAS

In summer, regions near the ocean usually have cooler air temperatures than do inland areas. In winter, coastal areas usually have warmer air temperatures than do inland areas. The nearness of the ocean is responsible for the differences in temperature.

The ocean heats up slowly during summer and cools off slowly during winter. In mid-summer, the temperature of the Atlantic Ocean at the approximate latitude of New York, New Jersey, and Connecticut reaches a high of slightly over 20°C (70°F). The temperature of the nearby land areas during the day may be 30°C (90°F) or higher. Consequently, on summer days an air mass over the ocean usually is cooler than an air mass over the land.

On a hot summer day, the air temperature several kilometers inland may be about 30°C (90°F). Along the coast, the air temperature often is 5°C (10°F) cooler. You can think of the ocean as a huge air conditioner in the summer, cooling air masses along the coast.

In winter, the ocean functions like a gigantic radiator, heating the land along the coast. For example, during winter in the middle latitudes, the top layers of soil are frozen solid, and the air temperature often drops below freezing. The ocean temperature, however, is about 5°C (40°F). Consequently, the ocean warms air masses above it. The warmed air then moves over nearby land areas. As a result, the temperature over coastal areas in winter often is several degrees warmer than the temperature inland.

The heat energy is stored in the world's oceans in several different ways. Large air masses warmed by the ocean drift across the continents, giving up heat to the land. Ocean currents, such as the Gulf Stream, carry huge volumes of warm water from the Caribbean Sea and the Gulf of Mexico northward toward Europe. The Gulf Stream helps to produce milder winters in Scandinavia and in the British Isles than would be expected based on their latitude. In fact, some palm trees thrive in the British Isles.

## ALTITUDE AND TEMPERATURE CHANGES

The air temperature steadily decreases from Earth's surface through the upper part of the troposphere. The temperature drops with increasing altitude mainly because Earth, which is the major source of heat for the atmosphere, is farther away.

Suppose that a series of thermometers was placed 300 m apart, from the ground to the tropopause. Each thermometer would read

about 2°C less than the thermometer placed 300 m below it. If the thermometer on the ground reads 15°C, the thermometer 300 m above it will read 13°C. The next thermometer would read 11°C, and so on (see Fig. 15-14). Sometimes the air temperature increases for several hundred meters and then begins to decrease again. This condition of inverted, or upside-down air temperature, is known as a **temperature inversion.**

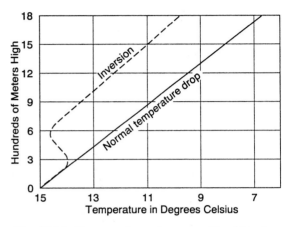

**Fig. 15-14. Temperature changes with altitude.**

## MEASURING AIR TEMPERATURE

Most people have no difficulty sensing large changes in air temperature. However, you must use a *thermometer* to determine the exact air temperature. A thermometer is a device that measures the temperature of gases, liquids, and solids. Some thermometers contain gas, others contain liquid, and some thermometers are made of metals. The kind of thermometer commonly used by meteorologists is called a *liquid thermometer.*

**Liquid Thermometers.** A **liquid thermometer** is a sealed glass tube with a bulb at one end. The bulb is a reservoir for a liquid. The sides of the sealed tube are marked with a scale from which the temperature is read. One type of liquid thermometer contains red-colored alcohol. Another type contains mercury, a silver-colored liquid metal. Alcohol and mercury are used in thermometers because these liquids expand and contract at a constant rate when heated and cooled. Alcohol thermometers are useful for measuring extremely cold temperatures because alcohol does not freeze until it reaches about −115°C (−175°F), whereas mercury freezes at −39°C (−38°F).

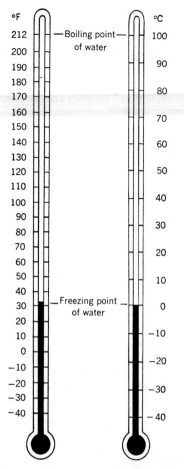

Fig. 15-15. Temperature scales.

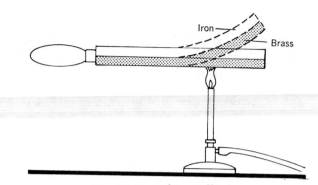

Fig. 15-16. A bimetallic bar.

freezing point of water is 32 degrees, and the boiling point of water is 212 degrees. On the Celsius scale, the freezing point of water is 0 degrees, and the boiling point of water is 100 degrees (see Fig. 15-15).

**Thermograph.** A device that writes a continuous record of temperature changes over time is called a **thermograph.** Instead of a liquid in a sealed glass tube, a thermograph commonly uses a *bimetallic bar* that reacts to changes in temperature. A **bimetallic bar** is made of two different metals welded together into one strip. When heated, the metals expand at different rates (see Fig. 15-16).

Inside the thermograph, a bimetallic strip is coiled like a watch spring. As the temperature changes, each metal expands or contracts at a different rate, thus causing the bimetallic strip to coil or uncoil. As the temperature increases or decreases, a pen attached to the bimetallic strip moves along a chart attached to a revolving cylinder. The chart keeps a written record of the temperature changes (see Fig. 15-17).

The scales commonly used on thermometers are the *Fahrenheit* scale and the *Celsius* scale. The Celsius scale is also called the *centigrade* scale (*cent* = 100, *grade* = divisions) thermometer. The Fahrenheit and Celsius scales are based on the freezing and boiling points of water. These are known as the *fixed points* of a thermometer. On the Fahrenheit scale, the

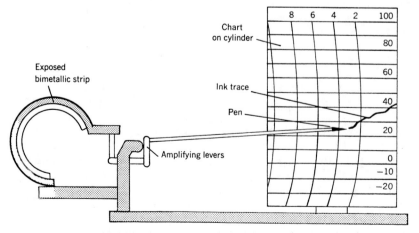

Fig. 15-17. A thermograph (in degrees Fahrenheit).

**Thermometer Location.** A thermometer placed directly in the sun will give a temperature reading much higher than the actual air temperature. To get an accurate temperature reading, a thermometer should be placed in the shade, where it will not be heated directly by solar radiation. Meteorologists keep thermometers in white-painted shelters with slatted sides. The slats allow air to circulate freely through the shelters, and the white paint reduces the amount of solar radiation absorbed by the shelters.

# CHAPTER REVIEW

## Science Terms

*The following list contains all of the boldfaced words found in this chapter and the page on which each appears.*

bimetallic bar (p. 220)
conduction (p. 214)
convection (p. 215)
convection currents (p. 215)
greenhouse effect (p. 216)
heat energy (p. 211)

liquid thermometer (p. 219)
radiant energy (p. 211)
radiation (p. 211)
specific heat (p. 212)
temperature inversion (p. 219)
thermograph (p. 220)

## Matching Questions

*On the blank line, write the letter of the item in column B that is most closely related to the item in column A.*

### Column A

_____ 1. describes motion of particles in a substance

_____ 2. they absorb and scatter solar radiation

_____ 3. form in which the sun emits heat

_____ 4. they absorb infrared rays *and* Earth's heat

_____ 5. transfer of heat between molecules

_____ 6. oxygen that absorbs UV radiation

_____ 7. transfer of heat by currents

_____ 8. trapping of heat in Earth's atmosphere

_____ 9. keeps written record of temperature

_____ 10. "upside-down" air temperature in troposphere

### Column B

*a.* ozone
*b.* thermograph
*c.* conduction
*d.* dust particles
*e.* convection
*f.* radiant energy
*g.* heat energy
*h.* water vapor and carbon dioxide
*i.* inversion
*j.* thermometer
*k.* greenhouse effect

# Multiple-Choice Questions

*On the blank line, write the letter preceding the word or expression that best completes the statement.*

1. Of the solar radiation entering Earth's atmosphere, the amount that reaches Earth's surface is about
   *a.* 10%  *b.* 50%  *c.* 75%  *d.* 98%                          1 _____

2. Which land area absorbs the most solar radiation?
   *a.* forest  *b.* sandy beach  *c.* snowfield  *d.* desert        2 _____

3. When the sun is directly over the ocean, the amount of solar radiation that the ocean will reflect back into the atmosphere is about
   *a.* 3%  *b.* 40%  *c.* 80%  *d.* 97%                            3 _____

4. Which of the following factors is *least* important in distributing heat throughout the ocean?
   *a.* waves                  *c.* the transparency of the water
   *b.* currents               *d.* the flow of rivers into the ocean    4 _____

5. On a hot summer day, the ground is heated to a depth of about
   *a.* 10 cm  *b.* 1 m  *c.* 3 m  *d.* 6 m                          5 _____

6. The atmospheric gases that absorb the most heat are
   *a.* carbon dioxide and oxygen        *c.* nitrogen and water vapor
   *b.* carbon dioxide and water vapor   *d.* nitrogen and carbon dioxide    6 _____

7. The sky's blue color is caused mainly by the
   *a.* absorption of sunlight by particles in Earth's atmosphere
   *b.* refraction of sunlight by Earth's ionosphere
   *c.* reflection of sunlight from the ocean
   *d.* scattering of sunlight by gases in Earth's atmosphere          7 _____

8. Compared to other parts of Earth, the equatorial region is heated most by solar radiation because the equator
   *a.* is closer to the sun
   *b.* has very large landmasses along it
   *c.* has very large sea areas along it
   *d.* receives the sun's most direct rays                           8 _____

9. The increased motion of molecules that have absorbed solar radiation is measured in terms of
   *a.* radiation  *b.* heat  *c.* convection currents  *d.* evaporation    9 _____

10. Sand is a better conductor of heat than air because the molecules in sand are
    *a.* closer together               *c.* heavier than those of air
    *b.* opaque to radiation           *d.* in more rapid motion         10 _____

11. The sun transfers its heat to other objects by
    *a.* conduction  *b.* convection  *c.* radiation  *d.* expansion     11 _____

12. When an air mass is heated, it expands because the gases in the air mass
    *a.* become lighter                *c.* become heavier
    *b.* move faster                   *d.* exert less pressure on each other    12 _____

13. Heat is transferred from Earth's surface to high altitudes by
    *a.* convection currents  *b.* dust particles  *c.* winds  *d.* cloud reflections    13 _____

14. The way that atmospheric gases trap the heat energy given off by Earth is called the
    a. blocking effect          c. greenhouse effect
    b. heat sink                d. radiation trap          14 _____

15. As an air mass expands, it becomes
    a. cooler  b. warmer  c. compressed  d. denser          15 _____

16. The coldest time of the day is usually around
    a. sunrise  b. sunset  c. midnight  d. 2 A.M.          16 _____

17. The coldest month of the year is usually
    a. December  b. January  c. February  d. March          17 _____

18. As an airplane climbs into the sky, a thermometer on the airplane recording the outside air temperature should show that the air temperature decreases at a rate of about
    a. 2°C per 300 m          c. 20°C per 300 m
    b. 8°C per 300 m          d. 30°C per 300 m          18 _____

19. A thermometer on which the freezing point of water is 0° and the boiling point is 100° uses the
    a. Fahrenheit scale          c. Celsius scale
    b. Absolute scale          d. Kelvin scale          19 _____

20. An instrument that makes a written record of the air temperature is called a
    a. thermograph          c. bimetallic thermometer
    b. hypograph          d. alcohol thermometer          20 _____

# Modified True-False Questions

*In some of the following statements, the italicized term makes the statement incorrect. For each incorrect statement, write the term that must be substituted for the italicized term to make the statement correct. For each correct statement, write the word "true."*

1. The types of solar radiation that are most important in heating Earth are visible light and *ultraviolet* light.          1 _____

2. Bodies of water are usually heated to a much greater depth than are land areas because water is *less dense.*          2 _____

3. On a hot summer day, heat can easily penetrate *a few centimeters* below the top of a paved surface.          3 _____

4. Atmospheric water vapor, *nitrogen,* and dust help to heat Earth's atmosphere because they absorb solar radiation.          4 _____

5. The difference between summer temperatures and winter temperatures in most places on Earth is determined mainly by the *angle* at which the sun's rays strike Earth at different times of the year.          5 _____

6. The process by which heat travels throughout a solid object is called *conduction.*          6 _____

7. Cold air in contact with heated land is warmed by *convection.*          7 _____

8. Daily air temperatures are warmest at about *midmorning*.  8 _____

9. The hottest days of the year usually occur at the end of *July*.  9 _____

10. The freezing point of water is 0° on the *Fahrenheit* scale.  10 _____

# Testing Your Knowledge

1. Of all the solar radiation that enters Earth's atmosphere, only about half reaches Earth's surface. Explain. _____

_____

2. Why do large bodies of water heat up more slowly than large land areas? _____

_____

_____

3. Name and briefly describe the three methods by which heat may be transferred through solid, liquid, and gaseous objects.

   *a.* _____

   *b.* _____

   *c.* _____

4. Containers filled with a gas under pressure, such as carbon-dioxide fire extinguishers and aerosol containers, get cold when the compressed gases are sprayed out of them. Explain.

_____

_____

5. A weather report for a coastal area predicted: *Rain mixed with wet snow along the coast, snow inland.* Explain the different types of precipitation along the coast and inland. _____

_____

_____

6. Suppose two thermometers were installed on opposite sides of a building. Both thermometers were accurate. At the same time one day, one thermometer reads 92°F, and the other thermometer reads 107°F. Explain. _____

_____

# The Greenhouse Effect: A Hot Topic Getting Hotter?

The 1990s set a record! Usually, records are exciting and welcome events, like those set by athletes in the Olympic Games and in sports such as baseball, football, and basketball. We marvel at them and share the glory of the record setters. But the record set in the 1980s should give us no cause for celebration.

The 1990s, it turns out, broke the record for the warmest decade ever recorded!

So what's the matter with a little warmer weather? Who needs icy winters anyway? And even the sting of hotter summers can be softened by a dip in the ocean or lake, or by turning up the air conditioner a bit.

That may be true, but an increase in average worldwide surface temperatures of, say, 2°C to 6°C (which is just what some scientists predict will happen over the next century) would have devastating effects. For one thing, much of the ice cap in the Antarctic would melt. This water would pour into the oceans, raising their levels from 0.5 to 1.5 meters. Many of the world's great coastal cities would become flooded. Thousands of acres of farmland would be covered by salt water. Clearly, the world will not be a better place to live in if the "warmest decade" record continues to be broken each decade!

But what is causing such record-high temperatures and what, as concerned citizens, can we do about it—if anything?

A clue to the answer to the first question lies in a simple fact: The amount of carbon dioxide in the atmosphere has risen about 25 percent from the level it was in the mid-1800s. So what began at that time that might account for the upsurge in this common gas? People in the industrialized nations of the world began to burn more and more fossil fuels, like coal and oil.

Today, the world's industries and people burn huge amounts of these fuels to run factories, heat homes, produce electricity, and operate cars, trucks, and buses.

As any chemist will tell you—and as you might already know—fossil fuels consist mostly of carbon. And when carbon is burned it produces the gas carbon dioxide, or $CO_2$ for short. Carbon dioxide mixes into the atmosphere where it does something interesting. It lets the radiation of the sun pass in but it doesn't let the hot infrared radiation released from Earth's surface pass back out into space.

Simply put, $CO_2$ in the atmosphere traps heat just as the glass of a greenhouse does. That's why the rise in global temperatures resulting from the increase of $CO_2$ in the atmosphere is called the greenhouse effect. One way to try to stop this "global warming" is to reduce the use of fossil fuels. But even this would not bring an immediate cooling effect, because $CO_2$ remains in the atmosphere for about 100 years.

Some scientists have suggested that other pollutants in fossil fuels may actually help cool the planet. These pollutants are sulfates, produced when the sulfur in coal and oil combines with oxygen during burning. Besides causing acid rain and respiratory problems, the sulfates rise into the atmosphere where they block sunlight and trigger the production of clouds, actions which tend to lower Earth's surface temperatures.

However, since sulfates remain in the atmosphere for only a few weeks and are more unevenly distributed than $CO_2$ most climate experts believe sulfate pollution will do little to offset the warming caused by $CO_2$. So, we're back to looking for ways to reduce $CO_2$ emissions. This would involve developing

alternative energy sources and finding ways to improve energy conservation. But getting the whole world to agree on how to go about doing this won't be easy.

In 1997, a treaty called the Kyoto Accord was drafted at a climate summit in Japan. The treaty would require developed nations like the United States, Russia, and European countries to reduce their $CO_2$ emissions to below 1990 levels by 2012, but would place no limits on $CO_2$ production by developing nations like India and China. The U.S. has refused to sign the Accord, objecting that it would harm the economy, and several other nations have also expressed reservations. Stopping the greenhouse effect without harming economic growth may prove to be a delicate balancing act. What do you think should be done?

**1.** *Some scientists predict that during the next century, average worldwide surface temperatures may*

   a. fall 2°C to 6°C.
   b. rise 2°C to 6°C.
   c. rise 1°C to 2°C.
   d. remain unchanged.

**2.** *Since 1850, the amount of carbon dioxide in the atmosphere has*

   a. fallen 10 percent.
   b. risen 10 percent.
   c. risen 25 percent.
   d. remained unchanged.

**3.** *Sulfates in the atmosphere can*

   a. increase the greenhouse effect.
   b. decrease the greenhouse effect.
   c. neither increase nor decrease the greenhouse effect.
   d. cause ocean flooding.

**4.** *What began happening in about 1850 that affected the carbon-dioxide level in the atmosphere, and how was it affected?*

_____

_____

_____

**5.** *On a separate sheet of paper, present your opinion of what should be done to help keep Earth's temperature constant. Support your opinion with references to the writings or statements of various scientists.*

# *Wind*

## LABORATORY INVESTIGATION

### *CONVECTION CURRENTS*

**A.** Set up the candle and glass chimneys as shown in Fig. 16-1. Light the candle, and close the sliding glass front.

**1.** Hold one of your hands about 15 centimeters (cm) above each of the glass chimneys. Do

you feel any differences? _____

_____

_____

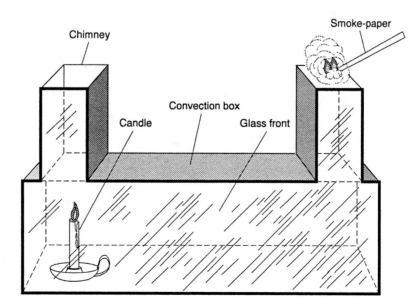

**Fig. 16-1.**

**B.** Light a piece of smoke paper, and hold it just above one of the chimney openings for about 10 seconds. Then, do the same at the other chimney opening.

**2.** In what direction does the smoke move? _____

_____

**3.** Draw the path followed by the smoke, using arrows on the picture of the convection box.

**C.** Blow out the candle, and place it on the opposite end of the convection box. Repeat steps A and B.

**4.** In what direction does the smoke move now? Draw a picture of the convection box on a separate piece of paper. Using arrows, show the new path followed by the smoke.

**5.** Would air flow through the convection box in a particular way without a candle burning?

_____

_____

Check your answer by repeating the experiment without lighting a candle.

**6.** Based on your observations, explain the way that (*a*) smoke moves in a convection box *with* a burning candle and (*b*) smoke moves in a convection box *without* a burning candle.

_____

_____

## AIR IN MOTION

There are two kinds of air movement. Air moving toward or away from Earth's surface is called an **air current.** Air moving parallel to Earth's surface is called **wind.** Winds and air currents influence the weather by carrying large amounts of heat and moisture from one part of the atmosphere to another.

In cities, where air pollution is a constant threat, winds help carry away polluted air and bring in fresh, clean air. If winds did not blow, harmful gases would accumulate near Earth's surface, endangering humans and other living things. Fortunately, differences in air temperature, pressure, and moisture in Earth's atmosphere continually produce wind.

## CAUSES OF WINDS

In Chapter 15, you learned how solar radiation heats Earth. You know that some places on Earth's surface receive more heat than do other places. The atmosphere mainly receives heat from Earth's surface. For this reason, the atmosphere is heated unequally. The unequal heating of the atmosphere produces differences in air density, which result in differences in air pressure. The differences in air pressure cause winds.

**Convection Cells.** In the laboratory investigation at the beginning of this chapter, you observed how unequal heating can result in air movements. On a gigantic scale, the unequal heating of Earth produces air currents and winds in the atmosphere.

Air circulation through a convection box is an example of a *convection current,* which was described in Chapter 15. Another example of a convection current described in Chapter 15 is air circulation in a closed room.

In a closed room, air follows a circular path. Heated air rises from a radiator and flows across the ceiling toward the far side of the room. The air cools, settles to the floor, and then flows along the floor to the radiator, where the air is reheated. The circular movement of air in convection currents is called a **convection cell.**

In the warmer part of the convection cell, air molecules are farther apart, making the air less dense. This area of the convection cell becomes a region of low pressure. In the cooler part of the convection cell, the air molecules are closer together, making the air more dense. This area of the convection cell becomes a region of high pressure. Air always moves from a region of higher pressure to a region of lower pressure.

**Earth's Convection Cells.** At the equator, the heat energy absorbed by Earth warms air above the surface. The warmed air becomes less dense and is lifted by heavier air that moves in to displace it. The rising air produces a region of permanent low atmospheric pressure at the equator. If Earth did not rotate, the rising air at the equator would circulate in two huge convection cells: one convection cell in the Northern Hemisphere and the other in the Southern Hemisphere (see Fig. 16-2).

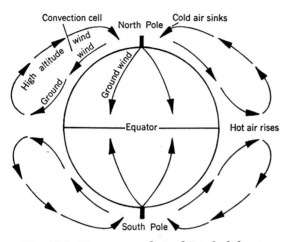

**Fig. 16-2. Movement of air if Earth did not rotate.**

The warm, low-pressure air mass would divide into two parts as it moved through Earth's troposphere. One part of the air mass would flow south, and the other part would flow north. Gradually, both of the air masses would lose their heat. As the cooling air masses traveled toward the poles, they would become denser and sink toward the ground.

In the Northern Hemisphere, the cold air mass would flow *south*, back toward the equator, gradually absorbing heat from the ground. When the air reached the equator, it would be reheated and rise into the atmosphere. The same kind of circulation would occur in the Southern Hemisphere. In the Southern Hemisphere, however, the cold air mass would flow *north* to the equator.

If Earth did not rotate, low-altitude winds would always blow from the poles toward the equator, and high-altitude winds would always blow from the equator to the poles. This kind of air movement is another example of a convection cell and similar to the movement you observed in the convection box.

Because Earth rotates, this simple system of wind circulation does not exist. In fact, the winds are deflected from their north-south paths, making the simple convection cells even more complicated.

**How Winds Are Deflected.** You can conduct a simple demonstration that illustrates how winds are deflected by Earth's rotation. You will need a mounted globe that can spin.

When the globe is not spinning, you observe that water droplets falling on the North Pole run straight down the globe (see dotted line in Fig. 16-3). The path followed by the water droplets is similar to the path that would be followed by low-altitude winds in the Northern Hemisphere if Earth did not rotate. When the globe is spun counterclockwise, water droplets falling on the North Pole curve in a direction opposite to the globe's rotation (see solid line in Fig. 16-3).

Air moving over Earth's surface moves in a similar way. Fig. 16-3 shows what happens to low-altitude winds as they travel over Earth. As Earth rotates from west to east, the winds in the Northern Hemisphere do not follow direct north-south paths. Instead, the winds curve, or are *deflected*.

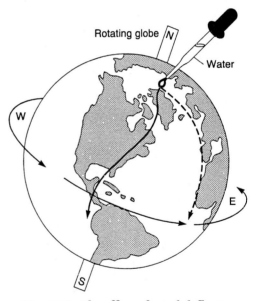

**Fig. 16-3. The effect of wind deflection.**

**Wind Direction and the Coriolis Effect.** Wind direction is described by the direction from which the wind blows. For example, a northerly wind blows from the north to the south. A westerly wind blows from the west to the east.

The influence of Earth's rotation on the direction of winds was first studied by a French mathematician, *Gaspard-Gustave de Coriolis* (1792–1843). Consequently, it is called the **Coriolis effect.**

Coriolis discovered that winds are deflected because they move at speeds different from the velocity of Earth's rotation. The velocity of Earth's rotation decreases from 1600 kilometers (km) per hour at the equator to 0 km per hour at the poles. Winds are deflected to the right in the Northern Hemisphere and to the left in the Southern Hemisphere because Earth is rotating underneath them. Instead of blowing directly north or south, winds curve according to *Ferrel's law*.

**Ferrel's Law.** Meteorologists use Ferrel's law to determine the direction toward which winds on Earth's surface are deflected. The law states: In the Northern Hemisphere, winds are deflected to the right of their original path; in the Southern Hemisphere, winds are deflected to the left of their original path.

# MEASURING WIND DIRECTION AND WIND SPEED

**Wind Direction.** An instrument commonly used to indicate wind direction is a **wind vane.** Wind vanes are often found atop church steeples and other tall buildings. A wind vane (sometimes called a *weather vane*) is an arrow-shaped device with a large tail. It is mounted so that it can turn easily when the wind blows. As the wind pushes against the large tail of a wind vane, the arrowhead turns and points into the wind, indicating the direction from which the wind is coming (see Fig. 16-4).

· The wind vane of a weather station usually is placed high above the ground and connected electrically to an indicator inside the

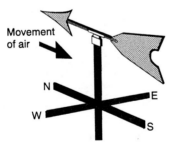

**Fig. 16-4. A wind vane.**

weather station. In this way, meteorologists can determine the wind direction without going outside.

**Wind Speed.** Wind speed is measured with an instrument called an **anemometer.** An anemometer consists of several small cups mounted on arms. The arms are connected to a single shaft (see Fig. 16-5).

The cups catch the wind, causing the shaft to rotate. The faster the wind blows, the faster the shaft rotates. The speed at which the shaft

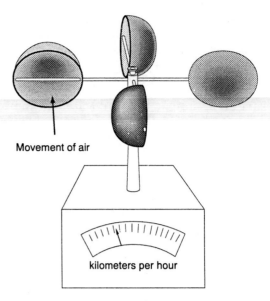

**Fig. 16-5. An anemometer.**

rotates is registered on a dial or on a digital display marked off in kilometers per hour or in knots (nautical miles per hour). The wind speed also may be printed out on a chart.

**Weather Balloons.** Wind speeds near Earth's surface can be measured with an anemometer. However, meteorologists use *weather balloons* to measure wind speeds high above Earth's surface. Helium-filled weather balloons are released and tracked visually with speed and direction-finding equipment on the ground. Observations of the balloons' movements are then used to calculate the speed and direction of the wind (see Fig. 16-6).

Weather balloons give accurate readings and are inexpensive. They cannot, however, be seen through dense clouds or over great distances. This drawback is overcome by using radar to track weather balloons. In addition, some weather balloons sent aloft regularly carry a package of instruments that measures air temperature, humidity, and air pressure at high altitudes. The package of instruments is called a *radiosonde*. A tiny radio transmitter included in the radiosonde sends the measurements back to weather stations on Earth.

# GLOBAL WIND BELTS

Fig. 16-7 shows Earth's major *wind belts.* The **wind belts** in each hemisphere consist of three convection cells that encircle Earth like belts. One convection cell circulates air between the equator and about 30 degrees (30°) latitude. A second cell circulates air between

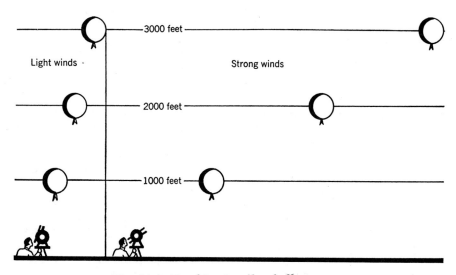

**Fig. 16-6. Tracking weather balloons.**

30° and 60° latitude, and a third convection cell circulates air between 60° latitude and the pole. In addition, another convection cell circulates air high in the troposphere between the equator and each pole.

Because of the complicated way that Earth's atmosphere is influenced by temperature, pressure, and Earth's rotation, each hemisphere has three convection cells instead of the single cell described earlier. Consequently, there are three major wind belts on Earth: (1) *trade winds*, (2) *prevailing westerlies*, and (3) *polar easterlies*. The formation of these wind belts depends on differences in air pressure that develop in the *doldrums*, the *horse latitudes*, and the *polar regions*.

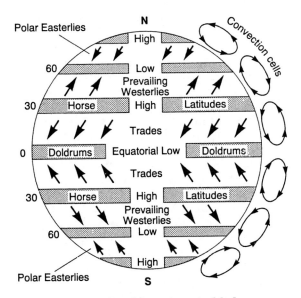

**Fig. 16-7. Earth's major wind belts.**

**Doldrums.** The region directly surrounding the equator is called the **doldrums**. In the doldrums, heated air usually rises straight up into Earth's atmosphere. In fact, solar heating of this region is similar to the heating of the air around the candle flame in the convection box. Warm air rising into the atmosphere creates a region of low air pressure (equatorial low) called a *low-pressure belt*.

**Horse Latitudes.** Air heated at the equator cools slowly as it rises higher into the atmosphere. Several kilometers above Earth, this air current divides into two parts. One part moves north; the other part moves south.

The warmed air cools gradually as it moves north and south of the equator. Because of Earth's rotation, the air is deflected toward the east. When it reaches 30° north (30° N) and 30° south (30° S) latitude, the air has cooled, and its density has increased to the point at which the air sinks toward Earth again. As the air sinks, it warms up slightly.

The sinking air produces a region of high air pressure at 30° latitude (see Fig. 16-7). This region of high barometric pressure, characterized by frequent calms and light winds, is called the **horse latitudes**. The horse latitudes are named from the days when sailing ships often remained motionless for weeks at a time in this region. At these times, horses that were carried on the ships were thrown overboard to conserve drinking water for the crew.

**Easterly Trades and Prevailing Westerlies.** In the horse latitudes, the sinking air divides into two parts. One part moves toward the equator, and the other part moves toward one of the poles. In each case, air moves from a region of high pressure (the horse latitudes) to regions of lower pressure north and south of the horse latitudes.

The winds moving toward the equator from the north and south are called the **trade winds.** They were named *trade winds* because they provided the energy for fifteenth- and six-teenth-century wind-powered trading ships to travel between Europe and the Americas. Be-cause of the Coriolis effect, the trade winds curve so that they actually blow from the northeast or southeast. In the Northern Hemi-sphere, these winds are called the *northeast trades,* and in the Southern Hemisphere, these winds are called the *southeast trades* (see Fig. 16-7). Together, they are known as the **easterly trades.**

The winds that move northward and south-ward from the horse latitudes are called the **prevailing westerlies.** In the Northern Hemi-sphere, the prevailing westerlies curve to the east so that they blow from the southwest. In the Southern Hemisphere, the winds that move southward from the horse latitudes curve so that they blow from the northwest.

The prevailing westerlies become stronger farther away from the horse latitudes. For ex-ample, the open ocean region near the tip of South America (at about 40° S latitude) was named the *roaring forties* by sailors because of the violent winds that blow there.

**Polar Easterlies.**   The dense, cold air moving away from both poles produces the **polar east-erlies.** These winds form part of the convection cell that extends from the poles to about 60° latitude in the Northern Hemisphere and the Southern Hemisphere.

In the Northern Hemisphere, the polar east-erlies blow toward the southwest, meeting the prevailing westerlies moving northward into Canada. The boundary between the polar east-erlies and the prevailing westerlies contin-ually shifts northward and southward. The shifting of these wind belts affects weather conditions in the United States and Canada (described in Chapter 18).

## WINDS CAUSED BY LOCAL TEMPERATURE CHANGES

**Sea Breezes.**   If you live near the ocean or one of the Great Lakes, you know that breezes blow off the water for many days of the year. These breezes are caused by the unequal heat-ing of the land and an adjacent, large body of water.

You learned in Chapter 15 that land heats up faster than water. On a hot, summer day, a sandy beach heats up faster than the adja-cent ocean or lake. Consequently, air over the beach becomes much hotter than air over the water.

Over the beach, the hotter air becomes less dense and rises, forming a low-pressure area. Over the water, the cooler, denser air forms a high-pressure area. The cooler, denser air over the water moves toward the land, forcing the warm, less dense air to rise.

The movement of cool ocean air toward the land is called a **sea breeze** (see Fig. 16-8a). Sea breezes usually begin blowing about mid-morning, ending about sunset. Sea breezes off the ocean can be felt about 25 km inland. Near the Great Lakes, a breeze develops in summer that can be felt about 5 km inland.

**Land Breezes.**   At night, land areas cool faster than do large bodies of water. Air over the land becomes denser than air over the water. As a result, the denser air flows toward the ocean or lake. A breeze that blows from the land to the ocean or a large lake is called a **land breeze** (see Fig. 16-8b). Land breezes usually are not as strong as sea breezes.

## MONSOONS

A huge land and sea breeze that reverses di-rection in different seasons is called a **mon-soon.** The monsoons of India and southern Asia are famous examples of these seasonal winds (see Fig. 16-9).

In summer, central India becomes much hotter than the adjacent areas near the Indian Ocean. Consequently, the interior of India be-comes a region of low pressure. Air over the Indian Ocean is much cooler than air over cen-tral India. Thus, this oceanic area becomes a region of high pressure. The denser, cooler air over the ocean moves inland, producing a steady wind called a *summer,* or *wet, monsoon.* The moisture-filled air of the summer mon-soon brings heavy rain to most of India. In southern India, up to 900 cm of rain may fall during the summer monsoon.

In winter, air over central India is much cooler than air over the Indian Ocean. Thus, the interior of India is a region of high pressure during winter. At the same time, the Indian Ocean warms the air over it. As a result, a re-gion of low pressure develops over the ocean. Cold, dense air then moves from central India toward the ocean. This cold, dry air brings

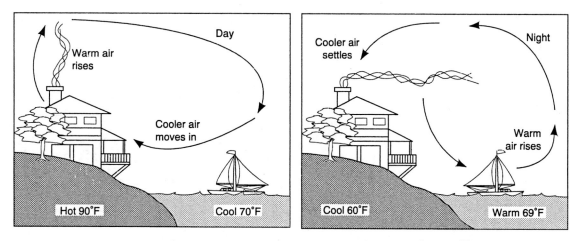

Fig. 16-8a. Sea breeze.    Fig. 16-8b. Land breeze.

India a long period of cold, dry weather called the *winter*, or *dry, monsoon.*

In the United States, people living in the central and eastern states also experience a seasonal change in the weather. In summer, air moving northward from the Gulf of Mexico brings warm, humid air to central and eastern states. In winter, air moving southward from the polar areas of Canada brings cold, dry air to these states.

**Mountain and Valley Breezes.** During the day, the sun heats mountain slopes faster than valleys. The heated air over mountain slopes becomes less dense. The cooler, dense air in the valleys moves up the mountainsides to displace the lighter air. As a result, a convection cell develops. The convection cell produces a wind that blows up the mountainside called a **valley breeze.**

At night, the air over mountain slopes cools faster than does the air in valleys. This occurs because the air at high altitudes contains less dust and moisture than does the air in valleys.

The heat gained by the mountain slopes during the day is rapidly lost at night by radiation into Earth's atmosphere. At night, the air over the mountain slopes becomes cooler and denser than the air in the valleys and moves down into the valleys. The movement of cooler, denser air into a valley from mountain slopes is called a **mountain breeze.**

Another kind of wind that blows down a mountainside is called a *chinook.* A chinook is produced by differences in air pressure on opposite sides of a mountain. Chinooks occur on the eastern slopes of the Rocky Mountains.

A chinook begins as a breeze over the Pacific Ocean. Winds blowing inland carry the moist air from the Pacific Ocean up the windward side, or western slopes, of the Rockies. As the air rises, it cools and gives up most of its moisture, which falls as rain or snow. As the now dry air descends the leeward side, or eastern slopes, of the Rockies, it is warmed by compression (described in Chapter 15). Thus, a **chinook** is a hot, dry wind that blows down a mountainside.

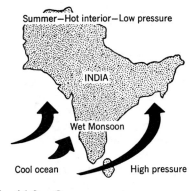

Fig. 16-9a. Summer wet monsoon.

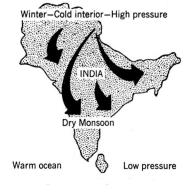

Fig. 16-9b. Winter dry monsoon.

## ROTATING WIND SYSTEMS

Huge air masses of low pressure and high pressure drift slowly over Earth's surface. These huge air masses are called *high-pressure systems* and *low-pressure systems*. The locations and shapes of high-pressure and low-pressure systems change continually. Masses of low-pressure air are called *cyclones;* masses of high-pressure air are called *anticyclones.*

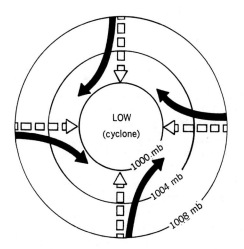

**Fig. 16-10a. Wind deflection in a cyclone.**

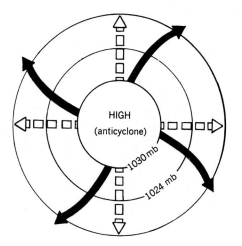

**Fig. 16-10b. Wind deflection in an anticyclone.**

**Cyclones.**   A low-pressure system in which wind blows *counterclockwise* is called a **cyclone.** In a cyclone, the air pressure is lowest at the center of the system and highest along its outer edge. Thus, higher-pressure air continuously moves toward the area of lower-pressure air in the center of the cyclone. On a weather map, you may have noticed that the wind arrows follow a counterclockwise pattern around a low-pressure system.

If Earth did not rotate, the air in a cyclone would move straight toward its center (see broken arrows in Fig. 16-10a). However, winds in the Northern Hemisphere curve to the right because of the Coriolis effect. Thus, the winds in a cyclone are deflected so that they spiral toward the center of the system, resulting in a counterclockwise circulation of the low-pressure air mass (see solid arrows in Fig. 16-10a).

**Anticyclones.**   A high-pressure system in which wind blows *clockwise* is called an **anticyclone.** In an anticyclone, air pressure is greatest at the center of the air mass and lowest along its outer edge. Because air in the center of an anticyclone has a higher pressure, it moves toward the outer edge of the air mass (see Fig. 16-10b). Again, the Coriolis effect causes the winds to curve to the right in the Northern Hemisphere. Consequently, the winds in an anticyclone are deflected so that as they spiral outward from the center of the system, the high-pressure air mass develops a clockwise circulation.

The movement of cyclones and anticyclones over Earth produces important weather changes. In general, cyclones bring stormy weather, and anticyclones bring clear or fair weather. The effect of cyclones and anticyclones on weather is described further in Chapter 18.

## THE JET STREAM

In the middle latitudes is a band of swiftly moving westerly winds called the **jet stream.** The jet stream is found near the tropopause, at an altitude that varies between 6 km and 11 km. Fig. 16-11 shows that the jet stream in the Northern Hemisphere is located farther south in the winter (between 20° and 30° N) than in the summer (between 35° and 45° N). Airplanes flying into the jet stream have been brought to a standstill, and sometimes even forced backward.

The jet stream was discovered during World War II. Pilots of high-flying B-29 bombers on long-distance missions westward over the Pacific Ocean sometimes were hampered by the extremely strong westerly wind. Their calculations indicated that the planes were flying against a wind that was blowing at about 480 km per hour (300 miles per hour).

A typical jet stream may be from about 40 km to 160 km wide and about 1.5 km high. Wind speeds of up to 643 km per hour have been recorded in the center of a jet stream. However, the speed of the jet stream usually ranges from 175 km to 370 km per hour.

The position and the speed of a jet stream is important to airline pilots and weather forecasters. Often, aircraft traveling from west to east can take advantage of this belt of strong winds to speed them to their destinations. On flights from east to west, pilots avoid the jet stream, which would prolong their flying time. During summer, however, the jet stream is less effective because it weakens and moves farther north.

Meteorologists have found that the jet stream influences weather throughout the world. They currently think that the jet stream steers, or directs, underlying low-pressure and high-pressure systems across the globe. When the jet stream is better understood, meteorologists should be able to make more accurate weather forecasts.

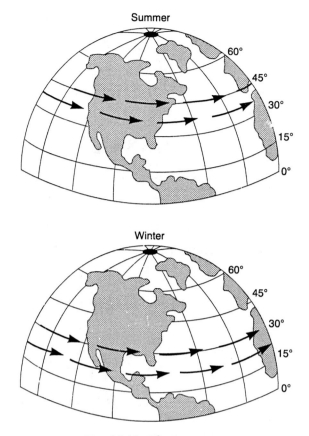

**Fig. 16-11. The jet stream.**

# CHAPTER REVIEW

## *Science Terms*

*The following list contains all of the boldfaced words found in this chapter and the page on which each appears.*

air current (p. 228)
anemometer (p. 230)
anticyclone (p. 234)
chinook (p. 233)
convection cell (p. 228)
Coriolis effect (p. 229)
cyclone (p. 234)
doldrums (p. 231)
easterly trades (p. 232)
horse latitudes (p. 231)
jet stream (p. 234)

land breeze (p. 232)
monsoon (p. 232)
mountain breeze (p. 233)
polar easterlies (p. 232)
prevailing westerlies (p. 232)
sea breeze (p. 232)
trade winds (p. 232)
valley breeze (p. 233)
wind (p. 228)
wind belts (p. 230)
wind vane (p. 230)

# Matching Questions

*On the blank line, write the letter of the item in column B that is most closely related to the item in column A.*

| Column A | Column B |
|---|---|
| ____ 1. low-pressure region all around the equator | *a.* horse latitudes |
| ____ 2. influence of Earth's rotation on winds | *b.* anemometer |
| ____ 3. seasonal winds that bring heavy rains | *c.* trade winds |
| ____ 4. circular movement of air in currents | *d.* wind vane |
| ____ 5. high-pressure region of calms and light winds | *e.* cyclone |
| ____ 6. instrument that shows wind direction | *f.* Coriolis effect |
| ____ 7. band of swiftly moving westerly winds | *g.* anticyclone |
| ____ 8. instrument that measures wind speed | *h.* convection cell |
| ____ 9. low-pressure, counterclockwise wind system | *i.* doldrums |
| ____ 10. high-pressure, clockwise wind system | *j.* jet stream |
| | *k.* monsoon |

# Multiple-Choice Questions

*On the blank line, write the letter preceding the word or expression that best completes the statement.*

1. The continuous replacement of warm, lightweight air by cooler, denser air produces atmospheric movements called
   *a.* whirlpools  *b.* drifts  *c.* convection cells  *d.* doldrums          1 ____

2. If Earth did not rotate, the winds high in the atmosphere over the Northern Hemisphere would be blowing toward the
   *a.* east  *b.* west  *c.* south  *d.* north          2 ____

3. Winds are deflected to the right in the Northern Hemisphere. This is summarized in a law stated by
   *a.* Ferrel  *b.* Galileo  *c.* Newton  *d.* Torricelli          3 ____

4. The region of Earth in which warmed air produces a low-pressure belt is called the
   *a.* horse latitudes  *b.* doldrums  *c.* Arctic  *d.* south temperate zones          4 ____

5. Air moving northward from the horse latitudes produces a belt of winds called the
   *a.* trade winds       *c.* southeasterlies
   *b.* northwesterlies   *d.* prevailing westerlies          5 ____

6. In the Northern Hemisphere, the trade winds blow from the
   *a.* southeast  *b.* southwest  *c.* northeast  *d.* northwest          6 ____

7. Dense, cold air flowing down from the polar regions produces wind belts called the
   *a.* polar northerlies   *c.* polar southerlies
   *b.* polar westerlies    *d.* polar easterlies          7 ____

8. Sea breezes are produced as
   a. land areas heat up more quickly than do adjacent water areas
   b. water areas heat up more quickly than do adjacent land areas
   c. warm ocean currents heat coastal areas
   d. warm land currents blow over the sea                    8 _____

9. Sea breezes begin blowing at about
   a. 6 A.M.  b. 10 A.M.  c. 5 P.M.  d. 8 P.M.                 9 _____

10. India's summer monsoon also is called the
    a. wet monsoon                c. hot monsoon
    b. dry monsoon                d. cold monsoon              10 _____

11. In winter, the interior of India becomes a high-pressure area because
    a. India is in the center of the doldrums
    b. air over India is cooler than air over the nearby Indian Ocean
    c. there is little snowfall in India
    d. warm ocean currents move offshore                      11 _____

12. A sinking air mass heats up because it is
    a. expanding                  c. losing water vapor
    b. being compressed           d. condensing               12 _____

13. The cause of the counterclockwise motion of air in a low-pressure system is Earth's
    a. revolution                 c. tilt on its axis
    b. rotation                   d. equatorial bulge          13 _____

14. Wind systems in which the wind blows outward in a clockwise direction are called
    a. hurricanes  b. cyclones  c. anticyclones  d. low-pressure troughs   14 _____

# Modified True-False Questions

*In some of the following statements, the italicized term makes the statement incorrect. For each incorrect statement, write the term that must be substituted for the italicized term to make the statement correct. For each correct statement, write the word "true."*

1. Sinking air masses produce two belts of *very calm* winds at 30° N and 30° S latitudes.                                      1 _____

2. The horse latitudes are regions of *high* air pressure.     2 _____

3. In the Northern Hemisphere, winds blowing from the horse latitudes toward the equator are called *westerlies*.              3 _____

4. Winds of the roaring forties are located near the bottom of *Alaska*.   4 _____

5. Wind speed is measured with an instrument called an *anemometer*.   5 _____

6. Sea breezes can be felt up to *100 kilometers* inland.      6 _____

7. The summer *monsoon* brings rains to India and other parts of Asia.   7 _____

8. Winds that blow upward along mountainsides are called *mountain* breezes.   8 _____

9. A low-pressure system in which the wind blows in a counterclockwise direction is called a *cyclone*.                        9 _____

10. Anticyclones often produce *clear, dry* weather.           10 _____

# Testing Your Knowledge

1. Make a drawing of your classroom on a separate sheet of paper showing the location of all radiators, windows, and doors. On this drawing, show the paths followed by heated and cold air when the radiators are operating.

2. If you placed a drop of water on top of a slowly rotating globe, the drop would curve in one direction. When the drop of water reached a point halfway between the top and bottom of the globe, the drop would curve in the opposite direction. Why does the drop of water follow two different curved paths on the globe? _____

_____

_____

3. On a separate sheet of paper, draw Earth's major wind belts on a circle representing Earth. Label each wind belt. Explain briefly how the trade winds and westerlies are produced.

_____

_____

4. How does a sea breeze form on a hot summer day? _____

_____

_____

5. Why does cool, dense air rise up a mountainside? _____

_____

6. Why are many deserts located on the leeward side of high mountain ranges? _____

_____

7. Name two methods used to measure wind speed. _____

_____

8. Identify the type of air pressure associated with each description below. Write either *high* or *low* next to each description.
   *a.* Air surrounding a hot radiator: _____

   *b.* Air at the equator: _____

   *c.* Air at the horse latitudes: _____

   *d.* Air over the ocean in summer: _____

   *e.* Interior of India in summer: _____

   *f.* Air on mountain slopes during the day: _____

   *g.* Pressure system in which winds blow in a clockwise direction: _____

# Clean Air: A Goal We Can Drive Toward?

The LEVs, ULEVs, and ZEVs are coming! What will they do to our cities?

No, these are not invading aliens from outer space set on gobbling up New York City, Chicago, and Los Angeles. As a matter of fact, they are earthlings—of a sort—that may help sweep clean the air over these cities.

The LEVs, ULEVs, and ZEVs are, in order, *low-emission vehicles, ultra-low emission vehicles,* and *zero-emission vehicles.* They are cars designed to reduce and, eventually, eliminate exhaust gases that pollute air, especially city air.

These air pollutants are gases produced by the burning of gasoline in internal combustion engines, the kind that in the early 1990s powered every car, truck, and bus in the United States. The gases include hydrocarbons, nitrogen oxides, and carbon monoxide. What's more, some of these gases, especially the nitrogen oxides, interact with sunlight to produce yet another air pollutant called *ozone.*

Taken together, all of these exhaust gases produce *smog,* a blanket of hazy, unhealthy air that, because of local weather conditions, is worst in the city of Los Angeles. No wonder, then, that in 1990 officials of the state of California created regulations that they hoped would reduce smog over the second most populated city in the country.

By the early 1990s, the states of New York, Massachusetts, Vermont, and Maine followed California's lead. Since then, many other states have joined the anti-air pollution crusade by adopting low-emission vehicle requirements.

However, there was a difference. California had a fast timetable for cleaning up its air. Other states decided to move more slowly.

Why the difference? For one thing, California's problem was more serious. The people of Los Angeles needed relief and they needed it fast. For another thing, the economic impact on consumers had to be considered, since nonpolluting—or less polluting—cars would probably cost consumers more money to buy or operate.

California had little choice. It was pay more or breathe less. Since people couldn't cut down on their breathing, they were ready to pay more. But could auto companies build less-polluting cars and, if so, how long would this take?

Some people said "never." Others said "a long time." California said "between now and the year 2003." In effect, California said to auto makers: "If you want to sell cars in our state, you're going to have to meet the strictest standards in the country." Here's where the LEVs, ULEVs, and ZEVs come in.

According to California regulations, as of 2003, auto makers in the state are required to sell a mix consisting of 75 percent LEVs, 15 percent ULEVs, and 10 percent ZEVs. Cars qualifying as LEVs and ULEVs were introduced in the state in 1997, and ZEVs made their appearance in 1998.

It's easy to understand what a ZEV is: It's a car that doesn't burn fuel at all—an electric car. But what are LEVs and ULEVs? There are several appoaches to making cars with extremely low emissions of air pollutants. One way is to design cars that run on alternative fuels—fuels that when burned, emit less pollution than gasoline does. Examples of such fuels are ethanol (also called wood alcohol) and natural gas. You may have already seen buses and cars with the label "NGV"—natural gas vehicle—on them. The Honda Civic GX is a car that runs on natural gas, which has been hailed as the

cleanest-running car on Earth with an internal combustion engine. Its tailpipe emissions are so low they're almost impossible to measure with conventional laboratory equipment!

Another way to make a low- or ultra low-emission vehicle is to design a car that can run on either gasoline or electricity. Such cars, called hybrid electric vehicles, have both a gasoline-powered internal combustion engine and an electric motor. The gasoline engine and the electric motor can be used together or separately to power the vehicle, which gets much higher gas mileage than an ordinary car while emitting far less pollution. The Honda Insight is a hybrid electric car that gets about 70 miles to the gallon and qualifies as an ULEV. Toyota's hybrid car, the Prius, gets about 50 miles per gallon and qualifies as a *super* ultra low-emission vehicle, or SULEV.

But even natural gas-powered and hybrid vehicles have some emissions; true ZEVs must be fully electric. And that's where the biggest problems with meeting California's requirements come in. After spending over a billion dollars on researching and developing batteries that can power electric vehicles, the best that auto makers have come up with are small cars with huge batteries that can only travel about 100 miles before needing to be recharged. What's more, the recharging can take up to six hours! Because of the high cost of producing such cars, they're more

expensive than ordinary cars, so in effect, the buyer gets less car for more money.

In light of such drawbacks, auto makers claim that there's no market for electric cars, and that California's goal of 10 percent ZEV sales by 2003 is simply unrealistic. Clean air advocates counter that there *is* a market for electric cars if they're promoted properly, and they point out that consumers can get tax rebates to help offset the higher cost.

As of this writing, several major auto manufacturers have gone to court to try to prevent the state of California from enforcing the 2003 ZEV requirement. Whatever the outcome of this court battle, it's clear that California's determination to reduce air pollution has forced auto makers to think of new and innovative ways to build cleaner running cars. But should people be forced to pay more for cleaner-running cars? And should government pick up part of the tab?

*1. An electric car is an example of a*
   a. LEV.       c. ZEV.
   b. ULEV.    d. SULEV

*2. "LEV" refers to a*
   a. low-energy vehicle.
   b. low-emission vehicle.
   c. light-energy vehicle.
   d. light-emission vehicle.

*3. Smog is worst in*
   a. Los Angeles.   c. New York City.
   b. Chicago.      d. Washington, D.C.

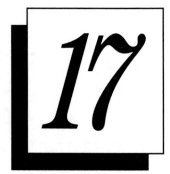

# Moisture in the Atmosphere

## LABORATORY INVESTIGATION

### WATER AND RELATIVE HUMIDITY

**A.** Dry the outside of a metal cup with a paper towel. Fill half the cup with water. Place a thermometer in the cup as shown in Fig. 17-1*a*.

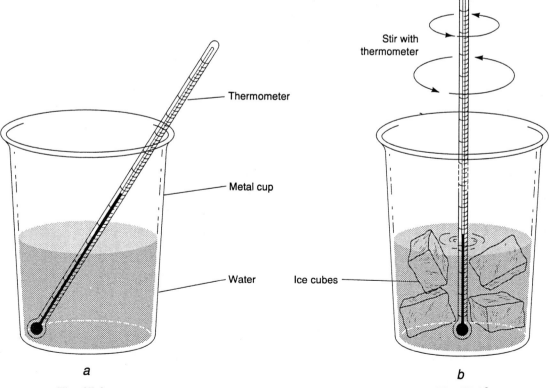

Thermometer

Metal cup

Water

Stir with thermometer

Ice cubes

*a*

**Fig. 17-1*a*.**

*b*

**Fig. 17-1*b*.**

**1.** What do you observe on the outside of the cup? _____

_____

**2.** What is the temperature of the water? _____

**B.** Add four ice cubes to the water, as shown in Fig. 17-1*b*. Stir the ice-water mixture slowly with the thermometer until the outside of the cup changes in appearance. As soon as you notice this change, read the thermometer.

**3.** What is the temperature of the water? _____

**4.** What do you observe on the outside of the cup? _____

_____

**5.** What is the relationship between the temperature change and the change in appearance

of the outside of the cup? _____

_____

**C.** Obtain a wide-mouthed jar and enough rubber sheeting to cover the mouth of the jar. Stretch the rubber sheet over the jar's mouth. Fasten the sheet in place by winding a piece of string tightly around the neck of the jar several times. Tie the loose ends of the string together (see Fig. 17-2).

Push the rubber sheet about 5 centimeters into the jar. Then, quickly release the rubber sheet while carefully observing the appearance of the air inside the jar. Repeat this procedure.

**6.** Describe your observations. _____

**7.** Does the volume of air inside the jar change? Explain. _____

_____

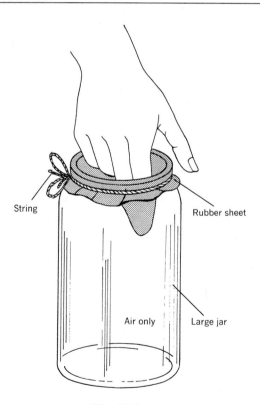

String      Rubber sheet

Air only      Large jar

**Fig. 17-2.**

**8.** What happens to the volume of air when you release the rubber sheet? _____

_____

**D.** Untie the string, and remove the rubber sheet. Pour 10 milliliters of water into the jar. Replace the rubber sheet, and tie it in place.
   Gently swirl the water around the bottom of the jar for about 1 minute. Then, repeat part C.

**9.** Describe your observations. _____

**E.** Remove the rubber sheet. Light a match, blow it out, and quickly drop it into the jar. Quickly replace the rubber sheet, and tie it in place. Repeat part C again.

**10.** Describe your observations. _____

**11.** Is there a relationship between the volume changes inside the jar and any other changes you observed inside the jar? Explain.

_____

_____

**12.** Did the smoke produced in part E affect the experiment? Explain.

_____

## WATER VAPOR AND THE WEATHER

A huge volume of water enters Earth's atmosphere every day as an invisible gas called *water vapor.* Under certain conditions, the water vapor changes into clouds, fog, dew, frost, rain, or snow.

The total amount of water vapor in the atmosphere does not change. That is because rain, snow, and other forms of water that fall to Earth in one place are replaced by water vapor that enters the atmosphere in other places.

Atmospheric water affects people in many ways. For example, heavy rain or snow, and dense fog, can affect transportation and outdoor events. It is less obvious how water vapor in the atmosphere causes the weather.

First, atmospheric water vapor prevents the daily air temperature from varying too much. Water vapor accomplishes this by absorbing heat energy during the day and releasing heat energy at night.

You learned that moist air is less dense than dry air. Thus, when the amount of water vapor in the air changes, the atmospheric pressure also changes. Changes in the air temperature, pressure, and amount of atmospheric water vapor are related to the weather. By studying these changes, meteorologists have learned how to predict weather changes.

## SOURCES OF ATMOSPHERIC MOISTURE

Most of the moisture in Earth's atmosphere is water vapor that evaporates from the ocean. Some water vapor also evaporates from the surface of lakes, streams, and the land. In addition, plants, animals, and other organisms release water vapor into the atmosphere.

**Evaporation.** The process by which a liquid changes into a gas is called **evaporation.** Most water enters Earth's atmosphere by evaporation. Like the molecules of all substances, water molecules are always moving. When water molecules collide at the surface of a body of water, some of the water molecules are pushed into the atmosphere and become water vapor (see Fig. 17-3).

As surface water absorbs heat energy from the sun, the water molecules begin to move faster. This results in more water molecules breaking loose and escaping into Earth's atmosphere as water vapor. Thus, as the water temperature increases, the rate of evaporation also increases.

As water molecules escape by evaporation, they remove heat energy. This causes the water temperature to decrease. If heat energy is removed faster than it is added, the water becomes cooler.

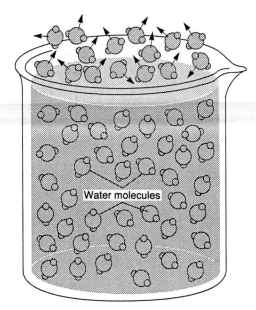

**Fig. 17-3. Evaporation of water.**

You can demonstrate the cooling effect of evaporating water by simply dipping your hand in water and then waving it through the air. Your hand feels cooler because the evaporating water molecules remove some of the heat from your skin. For the same reason, you may feel chilly when you emerge from a swimming pool or bathtub.

Moving air causes the rate of evaporation to increase by carrying away molecules of water vapor. As the wind removes molecules of water vapor from the surface of a body of water, room is made for other water molecules to escape into Earth's atmosphere. When the rate of evaporation increases, the cooling effect increases. For example, when you go swimming on a windy day, you feel much colder when you emerge from the water than on a calm day.

**Transpiration.** Most water evaporates from plants by **transpiration.** In this process, water that a plant absorbs from the soil evaporates through *pores,* or openings in the leaves. Transpiration adds a huge amount of water to the atmosphere. For example, a large oak tree releases about 114,000 liters (l) of water a year by transpiration. An acre of corn releases more than 1,000,000 l of water by transpiration in a growing season.

**Activities of Animals.** Animals add water vapor to Earth's atmosphere by their loss of perspiration to evaporation and by exhaling water vapor during breathing. Water vapor also is released into the atmosphere by humans when fuels are burned to heat homes, generate electricity, and power factories and motor vehicles.

## SPECIFIC HUMIDITY

The actual amount of water vapor in Earth's atmosphere is called the **specific humidity.** The specific humidity ranges from about 4% (by volume) to a low of near 0%. Where the sun shines strongly on a large body of water, such as the Gulf of Mexico, the specific humidity of the atmosphere is high. The specific humidity is low when the sun shines strongly on a place where there is little water, such as the Sahara Desert.

**Humidity and Temperature.** The amount of water Earth's atmosphere can hold depends mainly on the air temperature. As the air temperature increases, the atmosphere can hold more water. In contrast, as the air temperature decreases, the ability of the atmosphere to hold water decreases. You observed this relationship in your laboratory investigation.

The amount of water vapor in a given volume of air is measured in small units of weight called *grains*. There are 480 grains (gr) in 1 ounce (oz). (Sometimes, the amount of water vapor in a given volume of air—1 kilogram (kg)—is measured in grams. There are about 28 grams in 1 oz.) The amount of water vapor that a given volume, or *parcel*, of air can hold at a specific temperature is limited. The maximum amount of water vapor that a parcel of air can hold is called its *capacity*.

When a parcel of air is holding as much water vapor as it can, the air is said to be **saturated.** When the air is saturated, the specific humidity and the capacity are equal. When a parcel of saturated air is cooled, the maximum amount of water vapor it can hold decreases. As a result, excess water vapor *condenses* into tiny water droplets.

**TABLE 17-1. MAXIMUM
WATER VAPOR CONTENT
(1 CUBIC FOOT OF AIR)**

| Temperature (°C) | Water Vapor (grains) |
|---|---|
| −1.1 | 1.9 |
| 4.4 | 2.9 |
| 10.0 | 4.1 |
| 15.6 | 5.7 |
| 21.1 | 8.0 |
| 26.6 | 10.9 |
| 32.2 | 14.7 |
| 37.7 | 19.7 |

Table 17-1 shows the effect of temperature on the water vapor capacity of a parcel of air. At higher temperatures, the capacity of the air increases at a much greater rate than it does at lower temperatures.

**Measuring Humidity.** Humidity is commonly measured by comparing the amount of water vapor in the atmosphere with the maximum amount of water vapor that the air can hold *at a given temperature.* Meteorologists represent this comparison as a percentage and call it the **relative humidity.** For example, a parcel of air at 10° Celsius (C) can hold about 4 gr of water. When this parcel of air is holding

4 gr of water, the air is saturated, and the relative humidity is 100% (see Fig. 17-4). If the parcel of air at 10°C is holding 2 gr of water vapor, it is holding about one half of its capacity, or has a relative humidity of 50% (see Fig. 17-5).

**Determining Relative Humidity.** Relative humidity is measured by instruments called **hygrometers.** The *psychrometer* and *hair hygrometer* are two commonly used hygrometers. A psychrometer operates on the principle that evaporation is a cooling process. A hair hygrometer depends on the fact that human hair stretches when damp.

*Psychrometer.* A psychrometer is made of two thermometers that are mounted side by side on a narrow frame (see Fig. 17-6). One

Fig. 17-4. Saturated air has a relative humidity of 100 percent.

Fig. 17-5. Half-saturated air has a relative humidity of 50 percent.

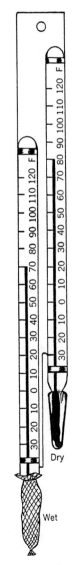

Fig. 17-6. A psychrometer.

thermometer is called a *dry-bulb thermometer.* This thermometer measures the existing air temperature. The other thermometer is called a *wet-bulb thermometer.* A wet cloth is wrapped around the bulb of the wet-bulb thermometer. Both thermometers are whirled through the air for several seconds. The motion causes water to evaporate from the cloth. This cools the cloth, resulting in a lower temperature reading on the wet-bulb thermometer.

The temperature of the wet-bulb thermometer depends on how fast water evaporates from the cloth, which in turn depends on how much water vapor the air is holding. If the air is very humid, little moisture will evaporate from the cloth. Then, the two thermometers will have similar readings. If the air is very dry, a lot of water vapor will evaporate from the cloth. Then, the wet-bulb thermometer will have a much lower reading than will the dry-bulb thermometer.

By comparing the difference in temperature between the dry-bulb thermometer and wet-bulb thermometer, a meteorologist can calculate the relative humidity (see Table 17-2).

Suppose that the reading of a dry-bulb thermometer is 30°C, and the reading of wet-bulb thermometer is 20°C. The difference between the two readings is 10°C. Using these data, you can find out that the relative humidity of the air is 39% (see Table 17-2). Table 17-3 shows typical dry-bulb and wet-bulb temperatures in a dry climate and in a humid climate.

**Hair Hygrometer.**  Human hair stretches when it is moist. The more moisture in hair, the more the hair stretches. Consequently, hair can show changes in the relative humidity of the atmosphere. A hair hygrometer is a long bundle of hairs with one fixed end and with the other end attached to an indicator (see Fig. 17-7). The indicator directly shows

### TABLE 17-2.  FINDING RELATIVE HUMIDITY

| Dry-Bulb Temperature (°C) | Relative Humidity (in Percents) Difference between wet-bulb and dry-bulb temperatures (°C) | | | | | | | | | | | | | | | | | | | |
|---|---|---|---|---|---|---|---|---|---|---|---|---|---|---|---|---|---|---|---|---|
| | 1° | 2° | 3° | 4° | 5° | 6° | 7° | 8° | 9° | 10° | 11° | 12° | 13° | 14° | 15° | 16° | 17° | 18° | 19° | 20° |
| 0 | 81 | 64 | 46 | 29 | 13 | | | | | | | | | | | | | | | |
| 2 | 84 | 68 | 52 | 37 | 22 | 7 | | | | | | | | | | | | | | |
| 4 | 85 | 71 | 57 | 43 | 29 | 16 | | | | | | | | | | | | | | |
| 6 | 86 | 73 | 60 | 48 | 35 | 24 | 11 | | | | | | | | | | | | | |
| 8 | 87 | 75 | 63 | 51 | 40 | 29 | 19 | 8 | | | | | | | | | | | | |
| 10 | 88 | 77 | 66 | 55 | 44 | 34 | 24 | 15 | 6 | | | | | | | | | | | |
| 12 | 89 | 78 | 68 | 58 | 48 | 39 | 29 | 21 | 12 | | | | | | | | | | | |
| 14 | 90 | 79 | 70 | 60 | 51 | 42 | 34 | 26 | 18 | 10 | | | | | | | | | | |
| 16 | 90 | 81 | 71 | 63 | 54 | 46 | 38 | 30 | 23 | 15 | 8 | | | | | | | | | |
| 18 | 91 | 82 | 73 | 65 | 57 | 49 | 41 | 34 | 27 | 20 | 14 | 7 | | | | | | | | |
| 20 | 91 | 83 | 74 | 66 | 59 | 51 | 44 | 37 | 31 | 24 | 18 | 12 | 6 | | | | | | | |
| 22 | 92 | 83 | 76 | 68 | 61 | 54 | 47 | 40 | 34 | 28 | 22 | 17 | 11 | 6 | | | | | | |
| 24 | 92 | 84 | 77 | 69 | 62 | 56 | 49 | 43 | 37 | 31 | 26 | 20 | 15 | 10 | 5 | | | | | |
| 26 | 92 | 85 | 78 | 71 | 64 | 58 | 51 | 46 | 40 | 34 | 29 | 24 | 19 | 14 | 10 | 5 | | | | |
| 28 | 93 | 85 | 78 | 72 | 65 | 59 | 53 | 48 | 42 | 37 | 32 | 27 | 22 | 18 | 13 | 9 | 5 | | | |
| 30 | 93 | 86 | 79 | 73 | 67 | 61 | 55 | 50 | 44 | 39 | 35 | 30 | 25 | 21 | 17 | 13 | 9 | 5 | | |
| 32 | 93 | 86 | 80 | 74 | 68 | 62 | 57 | 51 | 46 | 41 | 37 | 32 | 28 | 24 | 20 | 16 | 12 | 9 | 5 | |
| 34 | 93 | 87 | 81 | 75 | 69 | 63 | 58 | 53 | 48 | 43 | 39 | 35 | 30 | 28 | 23 | 19 | 15 | 12 | 8 | 5 |

## TABLE 17-3. SAMPLE PSYCHROMETER READINGS

| Types of Readings | Painted Desert, Arizona (dry) | Okefenokee Swamp, Florida (humid) |
|---|---|---|
| Dry-bulb temperature | 32°C | 32°C |
| Wet-bulb temperature | 18°C | 31°C |
| Difference between dry- and wet-bulb temperatures caused by evaporation (cooling effect) | 14°C | 1°C |
| Relative humidity | 24% | 93% |

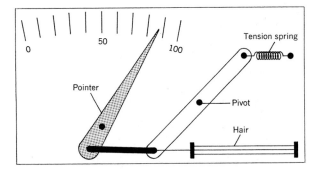

**Fig. 17-7. A hair hygrometer.**

the relative humidity. If the relative humidity is 60%, the indicator will point to 60.

## CONDENSATION

When air cools sufficiently, atmospheric water vapor undergoes **condensation** by changing from a gas into a liquid or solid. For example, when water vapor condenses at Earth's surface, the water vapor forms *dew* or *frost*. When water vapor condenses in Earth's atmosphere, it forms *clouds* made of water droplets or ice crystals.

**Condensation Near the Ground.** Water vapor that condenses on the ground or on solid objects near the ground is called **dew.** For ex-

ample, during warm weather, the water droplets that condense on a grass lawn at night are dew.

The air temperature at which water vapor begins to condense is called the **dew point.** The dew point depends on the amount of water vapor present in the atmosphere. Water vapor changes into a liquid or solid when the air temperature falls low enough for the air to become saturated. At the dew-point temperature, the water vapor capacity of the air has decreased to the point at which some of the water vapor condenses into a liquid. You observed this in your laboratory investigation.

Suppose humid air and dry air at a temperature of 25°C are cooled at the same rate by coming in contact with a flask of ice water. The humid air will become saturated first, and some of the water vapor will condense into dew (see Fig. 17-8). Table 17-1 shows that at 26.6°C, a sample of air has a capacity of almost 10.9 gr of water vapor. If a sample of air at 26.6°C is holding 10.9 gr of water vapor, it is saturated. If this saturated air cools to 21.1°C, its capacity would drop to 8 gr. As a result, 2.9 gr of water vapor would condense and collect on the flask as dew.

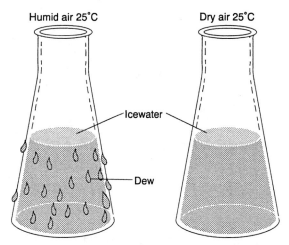

**Fig. 17-8. Formation of dew.**

In contrast, suppose a sample of dry air at 26.6°C contains 4.1 gr of water vapor. According to Table 17-1, the capacity of air at 10°C is 4.1 gr. The air would have to be cooled from 26.6°C to 10°C (or less) before any of the water vapor in this air will condense.

At night, the ground radiates most of the heat it absorbs during the day. For this reason, dew often forms after sunset. In addition, the ground radiates more heat on calm, clear nights. Then, the ground becomes cold enough so that the air above it can cool to the dew point.

Sometimes, the air may reach its dew point when the temperature falls below 0°C, or the freezing point of water. Then, atmospheric water vapor condenses to form ice crystals. This type of condensation is called **frost.** Frost forms when water vapor changes directly from a gas into a solid. Scientists call this change **sublimation.** When dew freezes, however, it does not form frost, but frozen dew. Like dew, frost forms mostly on calm, clear nights.

**Condensation in the Atmosphere.**   Most water vapor condenses in Earth's atmosphere on microscopic particles called **condensation nuclei** (see Fig. 17-9). Most condensation nuclei are salt particles from the ocean, wind-blown soil, and smoke and dust particles from forest fires and erupting volcanoes. As the water vapor condenses around the condensation nuclei, *fog* or *clouds* are formed.

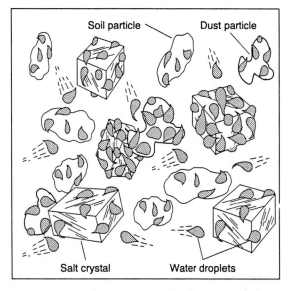

Fig. 17-9. **Condensation nuclei for water vapor.**

An accumulation of tiny water droplets light enough to float in the air is called **fog.** Like dew and frost, fog usually forms at night when the air temperature falls.

When air near the ground is cooled below its dew point and a light breeze or air current keeps the condensed droplets moving, the water droplets may remain suspended just above the ground. This type of fog is called *radiation fog* because it forms on clear nights after the ground has lost heat by radiation.

This type of fog also is commonly called *ground fog* because its thickness is usually less than 30 meters above the ground. Ground fog usually forms after sunset and disappears by late morning as heat from the sun causes the water droplets to evaporate.

Fog often forms when warm, moist air cools as it passes over a cold surface. This type of fog commonly occurs along seacoasts when warm, moist air from the ocean flows inland and is cooled as it meets colder land surfaces (see Fig. 17-10). This type of fog also forms when warm, moist air is cooled as it moves over cold ocean currents.

A combination of smoke and fog is called **smog.** Smog frequently occurs over cities because of large amounts of smoke that enter the air from chimneys and auto exhausts. If there is no wind to blow away the smoke and fog, it may accumulate and form a layer of polluted air that is unhealthy to breathe. Smog may irritate the lungs, causing serious respiratory disorders, especially for elderly or ill people.

Like fog, **clouds** form when water vapor in Earth's atmosphere condenses on condensation nuclei; but the process that forms clouds differs from the process that forms fog. Clouds form because (1) air cools as it expands, and (2) cool air cannot hold as much water as warm air can.

A common example of how an expanding gas becomes cool may be seen in the small carbon-dioxide cylinders used to charge bottles of water with carbon-dioxide gas. When released from the cylinder, the gas expands rapidly. As a result, the cylinder becomes very cold to the touch, sometimes so cold that frost forms on the cylinder.

In the laboratory investigation, you changed the air pressure inside the jar. You probably did not observe a change inside the jar. However, when you added water and smoke particles (condensation nuclei), the inside of the jar became cloudy.

In Earth's atmosphere, as a parcel of warm air rises, the air pressure surrounding the parcel decreases. Therefore, the parcel of air ex-

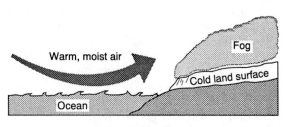

Fig. 17-10. **Fog.**

pands as it rises. As the parcel of air expands, it becomes cooler. When the temperature of the parcel of air falls to its dew point, the atmospheric water vapor condenses, and a cloud appears in the sky (see Fig. 17-11). If there are abundant smoke or salt particles in the air, the water vapor may condense even before the dew point is reached.

There are several ways in which air can rise, cool, and reach its dew point. Air may rise because (1) it passes over a hot surface; (2) it may be forced to rise over a mountain (see the section on chinook winds in Chapter 16); and (3) it may be lifted upward by a wedge of cooler air that forces its way beneath it. Each of these methods is responsible for the formation of different types of clouds.

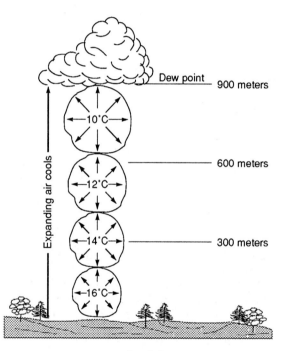

**Fig. 17-11. Cloud formation.**

## CLOUD TYPES

The type of cloud that forms depends mainly on the way an air mass rises into Earth's troposphere. If the air mass rises along a steep slope, the clouds will look like cotton balls. If the air mass rises along a gentle slope, the clouds will be flat or sheetlike.

Meteorologists classify clouds according to their shape and their distance above the ground. There are four main cloud types: *high clouds, middle clouds, low clouds,* and *clouds with vertical development.* If the base of a cloud is above 6 kilometers (km), it is called a *high*

*cloud.* If the base of a cloud is between 2 km and 6 km, it is a *middle cloud.* If the base of a cloud is between sea level and 2 km, it is a *low cloud.* Clouds with vertical development extend upward from about 0.5 km to more than 10 km.

Clouds are named according to their shapes. Some common cloud types are described below:

- *Stratus* clouds are sheetlike, or flat.
- *Cumulus* clouds are fluffy, resembling cotton balls.
- *Cirrus* clouds resemble tufts of hair, or feathers. Cirrus clouds occur at high altitudes, where the air temperature is far below the freezing point. Consequently, cirrus clouds are made of ice crystals.
- *Nimbus* clouds are rain clouds.

Some cloud types are a combination of two names. For example, a *cumulonimbus* cloud, also called a *thunderhead,* is a dark rain cloud that is piled up in a huge mound; an *altostratus* cloud is a flattened cloud at a middle altitude; and a *cirrostratus* cloud is a flattened cloud at a high altitude. These clouds and other common cloud types are shown in Fig. 17-12. Note how each name describes the cloud's appearance.

Clouds are important weather indicators. Some cloud types are indicators of fair weather, whereas other types indicate stormy weather. Table 17-4 provides brief descrip-

**TABLE 17-4. CLOUDS AND WEATHER**

| Cloud Types | Associated Weather |
| --- | --- |
| Cirrus | Fair, generally clear skies; little chance of rain or snow |
| Cirrostratus (covering a large part of the sky) | Possibly rain or snow within 24 hours |
| Altostratus (covering most of the sky) | Possibly rain or snow within a few hours |
| Altocumulus (covering a large part of the sky) | Fair, chance of rain or snow |
| Cumulus | Fair |
| Stratocumulus (covering a large part of the sky) | Chance of rain or snow |
| Cumulonimbus | Thunderstorms, hail |
| Nimbostratus | Rain or snow |

Fig. 17-12. Cloud types.

tions of the kinds of weather usually associated with different types of clouds.

## *PRECIPITATION*

The falling of water such as rain, sleet, hail, and snow from Earth's atmosphere to the ground is called **precipitation.** In general, water falls as precipitation when the very small droplets that make up clouds increase in size until they are heavy enough to fall to Earth's surface.

**Rain.** Meteorologists have proposed two main theories to explain how water droplets increase in size to become **rain.** According to one theory, water vapor condenses on condensation nuclei such as salt or smoke particles. As water continues to condense on a condensation nucleus, the droplet becomes larger. The water droplet attracts other droplets and merges with them. In this way, the water droplet increases in size until it is heavy enough to fall as a raindrop. Very small raindrops are referred to as *drizzle.*

According to the other theory, precipitation begins in the higher regions of thick clouds, where the temperature is below the freezing point. In these clouds, ice crystals form. When the ice crystals are large enough, they fall through the cloud. When the temperature near

the ground is below freezing, the ice crystals fall to the ground as snow. When the temperature near the ground is above freezing, the falling snow melts and changes into raindrops before it reaches Earth's surface.

***Cloud Seeding.*** Rain has been artificially produced by *cloud seeding.* This process involves placing either particles of dry ice or tiny crystals of silver iodide or other chemicals into clouds by aircraft. These chemical crystals function as condensation nuclei for the development of ice crystals. The ice crystals attract water droplets, which get large enough to fall through the atmosphere as precipitation.

Some common forms of precipitation are described in Table 17-5.

### TABLE 17-5. COMMON FORMS OF PRECIPITATION

| | |
|---|---|
| Drizzle | Fine drops of water (diameter less than 1/50 inch) |
| Rain | Drops of water larger than drizzle |
| Sleet | Frozen drops of rain |
| Hail | Frozen drops of water having a layered or onionlike structure |
| Snow | Ice crystals having a branched structure |

*Sleet, hail,* and *snow* are frozen forms of precipitation, each of which forms under a different condition.

**Sleet.** Pellets of ice that form when falling rain freezes are called **sleet.** This may happen when rain falls from a warmer region of Earth's atmosphere to a colder region at a lower altitude. Such a condition of reversed temperature in the atmosphere is called a *temperature inversion* (discussed in Chapter 15). If the temperature at the lower altitude is below freezing, the raindrops may freeze before reaching Earth's surface.

**Hail.** Strong updrafts of air in cumulonimbus clouds form frozen precipitation called **hail.** A hailstone begins as a raindrop, which is carried high into a cumulonimbus cloud by a strong updraft of air. The raindrop changes to ice when it reaches a level in the cloud where the temperature is below freezing. The frozen raindrop then falls through the cloud to a warmer level of Earth's atmosphere. As the frozen raindrop begins to melt, it acquires a coat of water. Another updraft then carries the raindrop higher so that the coat of water freezes.

This cycle of freezing and melting may occur several times. Each cycle of melting and freezing provides the hailstone with another layer of ice. Thus, if you cut a large hailstone in half, you will find several concentric layers of ice similar to the layers of an onion. Hailstones are usually less than half an inch in diameter, although some hailstones larger than grapefruits have been seen.

**Snow.** Like frost, **snow** forms when water vapor changes directly into a solid. Water vapor in clouds accumulates on condensation nuclei and changes directly into ice crystals, commonly called *snowflakes.* Large snowflakes are produced when several ice crystals clump together at temperatures close to the freezing point. As a rule, the colder the temperature, the smaller the snowflakes that form. The shapes of snowflakes are usually hexagonal (six-sided).

**Acid Precipitation.** The burning of fossil fuels and erupting volcanoes add sulfate and nitrate particles to the atmosphere. Both sulfate and nitrate particles make good condensation nuclei. As water condenses on these particles, they form droplets or ice crystals of sulfuric acid and nitric acid. When these fall to the ground, they are called acid precipitation. (Acid precipitation is discussed in Chapter 9.)

# MEASURING PRECIPITATION

Meteorologists measure the amount of rain that falls by catching the rain in a special container called a **rain gauge.** This device consists of a funnel leading into a cylinder (see Fig. 17-13). The diameter of the funnel is 10 times the diameter of the cylinder. Consequently, the cylinder magnifies the amount of rainfall 10 times.

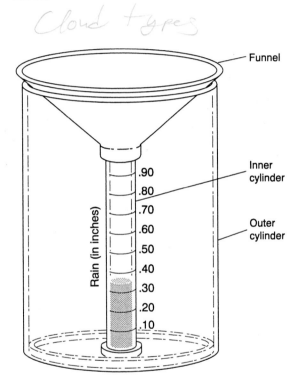

**Fig. 17-13. A rain gauge.**

When 0.1 inch (in) of rain falls into the funnel, the rain fills the cylinder to a depth of 1 in. In Fig. 17-13, the cylinder is filled to a depth of 3.2 in, which indicates 0.32 in of rainfall. Thus, with a rain gauge, small amounts of rainfall can be measured accurately.

Meteorologists usually measure snowfall by pushing a ruler into freshly fallen snow. Snowfall usually does not cover the ground evenly. Thus, measurements are taken in several places. The depth of the snow is the average of these measurements. By melting snow and allowing it to run into a rain gauge, the amount of water in the snowfall can be determined. In general, one foot of melted snow equals about one inch of rain.

# CHAPTER REVIEW

## *Science Terms*

*The following list contains all of the boldfaced words found in this chapter and the page on which each appears.*

clouds (p. 248)
condensation (p. 247)
condensation nuclei (p. 248)
dew (p. 247)
dew point (p. 247)
evaporation (p. 243)
fog (p. 248)
frost (p. 248)
hail (p. 251)
hygrometers (p. 245)
precipitation (p. 250)

rain (p. 250)
rain gauge (p. 251)
relative humidity (p. 244)
saturated (p. 244)
sleet (p. 251)
smog (p. 248)
snow (p. 251)
specific humidity (p. 244)
sublimation (p. 248)
transpiration (p. 244)

## *Matching Questions*

*On the blank line, write the letter of the item in column B that is most closely related to the item in column A.*

*Column A*

____ 1. humidity

____ 2. capacity

____ 3. saturation

____ 4. hygrometer

____ 5. dew point

____ 6. condensation nuclei

____ 7. radiation

____ 8. stratus

____ 9. sleet

____ 10. cirrus

____ 11. rain gauge

____ 12. hail

*Column B*

*a.* instrument that measures relative humidity
*b.* salt and smoke particles in Earth's atmosphere
*c.* method of heat loss that produces ground fog
*d.* sheetlike clouds; air mass rises diagonally
*e.* maximum amount of water vapor that a given parcel of air can hold
*f.* clouds composed of ice crystals
*g.* moisture in the air
*h.* form of precipitation composed of concentric layers of ice
*i.* temperature at which water vapor condenses
*j.* parcel of air holds all the water vapor it can at this point
*k.* instrument used to measure precipitation
*l.* this forms when raindrops freeze
*m.* also called a psychrometer

# *Multiple-Choice Questions*

*On the blank line, write the letter preceding the word or expression that best completes the statement.*

1. Evaporation involves the changing of liquid water into
   a. separate atoms of water
   b. solid water
   c. molecules of water vapor
   d. tiny drops of water

   1 _____

2. Water evaporates fastest on days when the air is
   a. hot, dry, and still
   b. muggy, hot, and windy
   c. hot, dry, and windy
   d. cold, dry, and still

   2 _____

3. On a summer day, the humidity is usually highest during the
   a. late morning   b. early afternoon   c. late afternoon   d. night

   3 _____

4. The water-vapor content of Earth's atmosphere is usually greatest over
   a. plains   b. tropical oceans   c. swamps   d. mountains

   4 _____

5. The temperature at which dry air weighs the most is
   a. 0°C   b. 10°C   c. 25°C   d. 35°C

   5 _____

6. When a sample of air that has a capacity of 6 gr of water vapor contains only 1 gr of water vapor, the relative humidity of the air is
   a. 17%   b. 60%   c. 83%   d. 92%

   6 _____

7. The principle of the wet-bulb and dry-bulb thermometers states that
   a. evaporation causes cooling
   b. evaporation causes heating
   c. moving air cools objects
   d. the dry-bulb thermometer absorbs water from the wet-bulb thermometer

   7 _____

8. Which of the following cities would probably have the highest humidity?
   a. Butte, Montana—dry bulb, 10°C, wet bulb, 9°C
   b. Biscayne, Florida—dry bulb, 25°C, wet bulb, 15°C
   c. Sacramento, California—dry bulb, 22°C, wet bulb, 12°C
   d. Gallup, New Mexico—dry bulb, 30°C, wet bulb, 27°C

   8 _____

9. An instrument that shows changes in relative humidity is the
   a. anemometer   b. hair hygrometer   c. barometer   d. thermometer

   9 _____

10. An important factor in the formation of ground fog is the loss of heat from Earth by
    a. evaporation   b. radiation   c. strong winds   d. frost

    10 _____

11. Clouds form when moist air cools by
    a. expansion   b. conduction   c. evaporation   d. condensation

    11 _____

12. The best description of a cirrocumulus cloud is
    a. thin, heaped up, consisting of ice crystals
    b. thin, flattened out, consisting of ice crystals
    c. thick, flattened out, consisting of water
    d. thin, feathery, consisting of water

    12 _____

13. The cloud type that is always associated with precipitation is the
    a. altostratus   b. cirrostratus   c. stratocumulus   d. nimbostratus

    13 _____

14. Which of the following is *not* considered a form of precipitation?
    a. hail   b. snow   c. sleet   d. frost

    14 _____

15. In a rain gauge, 5 in of water in the cylinder connected to the funnel is equal to a rainfall of

    *a.* 1/10 in  *b.* 1/2 in  *c.* 1 in  *d.* 5 in           15 \_\_\_\_

# Modified True-False Questions

*In some of the following statements, the italicized term makes the statement incorrect. For each incorrect statement, write the term that must be substituted for the italicized term to make the statement correct. For each correct statement, write the word "true."*

1. Most of the water vapor in Earth's atmosphere evaporates from the *ocean*.     1 *true*

2. The movement of water from the leaves of plants into Earth's atmosphere is called *transpiration*.     2 *true*

3. Heating water increases its rate of evaporation because individual water molecules *expand*.     3 *false*

4. The maximum amount of water vapor that a parcel of air can hold is known as *humidity*.     4 *false*

5. Atmospheric moisture that condenses on solid objects, such as grass, is called *dew*.     5 *true*

6. *Cumulus* clouds form when an air mass rises along a steep slope.     6 *true*

7. A cloud type usually associated with fair weather is *nimbostratus*.     7 *false — cirrus*

8. Particles in the atmosphere upon which water vapor condenses are called *precipitators*.     *false*  8 *true*

9. *Snow* is rain that freezes after falling from a cloud.     9 *false — sleet*

10. Hailstones form in *cirrus* clouds.     10 *false — cumulo nimbus*

11. Snow forms when water in the atmosphere changes directly from *liquid* to solid.     11 *false — gas / vapor*

12. One foot of melted snow is, on average, equivalent to *one inch* of rainfall.     12 *true*

# Testing Your Knowledge

1. Identify two conditions that cause water to evaporate rapidly.

    *a.* _____

    *b.* _____

2. Why does a wet cloth that freezes on a clothesline eventually dry even though the air temperature may remain below freezing? _____

_____

3. Explain the difference between capacity and saturation. _____

_____

**4.** Complete the table below.

| Amount of Water Vapor in the Air (in grains) | Capacity of the Air (in grains) | Relative Humidity |
|---|---|---|
| 2.0 | 6.0 | |
| 2.0 | | 25% |
| | 8.0 | 50% |
| 8.0 | 10.0 | |

**5.** On a hot, humid day, large drops of water appear on a cold-water pipe. However, a hot-water pipe nearby remains dry. Explain. _____

_____

**6.** What conditions are necessary for dew or frost to form on the ground? _____

_____

**7.** Describe how ground fog forms. _____

_____

**8.** Describe the three ways that an air mass may rise from ground level.

*a.* _____

*b.* _____

*c.* _____

**9.** How do the following cloud pairs differ?

*a.* stratus and cumulus _Stratus clouds are sheetlike or flat, and_ _cumulus clouds are fluffy._

*b.* cirrostratus and altostratus _Cirrostratus clouds mean there a possibility_ _for rain or snow in 24 hours and altostratus mean_

*c.* stratocumulus and cumulonimbus _with Stratocumulus clouds there's_ _a chance for rain and snow and with cumulonimbus thunderstorms_ _and hail._

**10.** What is the difference between the following precipitation pairs?

*a.* rain and snow _Rain falls as drops of water larger_ _than drizzle & snow falls as ice crystals having_ _a branched structure_

*b.* sleet and snow _Sleet falls as frozen drops of rain_ _& snow falls as ice crystals having a branched true_

*c.* hail and sleet _Hail falls a frozen drops of water with_ _layers & sleet falls as frozen drops of rain._

# Acid Rain: Is It Blowin' in the Wind?

Sprinkled by acid rain, many trees and animals in lakes in New England's six states are dying. Yet, New England produces less than 3 percent of the pollutants that trigger acid rain. So from where is the acid rain coming?

Before tackling this environmental detective story, you need to know something about acid rain. Acid rain, as you might guess, is precipitation that is more acidic than normal. The precipitation may fall as snowflakes, rain, or hail, or it may envelop cities, mountains, and forests as fog, mist, and clouds. Either way, it can harm or even destroy living things. It can also wear away statues, buildings, and the glossy finishes on expensive cars. Clearly, it's something we don't want around. So why have we had to put up with it?

Acid rain is a by-product of some of the most important industries in our country. It comes from the production of electricity, from all sorts of manufacturing processes, and from the cars, trucks, buses, and jet planes that carry people and products to all parts of the country.

Obviously, none of these things actually produce rain. That's the job of the atmosphere. But they do inject something into the atmosphere that transforms ordinary rain into acid rain.

That "something" comes in the form of two families of chemical gases that are produced from the burning of fossil fuels like oil, oil products, and coal. These families are oxides of sulfur and oxides of nitrogen. When these oxides get into the air from, say, the smokestacks of power plants or the exhausts of cars, they combine with water to form sulfuric and nitric acids. And when this water falls to the earth as rain, snow, hail, or sleet, what you have is acid precipitation.

Hold on, you might say, how do oil and coal produce oxides of sulfur and nitrogen? As you may know, oil and coal are the buried remains of living things that inhabited Earth many millions of years ago. Since living things contain many substances, including sulfur and nitrogen, so do their remains . . . like oil and coal.

Now, when oil and coal are burned to produce energy to make electricity or other products, or to get you from one place to another, the sulfur and nitrogen combine with oxygen. Presto! Oxides of sulfur and nitrogen.

Well, what has this to do with New England's low rate of air pollutant production, but high level of acid rain? Here's where seemingly dry statistics begin to yield key clues. You see, it turns out that among the top ten states for producing sulfur and nitrogen oxides are Ohio, Pennsylvania, Indiana, Illinois, Missouri, Kentucky, West Virginia, and Tennessee.

All these states are west and somewhat south of New England. So what? So take a look at a weather map over a few days. What do you notice? Storms usually move from west to east and frequently from south to north as well. *So the storm must carry the sulfur &c.*

What can New Englanders do about the problem that falls on their heads from the sky? Not much all by themselves. Clearly, the problem of acid rain is something that all Americans, and their government, must try to solve.

Regulations already exist to reduce the production of sulfur and nitrogen oxides by cleaning up gases coming from power plant smokestacks and automobile exhaust pipes. But many scientists say that the regulations aren't strict enough. In addition, scientists suggest that we use less high-sulfur coal to produce electricity. Let's use coal that holds smaller amounts of sulfur, they say.

Fine, reply utility companies. But they are quick to point out that low-sulfur coal is expensive. Using it will drive up the price of electricity. And what about all those coal miners who might lose their jobs if we stopped mining high-sulfur coal?

Obviously, there are no easy answers for New Englanders or anyone else. No one wants acid raining down on him or her. So what should be done? What do you think?

1. *New England produces approximately what percentage of the pollutants that trigger acid rain?*

   a. 0
   b. 0.3
   c. 3.0
   d. 30

2. *The substances that combine to produce acid rain are water and*

   a. dust.
   b. oxygen.
   c. oxides of sulfur and nitrogen.
   d. sulfuric acid and nitric acid.

3. *Oil and coal are the remains of organisms that lived*

   a. hundreds of years ago.
   b. thousands of years ago.
   c. hundreds of thousands of years ago.
   d. millions of years ago.

4. *What evidence is there to support the idea that acid rain in New England comes from the Midwest?*

   The states in the U.S that produce the most sulfur and nitrogen oxides are west and south of New England.

5. *On a separate sheet of paper, write a short essay discussing the pros and cons of trying to reduce the amount of acid rain that falls on New England. Refer to the arguments described above.*

# Weather and Weather Forecasting

## LABORATORY INVESTIGATION

### CHANGING THE CONDITIONS WITHIN A VOLUME OF AIR

**A.** Dip a strip of blue cobalt chloride paper in water, and exhale several times upon another strip of blue cobalt chloride paper.

    **1.** What do you observe in each case? _____

_____

The color change indicates that water is present.

**B.** Tape a thermometer to the inside of each of two large jars. Place the thermometers so that the bulbs are about 5 centimeters (cm) from the bottom of each jar. Make sure you can read the scale on each thermometer. Hang a strip of dry cobalt chloride paper from the rim of each jar, and cover each jar with a piece of cardboard. Label one jar A and the other jar B.

    **2.** What color is the strip of cobalt chloride paper in jar A? _____

What is the air temperature in jar A? _____

    **3.** What color is the strip of paper in jar B? _____

What is the air temperature in jar B? _____

Record your results in the top row of Table 18-1 below. Do not write anything else in the table until you have completed parts C and D.

**C.** Remove the strip of cobalt chloride paper from jar A. Cover the bottom of the jar with a thin layer of crushed ice cubes. Make sure the ice does not touch the thermometer. Replace the cobalt chloride paper, and again cover the jar with the cardboard (see Fig. 18-1).

**D.** Remove the strip of cobalt chloride paper from jar B. Pour warm water (about 40°C) into the jar to a depth of about 2 cm. Make sure the water doesn't touch the thermometer. Replace the cobalt chloride paper and again cover the jar with the cardboard.

    Read the thermometer in each jar, and note the color of the cobalt chloride paper at 1-minute intervals. Record your observations in the table for 8 minutes.

**TABLE 18-1.**

| | Jar A | | Jar B | |
| Minutes | Color | Temperature | Color | Temperature |
|---|---|---|---|---|
| 0 | | | | |
| 1 | | | | |
| 2 | | | | |
| 3 | | | | |
| 4 | | | | |
| 5 | | | | |
| 6 | | | | |
| 7 | | | | |
| 8 | | | | |

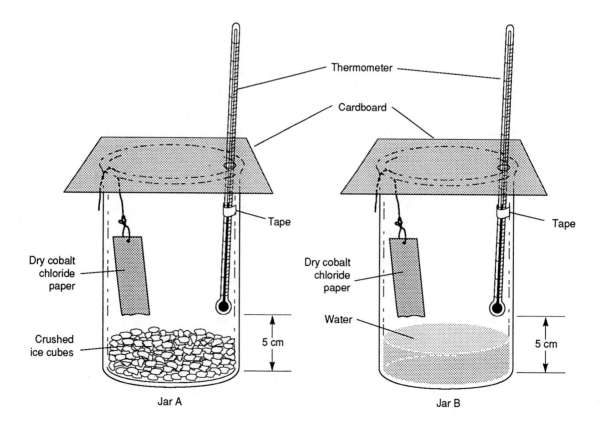

Thermometer

Cardboard

Tape

Dry cobalt chloride paper

Crushed ice cubes

5 cm

Jar A

Thermometer

Cardboard

Tape

Dry cobalt chloride paper

Water

5 cm

Jar B

**Fig. 18-1.**

**4.** How did the air temperature change in jar A? _____

_____

How did the air temperature change in jar B? _____

_____

**5.** What happened to the humidity in jar A? _____

_____

What happened to the humidity in jar B? _____

_____

**6.** Explain the difference in humidity between jar A and jar B. _____

_____

**7.** Where on Earth are conditions in the atmosphere similar to the conditions in jar A? Explain.

_____

_____

**8.** Where on Earth are conditions in the atmosphere similar to the conditions in jar B? Explain.

_____

_____

## PREDICTING THE WEATHER

People have long attempted to predict the weather accurately. Yet, until the mid-nineteenth century, most weather forecasting was either based on superstition or guesswork. Today, some people think weather changes can be predicted by observing the behavior of certain plants and animals. For example, if pigs return to the barn with straw in their mouths, some farmers believe that rain will fall shortly afterward. If the husks on ears of corn are thicker than normal, some farmers believe that this indicates the approach of a long, cold winter.

Weather forecasting based on the sky's appearance was more accurate. These forecasts were often expressed in short poems such as

*Red sky at night,*
*Sailors delight;*
*Red sky in the morning,*
*Sailors take warning.*

and

*When the wind is in the south,*
*The rain is in the mouth.*

As deputy postmaster general of the American colonies, Benjamin Franklin observed that storms usually travel from one city to another. Based upon letters Franklin received in Philadelphia from friends in other cities, he inferred that a storm occurring in one of those cities reached Philadelphia several days later. Given this information, Franklin concluded that the same storm affected different cities in succession as it moved across the country. As you will learn, modern weather forecasting is still based on Franklin's conclusion.

The invention of the telegraph in 1844 was a major factor responsible for improving the accuracy of weather forecasting. The telegraph made it possible for weather information to be collected and then communicated to distant areas as soon as the weather changes occurred. Today, the use of satellites and computers has enabled weather stations all over the world to communicate weather data instantly with one another.

Every day, weather stations across the country send weather data to the National Weather Service in Washington, D.C. Here, meteorologists compile weather maps from the data and send out copies of these weather maps by a *facsimile machine* to weather stations throughout the United States. Meteorologists at each weather station use these maps to prepare daily forecasts and extended forecasts for their region.

The key to understanding a daily weather forecast is knowing how to interpret a weather map. First, you need to know how weather maps are prepared to understand how meteorologists forecast the weather.

## PREPARATION OF WEATHER MAPS

A weather map of the United States gives a general description of the weather throughout the country at a particular time of day. The preparation of a weather map begins with a blank map of the United States. On the map, every weather station in the United States is represented by a small circle. At regular intervals throughout the day, meteorologists at each weather station observe and record local weather conditions such as the (1) *air temperature,* (2) *dew point,* (3) *air pressure,* (4) *wind speed and direction,* (5) *cloud cover,* and (6) *precipitation.*

Each weather station then sends a copy of its recorded observations to the National Weather Service in Washington, D.C. Here, the weather observations are assembled on a blank weather map as a group of numbers and symbols called a *station model* (see Fig. 18-2).

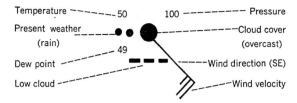

Fig. 18-2. Simplified weather station model.

Each number and symbol describes one weather factor observed at each local weather station.

On a station model, the numbers labeled *temperature* and *dew point* are in degrees Fahrenheit (see Fig. 18-2). The number labeled *pressure* is an abbreviated form of millibars. The millibar (mb) is the unit meteorologists use to measure atmospheric pressure (see Chapter 14). On a station model, only the last three digits of the millibar reading appear, and the decimal point is removed.

For example, the number 100 is an abbreviated form of 1010.0 mb in the station model. Likewise, a pressure of 1000.0 mb is shown as 000 on the weather map, 1000.5 mb is shown as 005, and so on. (Some of the symbols used on a weather map are explained in greater detail in Fig. 18-3).

When all the station models have been printed on a weather map, the map may show as many as 10,000 items. Copies of this map are then sent to weather stations all over the country. Fig. 18-4 shows a simplified weather map of the United States.

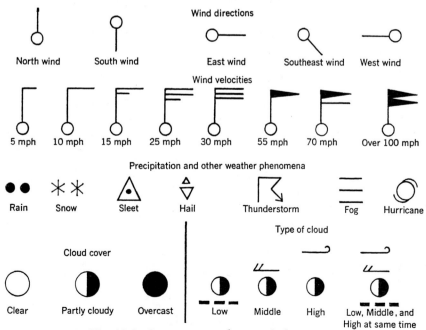

Fig. 18-3. Common weather symbols.

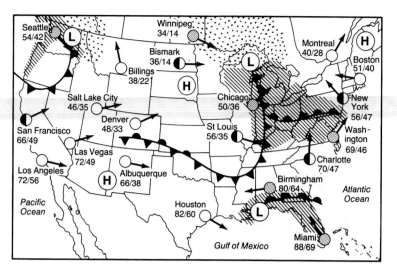

**Fig. 18-4. Weather map of the United States.**

# Air Masses and Storms

Since the early 1900s, data from many weather stations have been collected on a regular basis. However, accurate weather forecasts were rare because meteorologists were not aware of the combined effects of temperature, humidity, and air pressure on the weather changes.

Today, meteorologists know much more about how these weather factors influence weather changes. Meteorologists have discovered that Earth's atmosphere is divided into huge air masses—each having different characteristics. These huge air masses are made of either high-pressure or low-pressure systems.

The movement of these air masses and the changing positions of their boundaries cause weather changes in regions that they pass over. Because of different characteristics in each air mass, meteorologists can trace the paths of air masses as they move over Earth's surface. In fact, the boundary between pressure systems is often clear-cut and marked by distinct weather changes. This enables meteorologists to predict future weather.

**Formation of Air Masses.** A large section of Earth's troposphere in which the temperature and humidity at a particular altitude are similar is called an **air mass.** Air masses vary from hundreds to thousands of kilometers in diameter and may extend several kilometers into the atmosphere.

An air mass originates where surface conditions are relatively constant for a long time. Many air masses originate over the Gulf of Mexico and over central Canada. Air remaining over these regions for several weeks takes on the temperature and humidity characteristic of the surface below. For example, an air mass that remains over the warm water of the Gulf of Mexico for three or four weeks slowly becomes warm and moist. In contrast, an air mass that remains over central Canada for several weeks in winter becomes very cold and dry. On a small scale, you observed how the temperature and humidity of air can change in your laboratory investigation.

**Types of Air Masses.** Meteorologists name air masses according to the surface characteristics and location over which they originate. For example, air masses that form over land are called *continental* air masses, whereas those that form over large bodies of water are called *maritime* air masses. Air masses based on the latitude in which they form are called either *polar* or *tropical.*

You can identify the four main types of air masses by combining the latitude in which they originate with the words *continental* or *maritime.* For example, an air mass that originates over the Gulf of Mexico is *maritime tropical,* whereas one that originates over central Mexico is called *continental tropical.* And an

## TABLE 18-2. AIR MASSES AND THE WEATHER THEY BRING

| Name | Source Region | Weather Conditions |
|------|---------------|--------------------|
| Continental polar (cP) | Canada | Clear, cold, dry |
| Continental tropical (cT) | Southwestern United States and Mexico | Clear, hot, dry |
| Maritime tropical (mT) | Gulf of Mexico and Caribbean; Atlantic and Pacific oceans near the equator | Cloudy, warm, rain, thunderstorms |
| Maritime polar (mP) | Pacific and Atlantic oceans near the poles | Cloudy, cold, rain or snow |

air mass that originates over north-central Canada is *continental polar,* whereas one that originates over the ocean near either pole is *maritime polar.*

On weather maps, the names of air masses usually are abbreviated. For example, a continental polar air mass is abbreviated cP, a maritime polar one is abbreviated mP, and so on (see Table 18-2). Fig. 18-5 shows the air masses that affect North America and the paths they follow across the United States.

**Movement of Air Masses.** Daily weather depends mainly on the type of air mass that is passing over a region. For example, in winter a cP air mass moving south from central Canada brings very cold, dry weather to the regions of the United States that it passes over. In contrast, an mP air mass moving south brings cold, wet weather to the regions of the

United States that it passes over. Table 18-2 shows the weather conditions that some common air masses bring with them.

## WEATHER FRONTS

The boundary, or *interface,* between air masses with different characteristics is called a **weather front,** or simply a *front.* Meteorologists name fronts according to the relative positions and movements of the colliding air masses. Storms often form at the boundary between warm and cold air masses and move along with the front.

There are four main kinds of fronts: (1) the boundary along which a cold air mass overtakes a slower-moving warm air mass is called a *cold front;* (2) the boundary along which a warm air mass overtakes a slower-moving

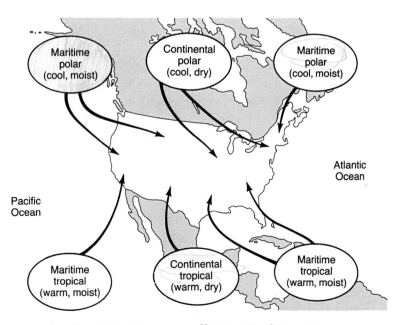

**Fig. 18-5. Air masses affecting North America.**

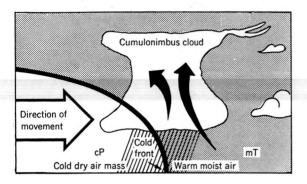

**Fig. 18-6. Weather front symbols.**

dry air is denser than warm, moist air. For this reason, the cold, dry (cP) air mass behind a cold front stays near Earth's surface as it moves, forcing its way under the warm, moist (maritime tropical, or mT) air mass.

As the warm, moist air mass rises and cools, the atmospheric water vapor condenses into clouds. Cumulonimbus clouds commonly form along a cold front because the warm, moist air is forced to rise along a steep slope by the dense air below (see Fig. 18-7).

A cold front is characterized by a sudden temperature change, high winds, and a relatively brief period of heavy rain or snow. Thunderstorms, accompanied by heavy rain, often form along a cold front. The greater the difference in temperature and pressure between the cold air mass and the warm air mass, the more severe is the weather as a cold front passes. A cold front usually passes in a few hours, followed by clear, dry weather and cooler temperatures.

cold air mass is a *warm front;* (3) the boundary between two air masses that are not moving relative to each other is called a *stationary front;* and (4) the boundary between a cold air mass, warm air mass, and a cool air mass is called an *occluded front.* Fig. 18-6 shows the symbols used to indicate each front on a weather map.

**Cold Front.** The air is cold and dry behind a **cold front.** Ahead of the cold front, the air usually is warm and moist. You learned that cold,

**Warm Front.** Along a **warm front,** mP air forces warm, moist (mT) air upward. Unlike a cold front, there is a gentle slope between the two different air masses. Along this gentle slope, the warm, moist mT air mass rises over

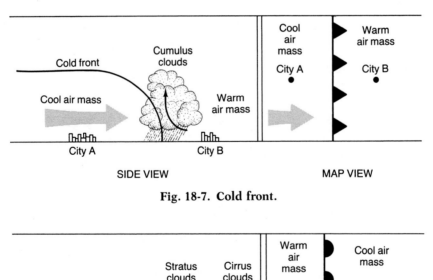

**Fig. 18-7. Cold front.**

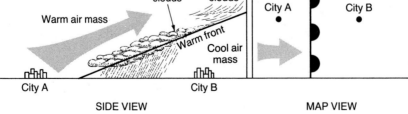

**Fig. 18-8. Warm front.**

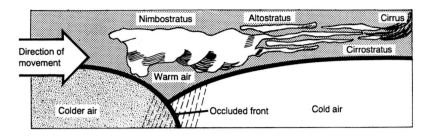

**Fig. 18-9. Occluded front.**

the moist, cool mP air mass. Because of the nearly horizontal slope between the two air masses, stratus clouds commonly form along a warm front (see Fig. 18-8).

Stratus clouds often appear high in the sky, more than 1750 kilometers (km) ahead of an approaching warm front. To a ground observer, cloudiness increases gradually as the warm front approaches. Gradually, the clouds ahead of the warm front become thick enough to block the sun or moon. At this point, steady precipitation may begin. The precipitation may continue for 24 hours or longer. After a warm front passes, the air temperature increases, wind direction changes, rain or snow ends, and the skies become partly sunny.

**Occluded Front.** Sometimes two masses of cold air move into a region occupied by warm, moist air. One of the cold masses overtakes the other mass and the parcel of warm, moist air becomes trapped between the two cold air masses. As a result, the warm, moist air is lifted completely off the ground. Now, the cold air mass ahead of the warm front can touch the colder air mass behind the cold front. The *interface* between two cold air masses and a warm air mass trapped above them is an **occluded front** (see Fig. 18-9). The weather along an occluded front is a combination of weather that occurs along a cold front and a warm front, characterized by several hours of thunderstorms with heavy precipitation, followed by steady rain that may last 24 hours or longer.

**Stationary Front.** Sometimes, a cold air mass and a warm air mass remain in the same relative position for several days. In this case, the two air masses move parallel to the front between them. When the front separating two air masses does not move, it is called a **stationary front.** Along a stationary front, warm air rises slowly into Earth's atmosphere. As a result, the weather along a stationary front is generally similar to the weather produced by the passing of a warm front.

## TYPES OF STORMS

Storms usually develop when different air masses meet. The most violent, or severe, storms are associated with mT air masses because they contain a lot of warm, moist air. Severe storms include *thunderstorms, tornadoes,* and *hurricanes.*

**Thunderstorms.** A **thunderstorm** is a brief, local storm produced by the rapid upward movement of warm, moist air within a cumulonimbus cloud. Thunderstorms always produce *lightning* and *thunder*, and often are accompanied by strong wind gusts, heavy rain, or hail.

Warm, moist air may rise rapidly into the atmosphere for two reasons. One reason is that the air mass may have received a lot of heat energy from Earth's surface. The other reason is that a mass of cold, dense air may have forced the warm, moist air upward.

The water vapor in the rapidly rising warm air may condense into towering cumulonimbus clouds called *thunderheads*. These huge cumulonimbus clouds range in height from about 7.5 km to about 24 km above the ground. Strong air currents, or *drafts*, rush upward and downward within a cumulonimbus cloud (see Fig. 18-10). When a thunderstorm passes overhead, you may feel these downdrafts as strong gusts of cool or cold wind. Often, heavy showers, lightning discharges, and thunder follow.

Meteorologists think that the strong updrafts and downdrafts in a cumulonimbus cloud split large raindrops into smaller droplets. This causes electricity to increase at different levels in the cumulonimbus cloud. A **lightning bolt** occurs when the difference in electrical charge becomes great enough between (1) the different levels of a cumulonimbus cloud, (2) two cumulonimbus clouds, or (3) the cumulonimbus cloud and the ground. A lightning bolt generates so much heat (about 28,000° Celsius) that the surrounding air suddenly expands, producing a loud explo-

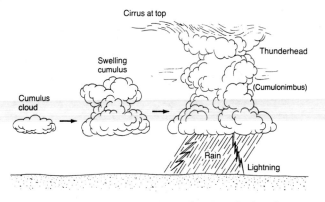

**Fig. 18-10. Development of a thunderhead.**

sion. The explosion caused by the expansion of air by a lightning bolt is commonly called **thunder.**

The sound waves generated by lightning take about 3 seconds (sec) to travel 1 km. Thus, you can calculate how far away a lightning bolt is from you by counting the number of seconds that pass between your sighting of the lightning and the sound of thunder. For example, if 6 sec pass between your sighting of the lightning and the sound of thunder, the lightning is about 2 km away from you.

During a thunderstorm, you should avoid open fields, hilltops, beaches, and bodies of water where the danger of a lightning strike is great. Because severe thunderstorms drop large amounts of precipitation in a short period of time, low-lying areas also should be avoided due to the danger of flash flooding. In fact, during a severe thunderstorm it is best to stay indoors.

Thousands of thunderstorms occur daily throughout the world. In the United States, most thunderstorms occur from April through September. In winter, thunderstorms rarely occur over the northern United States.

**Tornadoes.** A severe storm with swirling winds that may reach speeds of hundreds of kilometers per hour is called a **tornado.** Because of its rapidly swirling winds, many Americans refer to a tornado as a "twister." Tornadoes are formed by the rapid lifting and condensation of warm, moist air. For this reason, a tornado usually occurs when the sky is covered by large cumulonimbus clouds accompanied by thunderstorms. During an extremely violent thunderstorm, a funnel-shaped, swirling cloud may extend downward from a cumulonimbus cloud and reach the ground as a tornado. A swirling, funnel-shaped cloud that extends downward and touches a body of water is called a *waterspout.*

A tornado that touches the ground can be deadly and extremely destructive. The destructiveness of a tornado is caused by a combination of extremely strong winds and very low air pressure inside the funnel-shaped cloud. Wind speeds may exceed 500 km per hour inside a tornado.

A tornado is a very narrow storm. The diameter of a tornado's funnel averages less than 0.5 km. Consequently, the path of destruction left by a tornado is limited to a very narrow band on the ground. In some cases, a tornado's path has been so narrow that a house on one side of a street was undamaged, while a house across the street was completely destroyed (see Fig. 18-11).

When passing over a building, a tornado causes the air pressure outside the building to drop sharply. As a result, the much greater air pressure inside the building pushes outward

**Fig. 18-11. Movement of air in a tornado.**

against the walls with great force. At the same time, the violent, swirling winds of the tornado bombard the outside of the building. The combined forces may tear a building apart with explosive force. The safest place to be during a tornado is in an underground cellar or basement away from outside walls, windows, and doors.

Most tornadoes that form over the United States travel in a northeasterly direction, at about 55 km to 70 km per hour. Tornadoes may form anywhere and at anytime in the United States. Even New York State has had tornadoes. However, they mainly occur during late-spring afternoons in the Mississippi River valley and the Great Plains. Tornadoes regularly occur each spring in Oklahoma, Iowa, and Kansas.

***Severe Thunderstorm and Tornado Watches and Warnings.*** The National Severe Storms Forecast Center in Kansas City, Missouri, issues severe thunderstorm or tornado watches when atmospheric conditions appear favorable for the formation of either a severe thunderstorm or tornado. A *watch* covers an area of 100 km by 200 km or greater and gives the approximate time during which a severe thunderstorm or tornado may occur. A *warning* is issued when a severe thunderstorm or tornado has actually been sighted or detected by radar and predicts the approximate time during which the storm will strike an area.

**Hurricanes.** A large, rotating, low-pressure system accompanied by heavy precipitation and strong winds is called a *tropical cyclone,* or **hurricane.** In some places, such as the Pacific region, a hurricane is called a *typhoon.*

Hurricanes originate in hot, moist air masses over tropical oceans, mainly in the summer and early fall. The moist, hot air rises, cools, and condenses. The condensation of large amounts of water vapor releases huge amounts of additional heat. The heat causes the air to rise even more rapidly. As a result, an area of very low air pressure forms.

The surrounding air rushes in toward the center of this low-pressure area and is deflected by Earth's rotation. Consequently, the air begins to rotate in a counterclockwise direction. When the winds within this low-pressure system reach or exceed 120 km per hour, the storm is classified as a hurricane.

A large hurricane may cover an area with a diameter of about 600 km. From above, clouds spiral in toward the center of a hurricane. At the center is a small patch of clear sky called the *eye.* The eye of a hurricane generally ranges from 25 km to 65 km in diameter. Heavy rain showers and thunderstorms occur just outside the eye. The winds within the eye of a hurricane are relatively calm. However, just outside the eye, wind speeds are the most violent, reaching 320 km per hour. Like wind speeds, precipitation is greatest just outside the eye of a hurricane.

Most of the flooding and damage associated with hurricanes that strike coastal areas are caused by *storm surges,* which may raise sea level 3 to 5 meters (m) above normal. A storm surge is produced when a hurricane piles up seawater along the coast and then pushes the water inland. Most deaths associated with hurricanes are caused by storm surges that occur at times of high tide. Thus, people living along the coast and in flood-prone areas should leave their homes and move to higher ground before a hurricane strikes.

Hurricanes gradually weaken as they reach higher latitudes or when they move inland. Because the ocean is colder at higher latitudes, a hurricane moving northward increasingly loses the heat necessary to maintain its circulation. A hurricane that moves inland weakens rapidly because friction with the ground slows its winds and because land areas do not provide sufficient moisture to maintain a hurricane.

In the Northern Hemisphere, a hurricane first travels toward the west, and then usually curves toward the northeast. Fig. 18-12 shows the paths commonly taken by hurricanes in the Northern Hemisphere.

**Fig. 18-12. Hurricane paths.**

**Hurricane Watches and Warnings.** Meteorologists at the National Hurricane Center in Miami, Florida, use weather satellites, aircraft, and radar to forecast the path of a hurricane. After a hurricane has been detected, the National Hurricane Center issues a hurricane watch or a hurricane warning. A watch means that hurricane conditions are expected within 24 hours. A warning provides communities in the hurricane's path with 24 hours to take precautions against potential damage from strong winds and floods.

**Winter Storms.** Some severe storms occur only in winter. These storms, which feature precipitation in the form of snow and ice, are called **winter storms.** A violent winter storm, with strong winds, blowing snow, and frigid temperatures, is called a *blizzard*. Sometimes, falling rain freezes when it strikes the ground, coating solid objects with a layer of ice. This type of winter storm is called an *ice storm*. Traveling by car or on foot can be hazardous during blizzards and ice storms because of slippery roads and sidewalks. The National Weather Service issues winter storm watches and warnings 24 hours before blizzard or ice-storm conditions are expected.

## LOCATING FRONTS ON A WEATHER MAP

An important task of weather forecasting is predicting the movement of weather fronts. Consequently, a meteorologist first must locate and then chart fronts on a weather map before making a weather forecast.

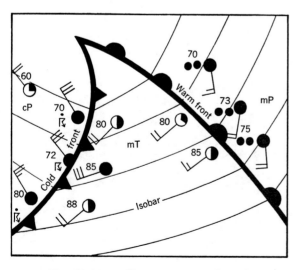

**Fig. 18-13. Differences in winds and temperatures mark the location of fronts.**

One way a meteorologist locates a front is by studying **isobars** on a weather map. Isobars are similar to the contour lines on a topographic map, which connect places on land with the same elevation above sea level. Isobars are lines on a map that connect places in Earth's atmosphere with the same air pressure. When isobars are drawn on a weather map, they show the position and extent of high-pressure and low-pressure air masses. Fig. 18-13 shows that isobars are smoothly curving lines that bend sharply where they cross a front.

Another way to locate a front on a weather map is to look for neighboring weather stations with very different wind speeds and air temperatures. For example, note the difference in air temperature and wind direction between weather stations on either side of the cold front and warm front in Fig. 18-13.

Fig. 18-14 shows part of a U.S. weather map that contains most features included on surface weather maps, such as weather stations, isobars, high-pressure areas, low-pressure areas, and fronts.

## WEATHER FORECASTING

Meteorologists make two kinds of weather forecasts: *short-range forecasts* and *long-range forecasts*. Short-range forecasts predict weather changes for the coming 24-hour to 48-hour period. Long-range forecasts predict the weather changes for the coming week, month, or season.

**Short-Range Forecasts.** Daily weather reports in the newspaper, over the radio, or on television are examples of short-range weather forecasts. A short-range forecast is based on a *computer model* of Earth's atmosphere. A computer model contains data gathered by satellites and radar on the winds, temperature, air pressure, humidity, cloud cover, and precipitation at Earth's surface and in the upper atmosphere. Using computer models, meteorologists prepare weather maps showing high-pressure systems, low-pressure systems, and fronts.

A meteorologist can make a short-range forecast after preparing and analyzing a series of surface weather maps. The maps show weather changes over continuous six-hour periods. From these weather maps, it can be estimated how fast and in what direction a particular front is moving. Once the speed and direction of the front is calculated, a meteo-

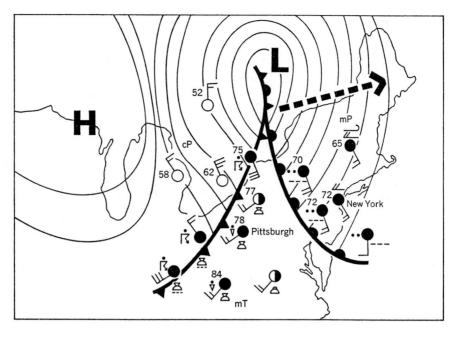

**Fig. 18-14. Section of a U.S. weather map.**

rologist can predict (1) where the front will be located in the next 24 to 48 hours, and (2) the kind of weather accompanying the front.

On the weather map in Fig. 18-14, it is shown that the low-pressure system (marked *L*) and the accompanying cold and warm fronts are moving east along the path shown by the dotted line. The weather stations ahead of this low-pressure system receive reports about its speed so they can determine when it will reach their region. Then, meteorologists at the weather station predict the wind speed, air temperature, sky conditions, and the probability of precipitation. The following is a sample forecast for New York City, which is in the path of the warm and cold front shown in Fig. 18-14:

June 6, 2003, 6:00 A.M.

New York City and vicinity. Increasing cloudiness, rain beginning in the early afternoon, ending by tomorrow morning. Partly sunny tomorrow with possibility of rain showers. The low tonight 70 to 75 and the high tomorrow in the upper 80s.

In preparing this forecast, a meteorologist must consider the following factors:

**1.** A warm front is approaching New York City.

**2.** The weather ahead of the warm front will pass over New York City before the front itself.

**3.** The weather ahead of the warm front is cloudy and rainy. This cloudy and rainy weather will take about 20 hours to pass over New York City.

**4.** After the warm front passes, mT air will pass over New York City, bringing partly sunny skies, higher temperatures, and scattered showers.

**5.** The warm air mass is expected to remain over the New York City area for the rest of the forecast period.

Farther west, in Pittsburgh, the warm front has already arrived. As a result, the forecast for the Pittsburgh area begins with the last two items of the New York City forecast:

June 6, 2003, 6:00 A.M.

Pittsburgh and vicinity. Partly sunny today with the possibility of scattered rain showers. Temperatures increasing to the high 80s or low 90s today. Probable thunderstorms late tonight, followed by rapid clearing. Mostly clear tomorrow with cooler temperatures. High tomorrow in the low 70s.

The meteorologist has considered the fol-

*Add detail*

lowing factors in preparing this forecast:

1. The mT air mass accompanied by high temperatures and scattered showers will be passing over Pittsburgh throughout the day.
2. At night, a cold front will pass over Pittsburgh. The passage of a cold front often produces thunderstorms.
3. A cP air mass will affect the Pittsburgh area after the cold front passes. The center of this cool, dry air mass is represented by the upper-case letter *H*, which stands for high pressure. A high-pressure air mass usually is accompanied by mostly clear skies and cooler weather.

**Long-Range Forecasts.** Using weather satellites and computer models, meteorologists can make fairly accurate weather forecasts for up to 10 days. Weather satellites provide continuous photographs of clouds, which make it possible to track low-pressure systems, hurricanes, and even thunderstorms. Because they can track storms over the ocean, satellites are especially valuable in making long-range forecasts.

General weather patterns can be predicted for several weeks or even months in advance. These predictions are based on the repeated observation that certain kinds of weather usually occur in cycles. For example, meteorologists are able to predict a colder than normal winter or a hotter than normal summer.

Using satellites and high-speed computers, a changeable Pacific Ocean current called *El Niño* has been forecast several months in advance. El Niño normally flows south along the coast of Ecuador each year at the end of December. However, about every 7 to 10 years, El Niño flows east, causing drought conditions along the west coast of South America. An El Niño was first predicted in 1975.

# CLIMATE

The general characteristics of the weather in a region are called the **climate.** Unlike the **weather,** which consists of hourly and daily changes in the atmosphere over a region, climate is the average of the weather conditions in a region over many years.

The main factors used to determine the climate of a region are *temperature* and *precipitation.* Climate varies from one place to another for two reasons. One reason is because Earth's surface is heated unevenly. The other reason is because the landmasses, oceans, and polar ice caps are not equally distributed over Earth's surface.

Based on average temperature, climates are classified into three general groups. One group includes the Arctic and Antarctic regions, in which air temperature is low all year. A second group includes the region near the equator, in which the temperature is very high all year. The third group includes regions in the middle latitudes, in which the temperature varies throughout the year.

The climate in Arctic and Antarctic regions is described as *polar,* whereas the climate in equatorial regions is described as *tropical.* Middle-latitude regions are said to have a *temperate* climate. These climate regions can be classified further by the amount of annual precipitation each receives. For example, the New York region is classified as a *temperate humid* climate because it has hot summers, cold winters, and receives about 40 inches (100 cm) of precipitation annually.

The type of climate in a region affects people in many ways. For example, jobs, outdoor activities, and home heating or cooling costs are often determined by the type of climate in a region. There are four main climates within the United States. Table 18-3 provides a description of the main climates of the continental United States.

**TABLE 18-3. MAJOR CLIMATES OF THE CONTINENTAL UNITED STATES**

| Location | Name | Temperature Range (approximate °C) | Annual Precipitation (approximate inches) |
|---|---|---|---|
| East Coast and Midwest | Humid continental | Below −18° to above 38° | 20 to 40 |
| Southeast | Humid subtropical | 0° to over 38° | 60 |
| Northwest Pacific Coast | Humid marine | Below 0° to 25° | 40 to 60 |
| Southwest | Desert | Below −18° to 45° | Less than 10 |
| West Coast | Mediterranean | About 25° to 45° | 20 to 30 |
| Great Plains | Steppes | Below −18° to 45° | Less than 20 |

# CHAPTER REVIEW

## Science Terms

*The following list contains all of the boldfaced words found in this chapter and the page on which each appears.*

air mass (p. 262)  lightning bolt (p. 265)  tornado (p. 266)
climate (p. 270)  occluded front (p. 265)  warm front (p. 264)
cold front (p. 264)  stationary front (p. 265)  weather (p. 270)
hurricane (p. 267)  thunder (p. 266)  weather front (p. 263)
isobars (p. 268)  thunderstorm (p. 265)  winter storms (p. 268)

## Matching Questions

*On the blank line, write the letter of the item in column B that is most closely related to the item in column A.*

### Column A

_____ 1. large section of troposphere with same temperature and humidity

_____ 2. boundary between two air masses *not* moving in relation to each other

_____ 3. warm air mass overtakes cold air mass

_____ 4. cold air mass overtakes warm air mass

_____ 5. brief, local storm with thunder and lightning

_____ 6. severe, narrow storm with fast, swirling winds

_____ 7. large, tropical cyclone with heavy rains and winds

_____ 8. severe storm with precipitation of snow and ice

### Column B

*a.* warm front
*b.* winter storm
*c.* occluded front
*d.* hurricane
*e.* stationary front
*f.* cold front
*g.* air mass
*h.* tornado
*i.* thunderstorm

## Multiple-Choice Questions

*On the blank line, write the letter preceding the word or expression that best completes the statement.*

**1.** The person who stated that storms move from place to place was
   *a.* Edison  *b.* Franklin  *c.* Jefferson  *d.* Morse          1 _____

**2.** A barometric pressure reading of 1005.0 millibars would appear on a weather map as
   *a.* 005  *b.* 050  *c.* 500  *d.* 1005          2 _____

**3.** The temperature and humidity within an air mass are
   *a.* fairly uniform at any particular altitude
   *b.* different between the center and edges of the air mass at any altitude
   *c.* likely to vary widely within the center of the air mass
   *d.* always different from the temperature and humidity within nearby air masses          3 _____

**4.** Air masses that form over large bodies of water are called
  *a.* polar  *b.* tropical  *c.* continental  *d.* maritime                 4 ____

**5.** An air mass that forms over the Gulf of Mexico would be represented on a weather map as
  *a.* cP  *b.* aM  *c.* mT  *d.* mP                                          5 ____

**6.** Cumulonimbus clouds may form when warm, moist air is lifted into the atmosphere by
  *a.* strong winds                 *c.* condensation
  *b.* expanding air                *d.* a mass of cold air                   6 ____

**7.** Meteorologists think that lightning is caused by
  *a.* friction between rain clouds
  *b.* friction between cumulonimbus clouds and layers of moist air
  *c.* the splitting of raindrops into smaller droplets
  *d.* the scattering of moist air within clouds by thunder                  7 ____

**8.** If you hear a clap of thunder about six seconds after seeing a flash of lightning, the distance of the lightning to the observer is about
  *a.* 1 km  *b.* 2 km  *c.* 6 km  *d.* 8 km                                  8 ____

**9.** Thunderstorms rarely occur during the
  *a.* summer  *b.* winter  *c.* spring  *d.* fall                           9 ____

**10.** The cloud type associated with tornadoes is the
  *a.* altostratus  *b.* cumulus  *c.* cumulonimbus  *d.* stratus           10 ____

**11.** Most tornadoes occur during
  *a.* afternoon hours in spring        *c.* early morning hours in summer
  *b.* afternoon hours in fall          *d.* late afternoon in winter        11 ____

**12.** A tropical storm is classified as a hurricane when the wind speed reaches about
  *a.* 60 km per hour               *c.* 80 km per hour
  *b.* 75 km per hour               *d.* 120 km per hour                      12 ____

**13.** Hurricanes weaken most rapidly when they pass over areas that are
  *a.* cold and dry                *c.* warm and wet
  *b.* cold and wet                *d.* warm and dry                         13 ____

**14.** The leading edge of a mass of warm air that overtakes a mass of cold air is a (an)
  *a.* warm front                  *c.* stationary front
  *b.* cold front                  *d.* occluded front                       14 ____

**15.** The type of weather that follows the passing of a cold front is
  *a.* cold and rainy              *c.* warm and dry
  *b.* warm and rainy              *d.* cold and dry                         15 ____

**16.** Which of the following statements is true?
  *a.* Cold fronts move slower than warm fronts.
  *b.* Precipitation may extend hundreds of kilometers in front of a warm front.
  *c.* Tornadoes are usually associated with the passage of a warm front.
  *d.* A stationary front forms where a cold front overtakes a warm front.   16 ____

**17.** On a weather map, a sharp bend in an isobar indicates a
  *a.* warm air mass               *c.* low-pressure system
  *b.* high-pressure system        *d.* weather front                        17 ____

18. Shortly after a warm front passes over your neighborhood, a change that does *not* occur is that
    *a.* the temperature rises        *c.* scattered showers occur
    *b.* the humidity remains high    *d.* the air becomes cold and dry     18 \_\_\_\_

# Modified True-False Questions

*In some of the following statements, the italicized term makes the statement incorrect. For each incorrect statement, write the term that must be substituted for the italicized term to make the statement correct. For each correct statement, write the word "true."*

1. An air mass that forms over the Gulf of Mexico would be warm and *dry*.      1 _____

2. An air mass that originates off the east coast of Canada is indicated on a weather map by the abbreviation *cP*.      2 _____

3. Hurricanes originate in *maritime tropical* air masses.      3 _____

4. The winds at the center of a *hurricane* may reach 480 km per hour.      4 _____

5. Most hurricanes originate over the *southeastern United States* during late summer.      5 _____

6. The leading edge of a cold air mass is called a *cold front*.      6 _____

7. The boundary between two different air masses that are not moving relative to each other is called a *stationary front*.      7 _____

8. Strong winds and heavy precipitation commonly occur along *cold* fronts.      8 _____

9. The weather associated with a cold front takes about *24 hours* to pass over a given weather station.      9 _____

10. Isobars are lines drawn on a weather map that indicate areas of equal air *temperature*.      10 _____

# Testing Your Knowledge

1. List five weather factors that a weather station records at regular intervals.

    *a.* _____      *c.* _____      *e.* _____

    *b.* _____      *d.* _____

2. Name the kind of region in which each of the following air masses originates. Describe the temperature and humidity you would expect to find in each air mass:

    *a.* maritime tropical _____

_____

  *b.* maritime polar _____

_____

  *c.* continental polar _____

_____

**3.** Why do strong gusts of wind and electrical discharges occur during a thunderstorm? _____

_____

**4.** Why does thunder usually occur after a lightning discharge? _____

_____

**5.** What steps should you take to avoid being injured during a tornado? _____

_____

**6.** What is the difference between weather and climate? _____

_____

**7.** Study the weather stations shown in Fig. 18-15. Write the letter (or letters) of the weather station(s) next to each description of weather conditions.

  *a.* Wind N at 40 miles per hour _____

  *b.* Wind SE at 15 miles per hour _____

  *c.* Overcast; rain falling _____

  *d.* Clear sky _____

  *e.* Thunderstorm _____ _____

  *f.* Closest to the warm front _____

  *g.* Closest to the cold front _____

  *h.* Location of an mT air mass _____

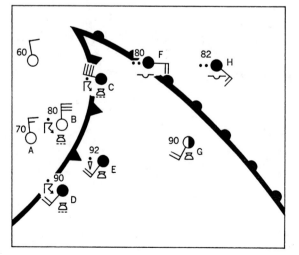

**Fig. 18-15.**

# Weather Forecasting: Can We Count on It?

I f meteorologists have their way, you may soon hear or read a weather forecast that sounds like this:

"The five-day outlook calls for a partly cloudy Wednesday with temperatures below average. Thursday will be warmer and clear, winds will pick up on Friday. Saturday will start foggy and end with drizzle, and Sunday will be cool and wet."

"Cancel the picnic," your mother says glumly to your father. But wait, the forecaster is still talking. *"The accuracy of this forecast is below 0.6."*

"That means the forecast has no value," you shout gleefully, as thoughts of smoke curling up from a grill and a shimmering lake go through your head.

That's right, scientists are not only trying to predict future weather, they are also trying to find out how to measure the chances that their predictions will come true. We already know that long-range weather predictions, and even short-range ones, often don't agree with what eventually happens. And the consequences can be more than disappointment over a washed-out weekend.

Resorts may lose thousands of dollars needlessly when rain is predicted but doesn't materialize. Lives and property may be threatened by an unexpected storm. The home heating industry may not have enough fuel for a surprisingly cold winter. And construction companies may have to lay off workers hired in anticipation of a dry spring that turns out to be unusually wet.

But what's the problem, you might ask? Weather forecasters use supercomputers that analyze more than 10,000 different measurements every six hours from satellites, balloons, and weather stations all around the world. Isn't that enough information to make an accurate prediction?

No, say meteorologists. How come? "Chaos" is their one-word reply.

The dictionary defines *chaos*, pronounced KAY-os, as a "confused and disorganized state of things." And that, say scientists, describes the winds, heat, moisture, and other properties of the atmosphere that produce weather. One noted meteorologist, Edward N. Lorenz, of the Massachusetts Institute of Technology, drives home the point this way: "A very small (change) . . . can make things happen quite differently from the way they would have happened if the small disturbance hadn't been there."

In 1972, professor Lorenz gave this idea a name: the *butterfly effect!* The name came from a question Lorenz posed during one of his lectures: "Could the flap of a butterfly's wings over Brazil spawn a tornado over Texas?"

What Lorenz was saying was that since all the factors influencing weather can't be measured at any given moment, it's impossible to predict with certainty what will happen in the next moment. And the further ahead you try to look into the future, the more inaccurate your prediction will be. There are simply too many things going on in the atmosphere—butterfly effects—that we can't identify or whose effect we can't determine.

Does this mean you shouldn't read weather forecasts in your local newspaper or listen to such forecasts on TV or the radio? No! Meteorologists are finding ways to make forecasts more reliable. They are also testing methods to find out just how reliable a forecast may be.

They do this by using computers to make a model of the atmosphere at a given time. But, since the atmosphere is chaotic, they can't know exactly what *is* happening in it

at a given time. To overcome this drawback, they make a number of models, each one slightly different from the others. They then let the models unfold, showing how each predicts upcoming weather. If all the models reveal similar weather patterns, the scientists feel a prediction of future weather patterns can be made with some confidence. But if the models don't produce similar results, the scientists will have less—or no—confidence in their predictions. That's why a weather forecaster of the future might qualify a forecast by saying that it is 0.6 (60%), or 0.7 (70%), or 0.9 (90%) reliable.

Will weather forecasting ever be 100 percent reliable? Should you make plans, or give them up, because of what a forecaster says? What do you think?

1. *Weather forecasters use supercomputers to analyze*

   a. about 1000 measurements four times a day.
   b. about 10,000 measurements four times a day.
   c. about 10,000 measurements twice a day.
   d. about 10,000 measurements once a day.

2. *A "confused and disorganized state of things" is defined as*

   a. chaos.
   b. weather.
   c. atmosphere.
   d. butterfly effect.

3. *If a weather forecaster of the future states that a forecast is 0.8 reliable, you can expect such a forecast to come true*

   a. 0.8% of the time.
   b. 8% of the time.
   c. 80% of the time.
   d. 88% of the time.

4. *Explain what meteorologist Edward N. Lorenz meant by the term butterfly effect.*

   _____

   _____

   _____

5. *On a separate sheet of paper, discuss why there may be limits on the accuracy of long-term weather forecasts and how such limits may affect your community.*

# Earth and the Moon

## LABORATORY INVESTIGATION

### *ANGLE OF SUNLIGHT*

**A.** Stick a thumbtack into the middle of a sheet of 1-centimeter (cm) graph paper. Attach one end of a 20-cm-long string to the thumbtack. Then, glue the other end of the string to a square cardboard cutout. Attach a cardboard tube to a flashlight as shown in Fig. 19-1. Place the cutout over the flashlight beam, and direct the beam onto the sheet of paper as shown in position A. Draw a line around the area of light produced on the graph paper. Repeat the procedure for positions B and C.

    **1.** In which position does the beam of light cover the largest area?

_____

    **2.** In which position does the beam of light cover the smallest area?

_____

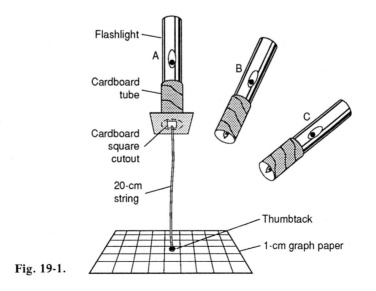

**Fig. 19-1.**

**B.** Place a *radiometer* on your desk. (A radiometer is a device that is used to measure the intensity of radiant light.) Using the flashlight with the cardboard tube attached, direct the beam of light as shown in position D in Fig. 19-2. The flashlight should be about 50 cm from the radiometer. Count how many turns the vanes of the radiometer make in 1 minute (min). Record the number. Repeat the procedure for positions E and F, keeping the flashlight at the same distance.

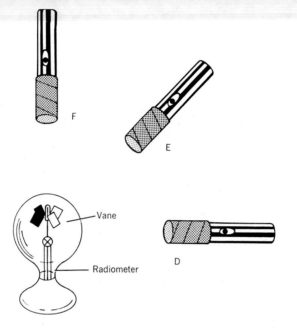

**Fig. 19-2.**

| Position | Number of Turns |
|:---:|:---:|
| D | |
| E | |
| F | |

3. Which position of the flashlight caused the radiometer vanes to turn the fastest? _____

4. Which positions in Figs. 19-1 and 19-2 have the beams of light most concentrated; least

concentrated? _____

**C.** Study Fig. 19-3, which shows rays of sunlight striking Earth. Then, answer the following questions.

5. Which ray of sunlight is most like beam D in Fig. 19-2? _____

6. Which rays of sunlight are most like beam F? _____

7. Which ray of sunlight striking Earth would you expect to be most concentrated? _____

8. From the observations you made in activities A, B, and C, explain how the angle of sunlight

affects the amount of heat received by different regions on Earth. _____

_____

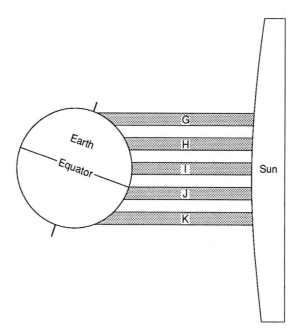

**Fig. 19-3.**

## EARTH'S SIZE AND SHAPE

Earth is shaped like a ball, or sphere. Unlike a true sphere, Earth is not perfectly round. There is a slight flattening of Earth at the North Pole and South Pole and a bulging of Earth at the equator. Because of Earth's polar flattening and equatorial bulging, the distance around the poles is shorter than the distance around the equator (see Fig. 19-4). Earth's diameter at the equator is about 12,755 kilometers (km). At the poles, Earth's diameter is about 12,715 km. Earth is the largest of the inner planets in our solar system, about twice the size of Mars (see Chapter 20).

Earth's spherical shape has been known for more than 500 years. Before Columbus sailed west across the Atlantic Ocean, he inferred that Earth was round based on two lines of evidence: (1) Earth makes a circular shadow on the moon during a lunar eclipse, and (2) the topmast on a sailing ship is always the last part of the ship to disappear and the first to appear over the horizon.

For centuries, people have *circumnavigated,* or circled, Earth in boats, and in this past century, in airplanes. Most recently, photographs taken from the moon and outer space during space flights have provided unquestionable evidence of Earth's spherical shape.

## EARTH'S MOTIONS

Earth spins like a top, turning about an imaginary line that extends through Earth between the North Pole and South Pole. This spinning motion is called **rotation,** and the imaginary line is called Earth's *axis of rotation* (see Fig. 19-5). As Earth rotates, you are carried around in a circle as if riding on a giant merry-go-round. At the same time, Earth also moves around the sun in a motion called **revolution.** The path Earth follows around the sun is called its **orbit.** The shape of Earth's orbit is an elongated circle, or *ellipse* (see Fig. 19-5).

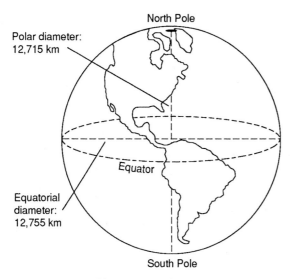

**Fig. 19-4. Earth's equatorial diameter is larger than its polar diameter.**

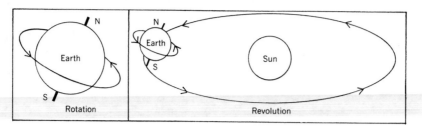

**Fig. 19-5. Earth's motions: rotation and revolution.**

Earth's rotation and revolution are important for several reasons. These motions cause the succession of night and day and the change of seasons. The lengths of the day and of the year also depend on rotation and revolution. In addition, the regular rise and fall of ocean tides is affected by Earth's rotation.

## NIGHT AND DAY

Because Earth is a sphere, only half of its surface is lighted by the sun at one time. As a result, one half of Earth's surface is always in sunlight, and the other half is always in darkness. Earth's rotation on its axis allows the sun to shine on every part of Earth in succession. If Earth did not rotate, the half facing the sun would have constant daylight, while the other half would never receive light.

There are 24 hours in a day. Thus, you might expect 12 hours of daylight followed by 12 hours of darkness every day throughout the year. However, this only occurs at the equator. In the middle latitudes, the amounts of daylight and of darkness change throughout the year. For example, in summer in New York, daylight hours may extend from about 5:30 A.M. until about 8:30 P.M. In winter, daylight hours may only extend from about 7:00 A.M. until about 5:00 P.M. The farther north or south of the equator you live, the longer the sun shines in the summer, and the longer are the winter nights. Above the Arctic Circle, the sun does not rise for several weeks in the winter.

**Why the Length of Daylight Changes.** There are two reasons why the lengths of days and nights change throughout the year. One is because Earth's axis of rotation is inclined, or tilted, 23.5 degrees (23.5°) from the vertical in relation to its orbit around the sun (see Fig. 19-6). The other is because the direction in which Earth's axis points doesn't change as Earth orbits the sun. This is called *parallelism* and is shown in Fig. 19-7. As you see, Earth's axis always points in the same direction in space, no matter where Earth is in relation to the sun.

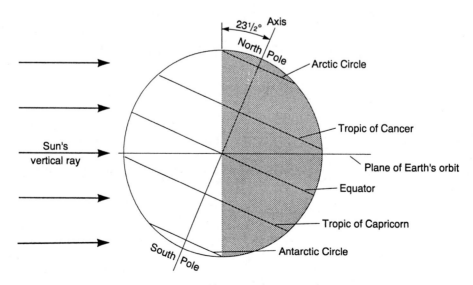

**Fig. 19-6. Earth's axis is tilted 23½° from its plane of orbit.**

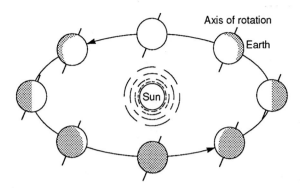

**Fig. 19-7. Parallelism of Earth's axis as it orbits.**

**Length of Daylight in Summer.** Fig. 19-8 shows the relation between Earth and the sun on the first day of summer in the Northern Hemisphere. On this date, Earth's axis is tilted toward the sun. You can see that more than half of the Northern Hemisphere's surface is exposed to the sun. As a result, the Northern Hemisphere has more hours of daylight than hours of darkness.

You can also see that the farther north you are, the greater the number of hours of daylight. For example, the equator has 12 hours of daylight and 12 hours of darkness. At the middle latitudes, daylight is about 15 hours long, and darkness is about 9 hours long. Still farther north, the area above the Arctic Circle has 24 hours of daylight on June 21. On this date, the sun remains in the sky throughout the 24-hour period.

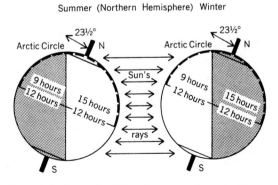

**Fig. 19-8. Length of day in summer and winter.**

**Length of Daylight in Winter.** On the first day of winter in the Northern Hemisphere, the northern tip of Earth's axis is tilted away from the sun (see Fig. 19-8). Consequently, the Northern Hemisphere has more hours of dark-

ness than hours of daylight. On the first day of winter, the area above the Arctic Circle has 24 hours of darkness. The middle latitudes have about 15 hours of darkness and 9 hours of daylight. At the equator, days and nights are of equal length.

**Length of Daylight in Spring and Fall.** On the first days of spring and fall, Earth's axis does not point toward or away from the sun. On these dates, the position of Earth's axis in relation to the sun causes the sun to shine for 12 hours everywhere on Earth. As a result, on the first day of spring and fall, there are 12 hours of daylight and 12 hours of darkness over all of Earth (see Fig. 19-9).

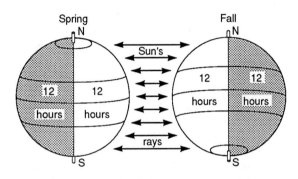

**Fig. 19-9. Length of day in spring and fall.**

## SEASONS

The tilt of Earth's axis also causes the change of seasons. Earth's axis is always inclined in the same direction in space, regardless of Earth's location in its orbit around the sun. Consequently, the northern tip of Earth's axis is tilted toward the sun during summer and tilted away from the sun during winter.

With each passing day, the Northern Hemisphere and Southern Hemisphere lean closer to or farther away from the sun. In the Northern Hemisphere, the tilt of Earth's axis causes the days to be about six hours longer in summer than in winter (see Fig. 19-8). As a result, the Northern Hemisphere receives more sunlight in summer than it receives in winter. Nights are relatively short in summer. Thus, there are fewer hours during which Earth's surface can radiate heat back into space.

The angle at which the sun's rays strike Earth's surface is especially important. The more directly the sun's rays strike Earth's surface, the more heat the surface absorbs. Fig. 19-8 shows that the sun's rays strike the Northern Hemisphere more directly in sum-

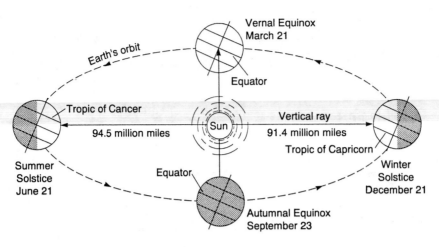

**Fig. 19-10. The tilt of Earth's axis causes the change of seasons.**

mer and more obliquely (at a slant) in winter. As a result of the longer days and the more direct sunlight in summer, the Northern Hemisphere accumulates more heat in summer than in winter.

The tilt of Earth's axis also explains why the seasons are reversed in the Northern Hemisphere and Southern Hemisphere. For example, when it is summer in North America, it is winter in South America.

Fig. 19-10 shows that Earth is about five million kilometers farther from the sun in summer than in winter. Thus, the distance from the sun to Earth does not influence seasonal changes. The seasons are actually caused by the position of Earth in its orbit.

In spring and fall, Earth's axis does not point toward or away from the sun. Consequently, the entire Earth has days and nights of equal length. Fig. 19-10 shows that the sun's rays fall directly on Earth only at the equator. This causes temperatures in the upper and middle latitudes of both hemispheres to fall between the normal winter and summer temperatures.

## SOLSTICES AND EQUINOXES

The moment that summer begins is called the *summer solstice*. The word **solstice** comes from two Latin words that mean "sunstop" and refers to the time when the sun seems to stop moving higher in the sky each day. In the Northern Hemisphere, the summer solstice occurs about June 21. On this date, the sun's most direct rays fall on the *Tropic of Cancer*, which lies 23.5° north of the equator. At the summer solstice, the North Pole is tilted farther *toward* the sun than at any other time of year. Thus, the first day of summer also is the longest day of the year.

The moment that winter begins is called the *winter solstice*. In the Northern Hemisphere, this occurs about December 21. On this date, the most direct rays fall on the *Tropic of Capricorn*, which lies 23.5° south of the equator. At the winter solstice, the North Pole is tilted *farther away* from the sun than at any other time of the year. Thus, the first day of winter also is the shortest day of the year. The moment that winter begins in the Northern Hemisphere, summer begins in the Southern Hemisphere (see Fig. 19-9).

The word **equinox** comes from two Latin words that mean "night of equal length." Equinoxes occur twice a year, when the most direct rays of the sun fall on the equator.

The *fall*, or *autumnal*, *equinox* occurs when the most direct rays of the sun cross the equator as the sun moves south toward the Tropic of Capricorn. The autumnal equinox takes place on or about September 23, the first day of fall in the Northern Hemisphere. The *spring*, or *vernal*, *equinox* occurs when the most direct rays of the sun cross the equator moving from south to north. The vernal equinox occurs on or about March 21, the first day of spring in the Northern Hemisphere (see Fig. 19-9).

## UNITS OF TIME

Before 1687, clocks only had an hour hand. The dials of these early clocks were divided into hours and quarter hours. In contrast, modern clocks and watches are accurate to within a few seconds or less in a day, and the most accurate timekeeping devices, *atomic clocks*, are accurate to within 1 second (sec) in 300 years.

Clocks are based on the amount of time it takes Earth to rotate on its axis and to orbit the sun. Earth's rotation and revolution are relatively constant. Thus, when you set a clock, you're actually adjusting the clock to Earth's position in space.

**Year.** The amount of time it takes Earth to orbit the sun is called a **year.** Earth rotates on its axis about 365.25 times in the period it takes Earth to complete one orbit of the sun. For this reason, a year is about 365.25 days long.

**Day.** The amount of time it takes Earth to rotate once on its axis is called a **day.** Astronomers can determine the exact length of a day by measuring when a particular star passes twice in succession over a specific point on Earth. A day based on the reappearance of the same star in a telescope is called a *sidereal day.*

*Sidereal Day.* Scientists who need to know the exact length of a day base their calculations on the **sidereal day.** That is because the successive passage of a star across meridians of longitude is not affected by Earth's orbit around the sun. Stars are so far away from Earth that scientists consider them to be motionless, or fixed, reference points that can be used to measure Earth's rotation.

A sidereal day, or one complete rotation of Earth, takes about 23 hours, 56 min, and 4 sec. The length of a sidereal day remains nearly constant for long periods of time. However, measurements have shown that the rate of Earth's west-to-east rotation has been decreasing very slowly over the past thousand years. Scientists think that the slowing is caused by ocean tides, which move (east to west) in an opposite direction. (Tides are discussed later in this chapter.)

*Solar Day.* From sunrise to sunset, the sun appears to take a curved path across the sky (see Fig. 19-11). When the sun is halfway between sunrise and sunset, it reaches the highest point in the curve. At this point, the sun is said to cross the *meridian,* and the time is "exactly noon, local time." The time it takes the sun to return to this point is called a **solar day.**

The 24-hour day commonly used is based on the solar day. Each day, the sun appears to rise in the east, travel across the sky, and set in the west. Actually, Earth's rotation is causing the apparent motion of the sun. The sun appears to rise above the eastern horizon because Earth rotates from west to east, as it moves toward the sun. To an observer, the sun appears to cross the sky and set below the western horizon because the rotating motion is carrying the observer sideways, away from the sun.

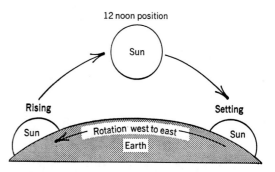

**Fig. 19-11. The sun is at its highest point at noon.**

Based on sidereal time, two successive passages of the sun to its highest point are rarely 24 hours apart. That is because (1) Earth travels about 2,400,000 km in its daily orbit around the sun, and (2) Earth's speed varies as it orbits the sun. Consequently, the length of the solar day can vary by as much as 16 minutes.

The average length of all solar days throughout a year is called *mean solar time.* To avoid confusion, clocks run on mean solar time. Based on mean solar time, all days are considered of equal length, regardless of the sun's position.

Each day, the 12-hour period before the sun reaches your local meridian is called A.M. This is an abbreviation for the Latin term, *ante meridiem,* which means "before the meridian." The 12-hour period after the sun crosses your local meridian is called P.M., for the Latin term *post meridiem,* which means "after the meridian." To differentiate between the two 12 o'clocks, it is customary to refer to the midday (meridian) as *noon* and to the end of the day as *midnight.*

## TIME ZONES

Until the end of the nineteenth century, people set their clocks by the noonday sun, or when the sun was at its highest point in the sky. This timekeeping method based on solar time was satisfactory until railroads spread across the United States. Railroad companies could not establish accurate schedules because towns and cities only 160 km apart often had official times that varied by as much as 8 min. In 1883, the railroad companies agreed to divide the United States into four standard **time zones.** Within each time zone, all clocks were set to the same *standard time.*

Eventually, all nations adopted standard time, and Earth was divided into 24 time zones. Each time zone extends through 1/24 of a circle, or 15° of longitude: 360°/24 time zones = 15° per time zone.

In theory, the line between two time zones should be drawn directly north and south along a meridian of longitude. However, the line separating two time zones is often shifted east or west to avoid splitting a city into two time zones. This avoids the confusion that would occur if the time in one part of a city was an hour earlier than the time in another part of the same city. For this reason, time zone boundaries on land are seldom straight lines.

**Time Zones of the United States.** Excluding Alaska and Hawaii, the United States is divided into four standard time zones. These time zones are called the *Eastern, Central, Mountain,* and *Pacific* standard time zones (see Fig. 19-12).

Fig. 19-12. **Time zones of the United States.**

As you travel west from New York City to Los Angeles, the time in each zone you enter is one hour earlier than the time in the zone you leave. Thus, when it is 11 A.M. in the Eastern standard time zone, it is 10 A.M. in the Central standard time zone, 9 A.M. in the Mountain standard time zone, and 8 A.M. in the Pacific

standard time zone. There is a three-hour difference in time between the east coast and the west coast of the United States. Thus, when it is 11 A.M. in New York City, it is 8 A.M. in Los Angeles.

**Daylight Savings Time.** Many communities in the United States do not follow standard time throughout the year. In late April, these communities set their clocks one hour ahead of standard time. By doing this, they gain one hour of daylight in the evening. In late October, these communities set their clocks one hour back to follow standard time again.

**International Date Line.** You learned that a person traveling westward gains one hour for each 15° of longitude. A person circling Earth gains 24 hours. To avoid confusion, the date is changed at the 180th meridian, commonly called the **International Date Line.** This is an imaginary line that runs in a north-south direction through the middle of the Pacific Ocean.

A new day always begins on the western side of the International Date Line. For example, suppose you were aboard a ship sailing west that crosses the International Date Line at 12 noon, Tuesday. Aboard the ship, the day automatically becomes 12 noon, Wednesday. In contrast, if you were aboard a ship crossing the line in the opposite direction (from west to east), you would have to set the calendar back one day—from 12 noon, Wednesday, to 12 noon, Tuesday.

## MONTHS

Long ago, people observed that a *full moon* reappears in the sky about every 30 days. For this reason, people divided the year into units of time called **months.** After the appearance of a full moon, the lighted part of the moon slowly *wanes,* or grows smaller. About 15 days later, the moon resembles a thin, curved sliver of light, before disappearing completely. Shortly afterward, the moon reappears as a thin sliver of light. The moon's lighted part continues to *wax,* or grow larger, for about 15 days until the moon is full once again. These daily changes in the moon's appearance are called **phases.**

**Phases of the Moon.** As the moon orbits Earth, its position in relation to Earth and the sun changes constantly. The amount of sunlight reflected from the moon's surface to Earth depends on the moon's position in re-

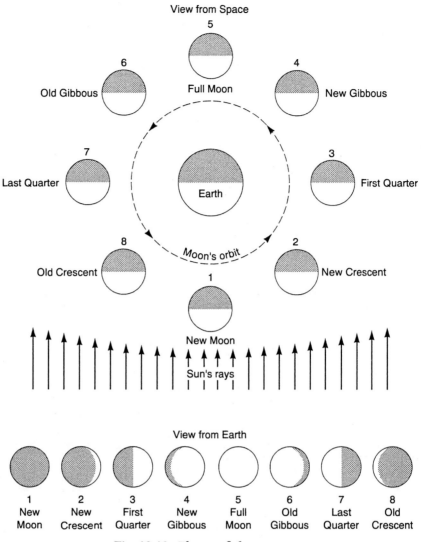

**Fig. 19-13. Phases of the moon.**

lation to Earth and the sun. For example, when the sun, moon, and Earth are lined up with one another, the moon is either fully lighted (a *full moon*), or completely darkened (a *new moon*). A *full moon* is visible only to an observer on Earth's darkened side (see Fig. 19-13).

When the moon is on Earth's lighted side, the moon's darkened side faces Earth. Then, the moon is not visible from Earth. This phase is called a *new moon*. A few days after a new moon, you can see a thin, curved sliver of the moon's surface. This thin, lighted part is called a *new crescent*. About seven days after a new moon, you can see half of the moon's lighted surface. This phase is called the *first quarter* because you can see half of the moon's lighted half. As the moon continues to orbit Earth, you can see more of its lighted surface

in the phase called *new gibbous*. This phase occurs between the first quarter and the full moon.

After a full moon phase, you see less of the lighted surface in the *old gibbous* phase. A few days later, you again see a quarter of the moon, a phase called *last quarter*. After passing through the *old crescent* phase, the moon again becomes a new moon and disappears from view.

Although a new moon does occur every 29.5 days, the moon actually orbits Earth once every 27.3 days. A new moon returns every 29.5 days because both the moon and Earth are orbiting the sun at the same time that the moon is orbiting Earth. For this reason, when the moon completes one full orbit of Earth, the moon, Earth, and the sun are no longer in a direct line. It takes about two days more for

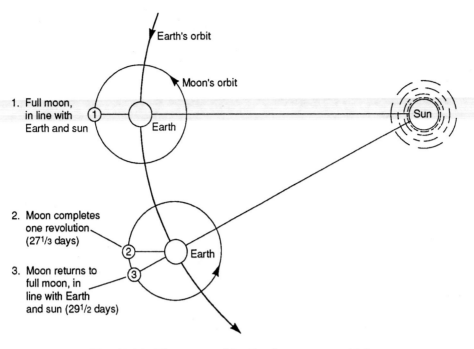

**Fig. 19-14. The moon orbits Earth once every 27.3 days, but returns to full moon once every 29.5 days.**

the moon to line up directly with the sun and Earth, as it was 29.5 days earlier (see Fig. 19-14).

It takes the moon 27.3 days to make one complete rotation on its axis. It also takes the moon 27.3 days to orbit Earth. For this reason, an observer on Earth always sees the same side of the moon.

## CALENDARS

A *calendar* is a way of marking off long periods of time, such as months or years. The modern calendar developed from lunar calendars used thousands of years ago, which were based on the phases of the moon. Lunar calendars lost about 10 days every year because 12 lunar months of 29.5 days added together are approximately 10 days shorter than a solar year of about 365 days.

**Julian Calendar.** In 46 B.C., Julius Caesar adopted what is now called the **Julian calendar.** The Julian calendar was based on a year of 365.25 days, or the amount of time for Earth to make one complete orbit of the sun. A year was divided into 12 months of varying length that totaled 365 days. Every fourth year, an extra day was added to the calendar. Thus, every fourth year is called a *leap year.*

The Julian calendar was more exact than the lunar calendars. Using it, people could predict the day of the month that a particular date would fall on. However, even the Julian calendar was slightly inaccurate. The Julian calendar was based on a year of exactly 365.25 days. Because a year is actually 11 min and 14 sec shorter, the Julian year was slightly long. After 128 years, the Julian calendar lagged behind the actual date by about one day.

**Gregorian Calendar.** By 1582, the vernal equinox arrived on March 11 instead of March 21, according to the Julian calendar. In that year, Pope Gregory XIII ordered the Julian calendar to be adjusted by dropping the 10 extra days. He also ordered that three leap-year days be eliminated every 400 years to make up for the slight lag between the Julian calendar and the progression of the seasons.

Leap-year days were eliminated by a simple rule. A century year (1600, 1700, 1800, 1900, and so on) becomes a leap year only if its first two digits are divisible by four. Thus, the year 2000 was a leap year. The next century year to be a leap year will be in the year 2400. This revised calendar, which is currently used, is called the **Gregorian calendar.**

## ECLIPSES

Because the moon and Earth are opaque objects, they block sunlight. As a result, they cast shadows into space. Both the moon and Earth cast two shadows, one inside the other. The darker, shorter, inner shadow that covers a cone-shaped space (completely shut off from the sun's rays) is called the **umbra**. The lighter, longer, outer shadow that covers a gradually widening space (partially shut off from the sun's rays) is called the **penumbra** (see Fig. 19-15).

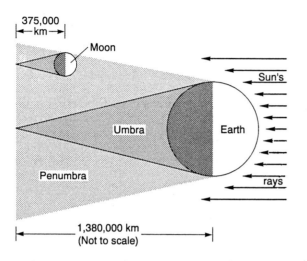

**Fig. 19-15. Shadows cast by Earth and the moon.**

Earth's umbra extends about 1,400,000 km into space. The moon's umbra extends about 375,000 km into space. Because the moon's diameter is about one quarter of Earth's diameter, the length of the moon's shadow is nearly one quarter the length of Earth's shadow.

Occasionally, as the moon orbits Earth, Earth comes between the sun and the moon. At this time, Earth's umbra is cast on the moon, and the moon is darkened. This is called a *lunar eclipse*. At other times, the moon passes between Earth and the sun. The moon's umbra is cast on Earth, and the sun is darkened. This is called a *solar eclipse*.

**Lunar Eclipses.** Once a month, Earth is located between the sun and the moon. Thus, you might expect that a lunar eclipse would occur each month. Actually, lunar eclipses occur only two to five times a year. The reason is that the plane of the moon's orbit is at an angle of 5° to the plane of Earth's orbit. As a result, the moon does not pass through Earth's shadow on every orbit (see Fig. 19-16).

When the moon is on the dark side of Earth (but not in Earth's umbra), a full moon appears. That is because all of its lighted half is visible from Earth. If the moon moves into Earth's umbra, the moon receives much less

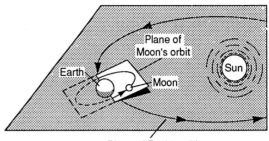

**Fig. 19-16. The moon's orbit is inclined 5° to the plane of Earth's orbit.**

sunlight, and a **lunar eclipse** occurs (see Fig. 19-17). A total lunar eclipse occurs when the moon is completely inside Earth's umbra. A partial lunar eclipse occurs when part of the moon is inside Earth's umbra.

The moon does not disappear completely from view during a total lunar eclipse. That is because the red rays of sunlight traveling through Earth's atmosphere are bent, or refracted, into Earth's shadow. For this reason, the moon turns a dull red or copper color during a total eclipse. Lunar eclipses may last for over three hours.

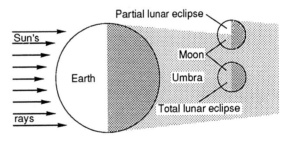

**Fig. 19-17. Lunar eclipses.**

**Solar Eclipses.** A **solar eclipse** occurs when the moon's umbra falls on Earth. In order for this to happen, the moon must (1) pass directly between the sun and Earth and (2) be close enough for its umbra to reach Earth. Solar eclipses occur two to five times a year at different locations on Earth.

The moon's distance from Earth varies as the moon orbits Earth. At its closest approach, the moon is about 360,000 km from Earth. At its farthest distance from Earth, the moon is about 400,000 km away. The moon's average distance from Earth is about 380,000 km.

You recall that the moon's umbra is about 375,000 km long. When the new moon is less than this distance from Earth, the moon's umbra is long enough to reach Earth. When

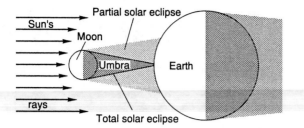

**Fig. 19-18. Solar eclipses.**

# TIDES AND THE MOON

the moon's umbra falls upon Earth, the sun cannot be seen from that part of the planet, and a solar eclipse occurs (see Fig. 19-18).

The moon's umbra covers an area on Earth of up to 270 km in diameter. Consequently, a total solar eclipse occurs only along this narrow band of darkness cast on Earth by the moon's umbra. However, a partial solar eclipse can be seen just outside of the moon's umbra.

During a total solar eclipse, a halo of light (the sun's corona) can be seen around the dark disk of the moon. The area on Earth within the moon's umbra becomes as dark as night for a few minutes. During this time, the stars can even be seen.

You learned that Earth completes one orbit of the sun in 365.25 days and rotates on its axis once every 24 hours. There is also a third motion of Earth that influences the **tides,** or the rise and fall of the oceans.

Earth and the moon revolve around a common center of gravity like opposite ends of a spinning bottle. This common center of gravity is called the **barycenter.** Because the narrow end of a bottle (the moon) is lighter, it spins in a larger circle than does the heavier, wide end (Earth). Both ends of a bottle spin around the bottle's center of gravity, which is a point inside the bottle, closer to the heavier bottom. Because Earth has about 81 times more mass than the moon, the barycenter of the moon-Earth system is located about 1600 km below Earth's surface (see Fig. 19-19).

The tides are mainly caused by two forces acting together. One force is the gravitational attraction of the moon, which pulls Earth's waters toward the moon. The other force is produced by the moon-Earth system's spinning about its barycenter. This causes the ocean on the side of Earth facing away from the moon to bulge outward.

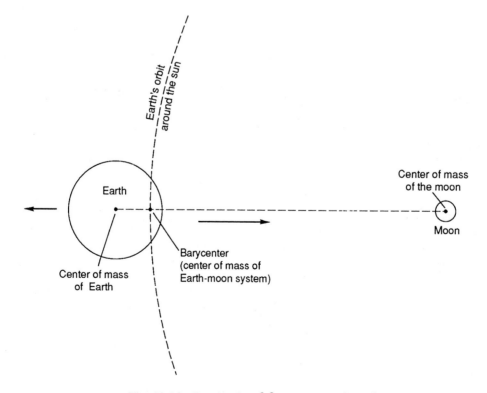

**Fig. 19-19. Gravitational forces cause the tides.**

The combined forces cause the ocean waters to swell up in two tidal bulges on opposite sides of Earth. Between the tidal bulges, the ocean level drops. Each bulge is called a *high tide*, and each depression is called a *low tide*. Fig. 19-20 shows the positions of high and low tides in relation to the moon.

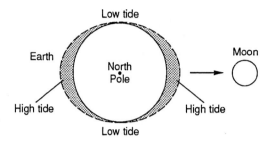

**Fig. 19-20. High tides and low tides.**

**High Tide and Low Tide.** When the moon is overhead on the open ocean, the waters beneath it bulge upward about one meter (m). As Earth rotates, coastal areas pass beneath the moon, and the bulging ocean moves onshore. This is called an *incoming tide*. As Earth continues to rotate, the tide reaches its peak and then begins to recede. The peak is called high tide. Several hours later, the tide falls to its lowest level, called low tide.

One type of tidal pattern—two highs and two lows each day—occurs along the east and west coasts of the United States. You might expect the tides to be 6 hours apart. Actually, the tides are about 6 hours and 13 minutes apart. The reason is the moon's revolution. As Earth rotates, the moon moves ahead in its orbit around Earth. After 24 hours, Earth needs another 52 minutes to "catch up" to the moon and thus complete its tidal pattern of two highs and two lows. By comparison, the Gulf of Mexico has a single daily tide. Although the east and west coast patterns are similar, the Pacific tides vary substantially in size.

The change in water level between high and low tides, or the *tidal range*, is influenced by the shape of the coastline and ocean floor, the depth of coastal waters, and the wind. These factors also influence when tides occur. The water level along a coastline usually rises about one meter (m) at high tide. Inland seas and lakes have little or no noticeable tide.

The tidal range can vary greatly. For example, along the coastline of the Gulf of Mexico, the tidal range may be only about 0.5 m. In the Bay of Fundy (along the coast of Nova Scotia),

high tide may rise 15 m or more above low tide because of the narrow V-shape of the bay.

**Neap Tides.** Twice a month, when the moon is at first or last quarter, the moon, sun, and Earth form a right angle. At these times, the sun's gravitational pull on Earth partially cancels out the effect of the moon's gravitational pull (see Fig. 19-21a). As a result, the tidal bulges produced by the moon are smaller than usual, so that there are lower high tides and higher low tides than normal. These weaker tides are called *neap tides*.

**Spring Tides.** Twice a month, at full moon and at new moon, the sun, moon, and Earth are in a direct line with each other. At these times, the sun's gravitational pull strengthens the moon's gravitational pull on Earth (see Fig. 19-21b). The combined pull of the sun and moon causes extreme high and low tides. These extra high and low tides are called *spring tides*.

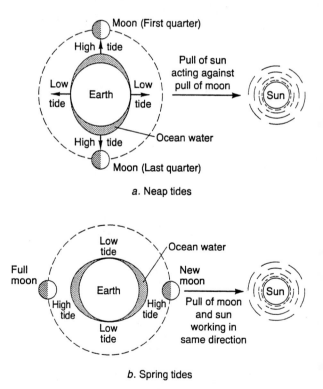

*a.* Neap tides

*b.* Spring tides

**Fig. 19-21. Neap tides and spring tides.**

## FEATURES OF THE MOON

The moon is a spherical body with a diameter of about 3475 km, a little more than one-fourth Earth's diameter. Because of its small mass (about 1/80 of Earth's mass), the moon's gravitational pull cannot hold an atmosphere. As a result, temperatures on the lunar surface are either extremely hot or cold. For example, on the moon's lighted side, temperatures may exceed 104° Celsius. On the dark side, temperatures may dip below −150°C.

In 1609, the Italian scientist *Galileo Galilei* (1564–1642) was the first person to view the moon with a telescope and record his observations. At the beginning of the seventeenth century, most people believed that the moon's surface was smooth. Galileo discovered that the moon's surface was rough and uneven. During the last 350 years, stronger telescopes, spacecraft equipped with cameras, and manned lunar landings have shown that many of Galileo's observations were correct.

**Highlands and Maria.** Galileo saw light and dark areas on the moon. The rugged, light-colored areas he saw were the mountainous, crater-filled regions called **highlands.** Based on radioactive dating, most highland rocks are between 4.0 and 4.3 billion years old. A few specimens from the highlands have been dated at 4.4 billion years, which may be slightly less than the age of the moon. The age of highland rocks supports the hypothesis that the highlands of the moon probably are made of the original lunar crust.

Galileo called the large, darker-colored areas **maria** because he thought that they were filled with water. *Maria* is a Latin word that means "seas." However, lunar probes and manned explorations have shown that there is no atmosphere or water on the moon. The maria are actually great basins and level plains, some of which are more than 1000 km in diameter—larger than Texas. The rocks inside the maria are the youngest lunar rocks, ranging in age from about 3.2 to 3.8 billion years. The age of maria rocks supports the hypothesis that the moon has been relatively inactive for about the last 3 billion years.

*Lunar Craters.* Scientists think that most lunar craters were caused by meteorites. There are two main types of craters on the lunar surface. One type is called an *impact crater.* Scientists think that impact craters were formed by the original impact of meteorites. The other type is called a *satellite crater.* These craters were probably formed by falling fragments ejected from impact craters.

Some craters have light-colored, thin streaks called *rays*, that extend outward from the crater walls. Rays are composed of shattered rock and fine-grained lunar surface material, or *regolith*, thrown outward by the explosive

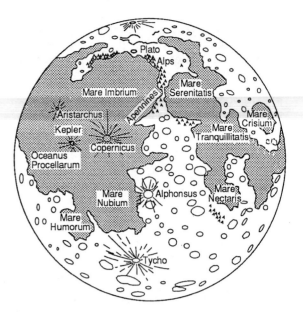

**Fig. 19-22. Features of the lunar landscape: highlands, maria, craters, and rays.**

impact of meteorites. Some rays are thousands of kilometers long and cross highlands, maria, and even other craters (see Fig. 19-22).

**Composition of Moon Rocks.** Scientists have identified three main types of lunar rocks. Most of the explored lunar highlands are composed of *anorthosite*, a light-colored igneous rock also found in the Adirondack Mountains of New York State. Anorthosite is made of a calcium-rich feldspar with a low specific gravity. It forms when a large body of magma cools slowly, allowing the light-weight calcium feldspar to float to the surface of the magma while heavier minerals sink to the bottom. Scientists believe that the newly formed moon was covered by an "ocean" of magma, and that the anorthosite rose to the surface of this ocean and cooled to form the lunar crust.

Most of the explored maria are filled with dark-colored igneous rocks similar to *basalt* and *gabbro*. Unlike basalt and gabbro on Earth, the lunar varieties contain more of the elements titanium and silicon, and they lack water. Lunar basalt and gabbro, which formed later than the anorthosite, probably hardened from lava that came from the moon's still molten interior. The lava rose to the surface through fractures in the crust caused by the impacts of huge meteorites.

Another lunar rock type (similar to a sedimentary rock on Earth called *breccia*) is composed of angular fragments of basalt, gabbro, and anorthosite. *Lunar breccia* was probably formed by the fragmentation and compaction of other lunar rocks when the moon's surface was struck by large meteorites.

**Lunar History.** Based on recent lunar data, the moon was probably formed about 4.5 billion years ago by a collision between Earth and a planet-like body about the size of Mars. The collision threw a mass of pulverized rocky debris into space. This material clumped together due to gravitational attraction and eventually became the moon.

During the moon's first billion years, huge meteorites blasted out great basins, while smaller ones formed craters. As the bombardment of the lunar surface by meteorites lessened, lava erupted from fractures in the great basins. As a result, the floors of the basins became dark and smooth as the lava hardened. For about the past 3 billion years, no new basins have been formed. However, new craters on the lunar surface have resulted from bombardment by small meteorites. In addition, very small meteorites, or *micrometeoroids*, have caused pitting of the lunar surface as well as powdering of surface rock into lunar dust.

# CHAPTER REVIEW

## Science Terms

*The following list contains all of the boldfaced words found in this chapter and the page on which each appears.*

barycenter (p. 288)
day (p. 283)
equinox (p. 282)
Gregorian calendar (p. 286)
highlands (p. 290)
International Date Line (p. 284)
Julian calendar (p. 286)
lunar eclipse (p. 287)

maria (p. 290)
months (p. 284)
orbit (p. 279)
penumbra (p. 287)
phases (p. 284)
revolution (p. 279)
rotation (p. 279)
sidereal day (p. 283)

solar day (p. 283)
solar eclipse (p. 287)
solstice (p. 282)
tides (p. 288)
time zones (p. 284)
umbra (p. 287)
year (p. 283)

## Matching Questions

*On the blank line, write the letter of the item in column B that is most closely related to the item in column A.*

*Column A*

_____ 1. solstice

_____ 2. equinox

_____ 3. sidereal

_____ 4. penumbra

_____ 5. lunar

_____ 6. solar

_____ 7. spring tides

_____ 8. neap tides

*Column B*

a. time measured by the stars
b. tides that are extra high or extra low
c. tides that are not very high or very low
d. beginning of summer or winter
e. beginning of spring or fall
f. lighter shadow of the moon or Earth
g. tides that occur only in the spring
h. type of eclipse produced when the moon passes through Earth's darker shadow
i. type of eclipse produced when Earth passes through the moon's darker shadow

# Multiple-Choice Questions

*On the blank line, write the letter preceding the word or expression that best completes the statement.*

1. Regions on Earth with about the same number of daylight and nighttime hours throughout the year are located
   *a.* within the Arctic and Antarctic circles
   *b.* along the tropics of Capricorn and Cancer
   *c.* at the North and South poles
   *d.* along the equator                                          1 ___

2. In the Western Hemisphere, parts of Earth with about 15 hours of daylight and 9 hours of darkness in summer are located
   *a.* above the Arctic Circle
   *b.* in northern Canada below the Arctic Circle
   *c.* in the northern parts of the United States
   *d.* along the Gulf of Mexico                                   2 ___

3. During spring and fall, the greatest concentration of sunlight falls on the
   *a.* equator                  *c.* Tropic of Capricorn
   *b.* Tropic of Cancer         *d.* Antarctic Circle             3 ___

4. The most direct rays of the sun shine on the Tropic of Capricorn on or about
   *a.* December 21  *b.* June 21  *c.* September 21  *d.* March 21   4 ___

5. The dates of March 21 and September 21 are called the
   *a.* equinoxes                *c.* seasonal progressions
   *b.* solstices                *d.* tropics of Cancer and Capricorn   5 ___

6. Earth has been divided into the following number of time zones:
   *a.* 4  *b.* 12  *c.* 24  *d.* 36                                6 ___

7. When it is midnight on the east coast of the U.S., on the west coast it is
   *a.* 9:00 P.M.  *b.* 10:00 A.M.  *c.* 3:00 A.M.  *d.* 5:00 A.M.   7 ___

8. The purpose of the International Date Line is to
   *a.* separate the Eastern and Western hemispheres
   *b.* separate one day from another day
   *c.* provide an international standard of time
   *d.* define the 180° of longitude                               8 ___

9. Lunar eclipses occur only two to five times a year because
   *a.* the moon comes close enough to Earth only two to five times a year
   *b.* Earth usually is between the sun and the moon
   *c.* the moon's orbit is at an angle to Earth's orbit around the sun
   *d.* the dark side of the moon is rarely between Earth and the sun   9 ___

10. Tides on the east and west coasts of the U.S. usually occur at intervals of
    *a.*  6 hours               *c.*  6 hours and 13 min
    *b.* 24 hours              *d.* 24 hours and 26 min            10 ___

11. When the sun, moon, and Earth are in direct line with each other, the tides that result are called
    *a.* neap tides  *b.* flood tides  *c.* ebb tides  *d.* spring tides   11 ___

12. In one month, the number of times that the moon and the sun are at right angles to each other in relation to Earth is
    *a.* one  *b.* two  *c.* three  *d.* four                      12 ___

13. Spring tides occur when the moon is in its
    a. quarter phases      c. full and quarter phases
    b. full and new phases      d. new phase only      13 _____

# Modified True-False Questions

*In some of the following statements, the italicized term makes the statement incorrect. For each incorrect statement, write the term that must be substituted for the italicized term to make the statement correct. For each correct statement, write the word "true."*

1. In early summer, the region above the Arctic Circle has 24 hours of *daylight*.      1 _____

2. The sun's rays shine almost vertically on the equator during the first week of *summer and winter*.      2 _____

3. The moment that summer or winter begins is called the *solstice*.      3 _____

4. When summer begins in the Northern Hemisphere, *fall* is beginning in the Southern Hemisphere.      4 _____

5. The apparent motion of the sun across the sky each day is caused by Earth's *revolution*.      5 _____

6. The 12-hour period that passes *before* the sun is at its highest point in the sky is called A.M.      6 _____

7. Earth is divided into 24 time zones, each time zone being *one degree* wide.      7 _____

8. During a *full* moon, the moon and the sun are on opposite sides of Earth.      8 _____

9. The moon makes one complete revolution around Earth every *27.3 days*.      9 _____

10. According to the Gregorian calendar, a century year becomes a leap year if the first two digits in the year are divisible by *four*.      10 _____

# Testing Your Knowledge

1. Why are there more hours of daylight during the summer than during the winter in the United States? _____

_____

2. Why are there 24 hours of sunlight in early summer at all places above the Arctic Circle?

_____

_____

3. Why does winter come in the Northern Hemisphere when Earth is at its closest position

to the sun? _____

_____

4. Identify the season during which the *vertical* rays of the sun are shining on the following
regions:

   *a.* Equator _____   *b.* Tropic of Capricorn _____   *c.* Tropic of Cancer _____

5. How are the following units of time derived from the motions of Earth in space?

   *a.* hour _____

   *b.* day _____

   *c.* year _____

6. Describe the difference between a solar day and a sidereal day. _____

_____

7. Why has Earth been divided into time zones? _____

_____

8. Draw the positions of the sun, Earth, and the moon during the following phases of the moon:

   *a.* new moon   *b.* full moon   *c.* first or last quarter

9. The moon revolves around Earth once every 27.3 days. Why is there a full moon every 29.5

days? _____

_____

10. Why do leap years occur? _____

_____

11. Draw the positions of the sun, Earth, and the moon during each of the following:

   *a.* solar eclipse   *b.* lunar eclipse

12. Lunar eclipses do *not* occur once a month. Explain. _____

_____

13. Identify the two forces that cause the ocean waters to bulge away from Earth.

   *a.* _____

   *b.* _____

14. What is the difference between a spring tide and a neap tide? _____

_____

15. What evidence indicates that life probably does not exist on the moon? _____

_____

# The Surface of the Moon: Should We Go Back?

On July 20, 1969, the first human being to set foot on an object beyond Earth hopped onto the dusty surface of the moon. His name was Neil Armstrong, a United States' astronaut who was part of the moon-exploring *Apollo* project.

During the next three years, other U.S. astronauts visited the moon, walked on its surface, rode over it in small carts, performed all sorts of scientific tests, and brought back samples of its soil and rocks. We learned much about the moon back then. So why go back again, as many scientists now argue?

In the late 1960s and early 1970s, we soared through space to explore the moon. Now, say those who propose a return of people to the moon, we should use the moon as a base from which to explore the rest of the universe.

For one thing, as a team of astronomers at New Mexico State University and the University of New Mexico wrote in the March 1990 issue of *Scientific American:* "The harsh, lifeless surface of the moon may well be the best place . . . for human beings to study the universe around them. The near absence of an atmosphere . . . the low levels of interference from light and radio waves . . . and the abundance of raw materials make the moon an ideal site for . . . astronomical observatories."

The moon base envisioned by scientists would be permanent. Engineers, scientists, technicians, laborers, and perhaps entire families would live in domed communities. These would be supplied with food from Earth, energy from the sun, and water and oxygen obtained from moon rocks. Raw materials, like aluminum and those used to make ceramics and special kinds of glass, would also be gotten from rocks mined on the moon. In moon factories, these raw materials would be turned into products, such as those used for constructing buildings and telescopes.

Hundreds of people would inhabit the moon for shifts ranging from a few months to a few years. Their tasks would vary depending on the projects undertaken by the nations of Earth, especially the United States.

One such project would be to build a super telescope with the tongue-twisting name of Lunar Optical-Ultraviolet-Infrared Synthesis Array, known more simply as LOUISA.

As you might guess from the word "array," LOUISA would not be a single telescope like the ones with which you may be familiar. On the contrary, it would be made up of two rings of telescopes: an outer ring of 33 "eyes" on the sky and an inner ring of nine more such devices.

The outer ring, 10 km in diameter, would be large enough to encircle an average-sized city on Earth. Data from all the telescopes would be transmitted to a kind of central clearinghouse for analysis.

How powerful would LOUISA be and what might it see? LOUISA would be 100,000 times more powerful than the most powerful telescope on Earth. To get an idea of what this means, scientists tell us that LOUISA could recognize a dime held in your hand here on Earth—384,000 km away. (Using your unaided eye, you probably couldn't tell a dime from a nickel if it lay on the floor across your room. Try it.)

LOUISA would, of course, not be aimed at Earth, but deep into space. Its 42 eyes would explore many unsolved mysteries of the universe. It would take a new look at star families called *galaxies*. It might give us a glimpse of streams of matter swirling into those most mysterious of objects—black holes. It would also hunt for "earthlike

planets orbiting nearby stars . . . the first step [in the search for] life elsewhere in our galaxy." In short, it would allow us to see the universe as we have never seen it before.

But a peek into outer space would not be the only reward we would get from a permanent moon base. The base would be a training ground and jumping off place for the human exploration of Mars, the "red planet" whose riddles have fascinated humans for years; riddles that may only be solvable by going there in person.

Unfortunately, setting up housekeeping on the moon has one important drawback. It could cost more than a hundred billion

dollars; money that many people feel might be better spent right here on Earth to improve our environment, our standard of living, and our health. Other people say we don't have to sacrifice one for the other. We can cut back, they say, on military spending, or on welfare programs, or on other kinds of government spending. We might even raise taxes to make more money available.

Which will it be? What do you think?

1. *Astronauts first walked on the moon in*

   a. 1968.
   b. 1969.
   c. 1970.
   d. 1971.

2. *LOUISA is the name of a(n):*

   a. astronaut.
   b. moon base.
   c. telescope.
   d. mission to Mars.

3. *A permanent moon base could be used as a jumping off place for the human exploration of*

   a. galaxies.
   b. black holes.
   c. stars.
   d. Mars.

4. *Explain why the moon might be "the best place for human beings to study the universe around them."*

   _____

   _____

   _____

5. *On a separate sheet of paper, express your views on whether government money should, or should not, be used to create a permanent moon base. Back up your views with facts and logical arguments.*

# The Solar System and the Universe

## LABORATORY INVESTIGATION

### CONSTRUCTING A TELESCOPE

**A.** You should have two convex lenses on your desk. Insert the thinner lens into a lens holder. Set the lens and the holder on a meterstick as shown in Fig. 20-1*a*. Aim the lens at a source of light, such as a light bulb, that is about 5 or 6 meters (m) away.

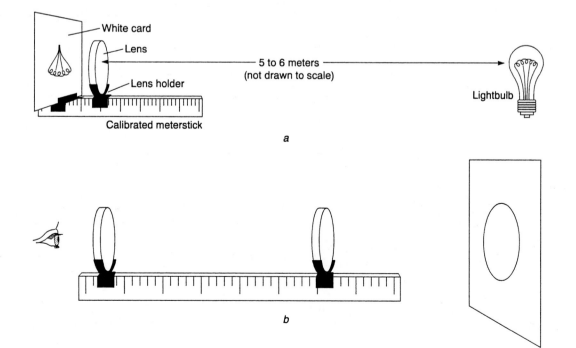

**Fig. 20-1.**

Insert a white card in another holder. Set the holder on the meterstick about 50 millimeters from the lens, with the lens between the white card and the light bulb. Slowly move the card away from the lens until a sharp image appears on the card.

**1.** The distance between the card and the thin lens is _____ centimeters (cm).

**B.** Remove the thin lens, and replace it with the thicker lens. Repeat step A with the thick lens.

**2.** The distance between the card and the thick lens is _____ cm.

**3.** The sum of the distances measured in steps 1 and 2 above is _____ cm.

**C.** See Fig. 20-1b. Set the thin lens on the meterstick so that the thin lens is closer to the source of light than the thicker lens. Set the thick lens a distance from the thinner lens equal to the distance you found in step 3 above. You have now made a simple telescope (which may need some adjustment).

**D.** Draw a small circle on the chalkboard. Without disturbing the positions of the lenses, carry the telescope to the opposite end of the room. Look through the thicker lens, and aim the telescope at the circle on the chalkboard. If you cannot see the circle clearly, move either lens slightly toward or away from the other lens. With a little adjustment, the circle will come into focus.

**E.** Measure the actual size of the small circle on the chalkboard. Note the large, measured circle your teacher has already drawn on the chalkboard. Compare the size of the small circle as seen through the telescope with the large circle on the chalkboard.

**4.** What is the magnifying power of your telescope? _____

**F.** Aim your telescope at a sheet of newspaper attached to the wall. Focus the telescope until you get a sharp image.

**5.** What do you see? _____

_____

**6.** Why does the position of a circle seen through a telescope look different from the position

of a letter of the alphabet seen through a telescope? _____

_____

**7.** Telescopes make distant objects appear closer. Explain why this is so.

_____

_____

# Study of the Solar System and the Universe

Looking into the night sky, you may have wondered, "What are the stars made of? How far away are they? Is there life on any other planets? How did the universe begin?" Ancient people could not answer these questions scientifically because they lacked the necessary instruments. Instead, they developed legends and myths to explain heavenly objects and the origin of the universe.

Many ancient civilizations worshiped the

sun, moon, and planets as gods. Long ago, people believed they could predict the future by studying the changing positions of these celestial objects. People also studied the movements of the sun and moon so that they could plant their crops and celebrate religious festivals at the proper times of the year.

## BEGINNINGS OF ASTRONOMY

The first **astronomers,** or scientists who studied the heavens, also observed the motions of the sun, moon, and planets. They noticed that some planets would disappear for several months and then reappear. Early astronomers also noticed that the planets appeared to slow down at times; some planets even seemed to move backward!

In 1543, the Polish astronomer *Nicolaus Copernicus* (1473–1543) claimed that Earth moved around the sun. Previously, most people believed that the sun moved around Earth. Planetary movements became better understood as people gradually accepted Copernicus's hypothesis that the sun instead of Earth was the center of the solar system.

## INVENTION OF THE TELESCOPE

An important date in astronomy was 1608. During that year, the *telescope* was invented. The invention and development of the telescope made it possible for people to observe thousands of new stars, as well as details of the moon and planets. In 1609, the Italian scientist *Galileo Galilei* (1564–1642) used a telescope to discover that the moon's surface was not smooth, but actually covered with craters and mountains. From his observations, Galileo also discovered that the planet Jupiter had several moons.

**Types of Telescopes.** The telescope that Galileo used was a tube with two convex glass lenses inside. A telescope that uses only convex lenses is called a *refracting telescope.* In a simple refracting telescope, light from a distant object passes through two lenses so that the distant object appears much larger and closer to the observer.

The first refracting telescopes had many defects. This led to the invention of *reflecting telescopes* by the English scientist *Sir Isaac Newton* (1642–1727). In a reflecting telescope, light from a distant object is reflected from a curved mirror into a lens. Because they can be made much larger than refractors, the reflectors can

gather light more efficiently, thus making more objects visible to an observer.

By making objects in outer space appear larger and brighter, refracting and reflecting telescopes have helped astronomers make important discoveries. During the past century, cameras and other devices attached to telescopes have helped astronomers make discoveries that form the basis of modern astronomy. And during the past half century, the use of powerful *radio telescopes* has greatly extended our knowledge of the universe. (See section on radio telescopes, below.)

**Refracting Telescopes.** In your laboratory investigation, you made a simple refracting telescope by placing two convex lenses in a straight line. The thicker lens was closest to your eye. By lining this lens up with the other lens and moving it back and forth, you found that a distant object came into view. The object appeared larger because the thicker lens magnified the sharp image of it that formed where the incoming light rays crossed. In a **refracting telescope,** light rays from a distant object (such as a planet or star) are first collected by an *object lens,* or *objective,* located at the far end of the tube. The objective then bends and concentrates light rays from the distant object to a single point inside the telescope. This point, where the light rays cross, is called the *focal point.* The image at the focal point is viewed through the *eyepiece lens,* which functions as a magnifying glass.

If the eyepiece is focused correctly on the focal point, an astronomer will be able to see a bright and larger image of a distant object such as the moon, another planet, or a distant star. Fig. 20-2 shows a model of a simple refracting telescope.

In a refracting telescope, the diameter of the objective lens is limited. If the lens is too large, the weight of the glass causes the lens to sag. This produces a blurred image. The world's largest refracting telescope is located at the Yerkes Observatory in Wisconsin. Its objective lens is 101 cm in diameter.

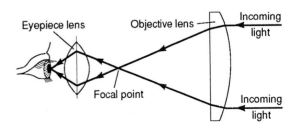

**Fig. 20-2. Refracting telescope.**

**Reflecting Telescopes.** In a **reflecting telescope,** light rays from a star are collected by one or more curved, or concave, mirrors. The two main kinds of reflecting telescopes are *single-mirror reflectors* and *multiple-mirror reflectors.*

In a single-mirror reflector, a large concave mirror is located at the far end of the telescope tube. The mirror is curved so that all of the incoming light rays are reflected toward a single focal point. A flat, smaller mirror inside the telescope then reflects the converging light rays to an eyepiece lens, which magnifies the image. The eyepiece can be mounted either at the side of the telescope or behind a hole in the center of the mirror (see Fig. 20-3). In a multiple-mirror reflector, several smaller concave mirrors are arranged in a ring. An image is formed when the light rays from each of the mirrors are combined and focused on a single focal point.

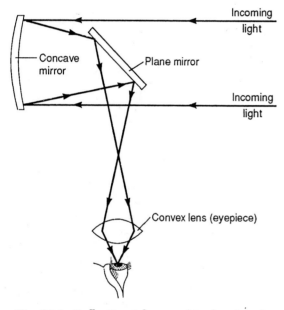

**Fig. 20-3. Reflecting telescope (single mirror).**

Large reflecting telescopes, especially multiple-mirror reflectors, are much easier to build than large refracting telescopes because it is simpler and less expensive to make a large, curved mirror than to make a convex lens from a huge piece of glass. In addition, the end of a reflecting telescope's tube supports the mirror's weight, which prevents the glass inside the mirror from sagging.

The largest and best-known single-mirror reflecting telescope in the United States is the *Hale telescope* on Mount Palomar in California. The diameter of its mirror is about 5 m. Other large single-mirror reflectors are located on Kitt Peak in Arizona and Cerro Tololo in Chile. Both have mirrors about 4 m in diameter.

One of the world's largest telescopes (the *Keck multiple-mirror telescope* in Hawaii) was completed in 1992. This multiple-mirror telescope has a combined power of a single-mirror reflector with a diameter of 10 m.

The Hubble Space Telescope, which is in orbit around Earth, is a single reflector with a 2.4 m mirror that can observe objects seven times farther and 50 times fainter than can be observed with any light-gathering telescope on Earth.

The next advance in space telescopes—the Webb—is being planned for launching in 2010. This telescope will have a light-gathering area six times larger than Hubble's. The Webb telescope should be able to detect light a 100 or more times fainter than that which the Hubble can see, hopefully enabling astronomers to analyze light from the theoretical beginning of the universe, about 14 billion years ago.

**Radio Telescopes.** Both refracting and reflecting telescopes are *optical telescopes* because they require visible light. A telescope that collects invisible radio waves is called a **radio telescope.** A radio telescope is a special receiver that collects radio waves generated by the sun and distant stars.

Most radio telescopes have huge dish-shaped collecting antennas (see Fig. 20-4). Like the curved mirror in a reflecting telescope, these collecting dishes gather and focus the incoming radio waves. The radio waves are then sent to a receiver, which converts them into electrical signals. The electrical signals are magnified and then changed into an image on an *oscilloscope* (which resembles a television screen). By studying radio waves with a radio telescope, astronomers have discovered information about the universe that they could not have learned with optical telescopes.

Radio telescopes have several advantages over optical telescopes. For example, radio signals can penetrate dust and gases in outer space, as well as clouds in Earth's atmosphere, which block the path of light rays. Radio telescopes can receive signals during the day as well as at night, whereas optical telescopes are ordinarily used at night. Also, radio telescopes can detect objects much farther away than the most distant objects that can be seen with optical telescopes.

The world's largest dish-type radio telescope is located in Arecibo, Puerto Rico. The collecting dish antenna of this telescope is more than 300 m in diameter. It cannot be

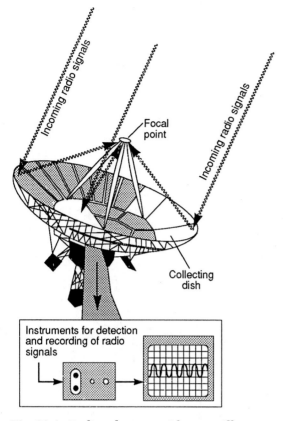

**Fig. 20-4. Radio telescope with an oscilloscope.**

than the eyes of an astronomer. For example, a camera equipped with special "slow" film can photograph stars in which the exposures last several hours. When developed, the film shows details and movements of stars that could not be seen by someone looking through a telescope.

***Spectroscope.*** One of the most useful telescope attachments is the **spectroscope**. A spectroscope is a device that separates white light into different colors. (Each color represents a different wavelength.) In one kind of spectroscope, a triangular-shaped piece of solid glass, or *prism*, bends and separates a beam of white light into its different wavelengths (see Fig. 20-5). The series of colors produced by a spectroscope is called a **spectrum**.

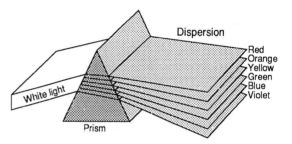

**Fig. 20-5. A prism separates white light into the colors of the spectrum.**

turned to aim at various parts of the sky. Another large radio telescope is located at Jodrell Bank, England. Its collecting dish is more than 75 m in diameter, and it is mounted on a machine that can be turned to aim the antenna at different parts of the sky.

Radio telescopes can be arranged in groups, or *arrays*. The Very Large Array (VLA) in Socorro, New Mexico, has 27 collecting dishes, each with a diameter of 25 m. Together, the VLA functions like a single collecting dish with a diameter of about 35 kilometers (km).

**Telescope Attachments.** Astronomers rarely look at objects in space directly through a telescope. Today, instruments attached to a telescope can collect much more data and details

The way in which a spectrum is produced by a glass prism is similar to the way in which water droplets produce a rainbow. When sunlight (white light) passes through water droplets in Earth's atmosphere, the water bends and spreads out the light into a rainbow, or the spectrum of colors that make up sunlight.

Astronomers often use spectroscopes to observe spectra of the sun and distant stars. Because each element has a different spectrum, astronomers can identify elements in a star's outer surface by studying its spectrum. Spectra also indicate the temperature, pressure, magnetic field of gases, and even the movement of stars toward or away from Earth.

# The Solar System

The sun and all of the natural objects in space (heavenly bodies) that orbit the sun make up the **solar system.** The gravitational attraction of the sun maintains the orbits of these heavenly bodies, which include Earth, the planets, asteroids, meteoroids, comets, and the dust and gases in the space between these larger bodies.

# THE SUN

In relation to Earth, the sun is the most important heavenly body in the universe. Besides keeping Earth in its orbit, the sun also makes life possible on Earth. By supplying the radiant energy necessary for plants to carry out photosynthesis, the sun indirectly produces food and oxygen for all animals.

The sun's radiant energy also causes water to evaporate from oceans, lakes, and rivers. Evaporated water forms clouds, from which rain and snow fall to Earth. Rain and snow change the landscape (as you learned in Chapter 4), and supply the water that plants and animals require for their life processes. Even the energy stored in coal and oil comes from prehistoric organisms that obtained energy from the sun.

**Size of the Sun.** The sun is a star. Although astronomers do not consider the sun a large star, it is huge compared to Earth. The sun's diameter is about 1,390,000 km, about 110 times the diameter of Earth (see Fig. 20-6). If the sun were a hollow ball, it could hold more than one million Earths.

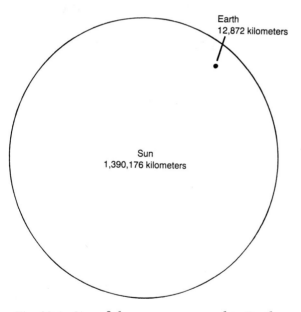

**Fig. 20-6. Size of the sun as compared to Earth.**

**Composition of the Sun.** You recall that astronomers can determine a star's composition with a spectroscope. By studying the sun with a spectroscope, astronomers have discovered that the spectral lines in the sun's spectrum match many of the spectral lines produced by elements on Earth. It has been determined that the sun consists of about 82% hydrogen

and 17% helium. The sun also contains tiny quantities of about 60 other elements, such as carbon, oxygen, and calcium.

Astronomers discovered helium on the sun in 1868, many years before they found it on Earth. Astronomers found that certain spectral lines in the sun's spectrum did not match the spectral lines of any element known at that time on Earth. They named this new element *helium,* from *helios,* a Greek word meaning "sun."

**Structure of the Sun.** At sunset, the sun often looks like a smooth, round disk with a sharply defined edge. The sun's surface and surrounding atmosphere are actually a turbulent mass of gases, which sometimes flares far out into space. The visible part of the sun, or its atmosphere, consists of the *photosphere, chromosphere,* and *corona* (see Fig. 20-7).

The innermost part of the sun's atmosphere, called the **photosphere,** is a layer of hydrogen gas about 500 km thick. Temperature in the photosphere is about 6000°C, causing its hydrogen atoms to glow brightly. As a result, the photosphere looks like a dazzling, bright yellow disk extending from the sun's surface.

Photographs of the sun taken through special filters show dark areas on the photosphere. These areas are called **sunspots.** Sunspots usually occur in pairs. A single sunspot may range from 100 to 80,000 km in diameter. A very large sunspot pair may cover an area of about 150,000 km in diameter. Sunspots look dark because they are more than 1000°C cooler than the surrounding photosphere.

Sunspots constantly appear and disappear on the photosphere. During a period of high sunspot activity, 100 or more sunspots can be seen on the photosphere. During periods of low sunspot activity, sunspots may not appear for several days. Sunspot activity occurs in a cycle, averaging about 11 years from one period of high sunspot activity to the next.

Most sunspots are visible for less than one day, but some sunspots have been observed for more than three months. Sunspots move across the sun's surface from left to right. This indicates that the sun is rotating on an axis. In fact, the sun makes one complete rotation on its axis about every 25 days at its equator and 27 days at its poles.

Occasionally, hot gases erupt from a sunspot, producing a **solar flare.** During a solar flare, electrically charged particles (ions) and radiation are ejected from the sun at high speeds. These electrically charged particles and radiations are known as **solar wind.** During periods of high sunspot activity, the solar wind produces electromagnetic disturbances

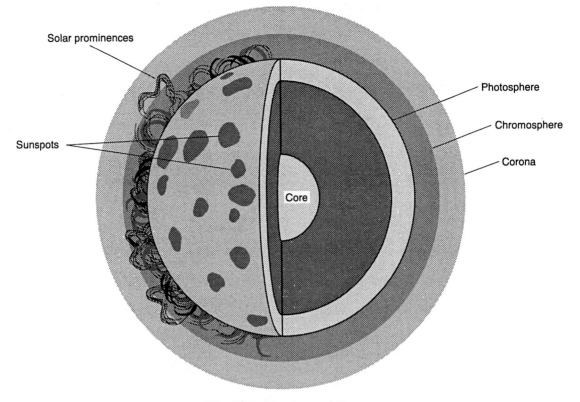

**Fig. 20-7. Structure of the sun.**

in Earth's atmosphere. These disturbances may disrupt long-distance telephone reception as well as cause static in radio receivers.

The electrically charged particles of the solar wind also excite, or *ionize*, gases in Earth's upper atmosphere. Consequently, the ionized gases glow with various colors (refer to the discussion of auroras in Chapter 13). The ionization of oxygen atoms in the upper atmosphere produces red, orange, and green light. A blue or violet light is produced when nitrogen atoms are ionized.

Extending thousands of kilometers outward from the photosphere is a region of glowing gas called the **chromosphere.** The red color of the chromosphere is caused by the ionization of hydrogen atoms.

Surrounding the chromosphere is a layer of very thin gas called the **corona.** The corona extends millions of kilometers beyond the chromosphere. At the beginning and end of a total solar eclipse, the corona appears as a faint halo of pearly light around the sun. Astronomers use a special instrument, called a *coronagraph,* to study the sun's corona. A coronagraph is a telescope with an opaque disk that blocks the brightness of the photosphere. This allows astronomers to observe the corona.

With a coronagraph, astronomers have ob-served streamers of hot gases arching thousands of kilometers above the photosphere. These streamers of gas are called *solar prominences.* Solar prominences may last for several hours and extend more than one million kilometers above the photosphere. Unlike solar flares, solar prominences do not disrupt communications on Earth.

**The Source of the Sun's Energy.** The energy that causes sunspots, solar flares, and solar prominences is related to the sun's composition and forces operating within the sun's core. Near the sun's core, hydrogen atoms are subjected to great heat and pressure. For example, the temperature near the sun's core is about 15 million °C. Under this tremendous heat and pressure, hydrogen atoms in the sun's core are changed into helium atoms. This process is called **thermonuclear fusion.**

During thermonuclear fusion, four hydrogen atoms are changed into one helium atom. In the process, some of the mass of the hydrogen atoms is converted into energy (see Fig. 20-8). The conversion of matter into energy takes place according to physicist *Albert Einstein's* (1879–1955) famous equation:

$$E = mc^2$$

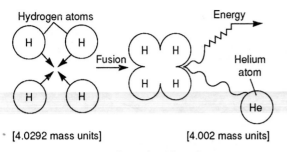

[4.0292 mass units]          [4.002 mass units]

**Fig. 20-8. Thermonuclear fusion.**

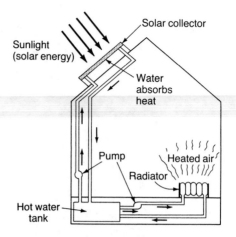

**Home with Solar Heating System**

**Fig. 20-9. Solar energy can be used to heat a home.**

In this equation, $E$ represents energy, $m$ is the mass of the matter that is converted into energy, and $c^2$ is the speed of light squared.

According to Einstein's equation, an enormous amount of energy is released when matter is converted into energy. For example, if the matter in three aspirin tablets, weighing one gram altogether, could be converted into energy, it would heat all of the water in a large swimming pool to the boiling point.

As a result of thermonuclear fusion, the sun converts about four million tons of matter into energy each second. Although the sun loses millions of tons of matter every second, its mass is so great that it can continue generating energy at its current rate for at least the next five billion years.

Only about 0.000000002 of the sun's energy reaches Earth. Fortunately, about one third of this energy is absorbed by the atmosphere before it reaches Earth's surface (refer to the discussion of the ozone layer in Chapter 13). If much of the sun's ultraviolet radiations were not absorbed by ozone in the upper atmosphere, most organisms would probably not have been able to evolve on Earth.

**Using the Sun's Energy.**   As you have learned, coal, oil, and food are indirect benefits of solar radiation. Scientists have also constructed devices that produce heat by using mirrors to concentrate sunlight. The heat produced by the concentrated sunlight can then be used to change water into steam. The steam, in turn, is used to turn turbines that generate electricity. As you can see in Fig. 20-9, *solar energy* also can be used to heat homes.

In addition, sunlight is used to provide electricity for artificial satellites. The artificial satellites are equipped with flat plates, or panels, that are made of light-sensitive substances. When sunlight strikes these *solar panels*, an electric current is produced. A number of solar panels joined together is called a *solar array.*

An experimental electrical car called *Sun-raycer* has been designed that uses solar energy. A solar array is located on the car's roof. Sunlight striking the car's roof produces enough electricity to power the car. Other solar-powered objects include telephones, radios, lights, and battery chargers.

## THE PLANETS

A large, solid body that shines by reflected sunlight and revolves in a stable orbit around the sun is called a **planet.** Long ago, people noticed that of all the stars in the sky, there were five bodies that continually changed their positions in relation to the others. The ancient Greeks used the word *planet*, from a Greek word that means "wanderer," to describe these bodies.

**Planetary Motion.**   From Earth, the stars, planets, and the sun appear to move across the sky from east to west each night. That is because of Earth's rotation. Each night, however, the stars appear in the same positions in relation to each other.

As time passes, the planets change position against the background of unmoving, or fixed, stars. The direction in which a planet moves does not remain constant. Over several weeks or months, a planet may slow down, stop, or even reverse its direction relative to the fixed stars. The reversed movement of a planet is called *retrograde motion.*

For more than a thousand years, people accepted the Greek astronomer *Ptolemy*'s explanation of planetary motion. According to Ptolemy (about A.D. 140), planets moved in great circles and smaller circles around Earth. The smaller circles were responsible for the back-and-forth planetary movements.

In 1543, the Polish astronomer Copernicus first claimed that the planets, including Earth, orbited the sun. The current theory of planetary motion was explained by the German astronomer and mathematician *Johannes Kepler* (1571–1630), in 1609.

**Kepler's Laws of Motion.** Kepler's explanation of planetary motion consists of three laws. These laws state that:

1. Planets travel around the sun in elongated circles, or *elliptical* orbits.
2. As a planet moves closer to the sun, its speed increases; as the planet moves farther from the sun, its speed decreases.
3. The time a planet takes to complete one revolution around the sun depends on the planet's distance from the sun.

Like Earth, all planets move in two ways. A planet (1) rotates around its axis and (2) revolves in an orbit around the sun (refer back to Fig. 19-5). The time a planet takes to complete one rotation is called its *period of rotation*. Earth's period of rotation is nearly 24 hours (one day). The time a planet takes to complete one revolution around the sun is called its *period of revolution*. Earth completes one revolution in its orbit about every $365\frac{1}{4}$ days.

**Characteristics of the Planets.** Astronomers classify planets into two groups based on their distance from the sun: the inner planets and the outer planets (see Fig. 20-10). The four inner planets are Mercury, Venus, Earth, and Mars. The five outer planets are Jupiter, Saturn, Uranus, Neptune, and Pluto. Excluding Mercury and Venus, all of the planets are or-

bited by natural satellites, or *moons* (see Table 20-1).

**Inner Planets.** The inner planets are alike in several ways. All of the inner planets consist mainly of rock and iron-nickel compounds. All have nearly the same density, about four to five times the density of water. Finally, none of the inner planets are more than 13,000 km in diameter. The smallest of the inner planets, Mercury, is about 5000 km in diameter. The largest of the inner planets, Earth, is slightly less than 13,000 km in diameter.

*Mercury* is the planet closest to the sun and is named after the Roman god who was the messenger of the other gods. The name is appropriate because Mercury orbits the sun faster than any of the other planets in the solar system.

Like our moon, Mercury does not have an atmosphere, and its surface has many impact craters. Some of the impact craters on Mercury are more than 1000 km in diameter.

Until 1965, Mercury's period of rotation was thought to be about 88 days, the same as its period of revolution around the sun. In 1965, astronomers using the large radio telescope at Arecibo, Puerto Rico, discovered that Mercury's period of rotation is actually about 58 days. This slow rate as well as Mercury's nearness to the sun cause a daytime temperature of more than 400°C. At night, on Mercury the heat quickly radiates into space, dropping temperatures to almost −200°C.

*Venus* is named after the Roman goddess of love. From Earth, Venus is the brightest heavenly object in the night sky besides the moon. Because Venus is located between Earth and the sun, Venus can be seen only for a few hours or so around sunrise or sunset. As a result, many people mistakenly refer to Venus as the "morning star" or "evening star."

Venus has a thick atmosphere, consisting of about 96 percent carbon dioxide. The atmo-

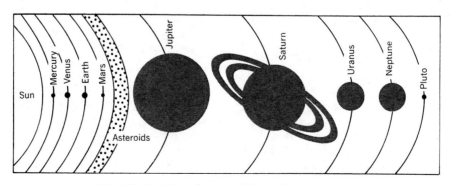

**Fig. 20-10. The planets of the solar system.**

**TABLE 20-1. PLANETS OF THE SOLAR SYSTEM: MAIN CHARACTERISTICS**

| Name | Average Distance From the Sun (millions of km) | Diameter (km) | Period of Revolution | Period of Rotation | Number of Moons |
|---|---|---|---|---|---|
| Mercury | 58 | 4880 | 88 days | 58 days | 0 |
| Venus | 108 | 12,104 | 225 days | 243 days | 0 |
| Earth | 150 | 12,756 | 365 days | 24 hours | 1 |
| Mars | 228 | 6794 | 1.9 years | 24.5 hours | 2 |
| Jupiter | 778 | 142,700 | 11.9 years | 10 hours | 16 |
| Saturn | 1427 | 120,000 | 29 years | 10.5 hours | 23 |
| Uranus | 2869 | 50,800 | 84 years | 17 hours | 15 |
| Neptune | 4486 | 48,600 | 165 years | 18.5 hours | 2 |
| Pluto | 5890 | 4000 | 248 years | 9.5 hours | 1 |

spheric carbon dioxide traps heat radiating from the surface of Venus. This is similar to the greenhouse effect that causes Earth's atmosphere to retain heat. As a result, the surface temperature of Venus is more than 500°C.

Because of the permanent cloud cover surrounding Venus, there are no photographs of the planet's surface. However, orbiting spacecraft using radar have detected surface features that include volcanoes, large impact craters, ridges, valleys, and high plateaus.

*Earth* is the third planet from the sun. The planet Earth was already discussed in Chapter 19.

*Mars* is the fourth planet from the sun and is named after the Roman god of war. Because Mars is farther from the sun than Earth, it is colder. At night, the temperature may drop to −125°C. The thin Martian atmosphere is about 95% carbon dioxide, 5% nitrogen and argon, with traces of other gases.

The surface of Mars features polar ice caps, impact craters, mountain ranges, several inactive volcanoes, deep canyons, and channels that emerge from both of the ice caps. The largest known volcano in the solar system (*Olympus Mons*) is located in the northern hemisphere of Mars. This huge volcano is 25 km high and has a diameter of about 600 km. Although there is no evidence of running water on the surface of Mars, it is thought that the deep canyons were formed by running water at some time in the past.

The northern polar ice cap of Mars consists mainly of frozen water, whereas the southern ice cap is predominantly frozen carbon dioxide. The reddish-brown color and large fields of sand dunes indicate that most of the Martian surface is probably a desertlike environ-

ment. Desert regions on Earth also appear reddish brown when viewed from high altitudes. To date, instruments on orbiting spacecraft and on the Martian surface have shown no signs of life.

**Asteroids.** A belt of thousands of rocky fragments is located between the orbits of Mars and Jupiter (the first of the outer planets). These rocky fragments are called **asteroids.** Astronomers think that some asteroids are leftover pieces from the birth of the solar system that never combined to form a planet (see again Fig. 20-10).

Asteroids revolve around the sun in circular orbits and in the same direction as the other planets. Some large asteroids have even been named. For example, the largest asteroid, about 1000 km in diameter, is named *Ceres.* Most asteroids, however, have a diameter of only about one km.

**Outer Planets.** With the exception of Pluto, the outer planets have several characteristics in common. The average density of the outer planets is about the same as water, so that they are about four times less dense than the inner planets. The outer planets consist mainly of frozen hydrogen and helium gas, which explains their low densities. Their atmospheres consist of methane and ammonia clouds that are thousands of kilometers thick. The outer planets are also much larger than the inner planets (see again Fig. 20-10). For example, the diameter of Jupiter is about 140,000 km, more than 11 times Earth's diameter.

*Jupiter* is named after the chief Roman god and is the largest planet in the solar system.

In fact, Jupiter has twice the total mass of all other planets combined. The most striking feature on Jupiter is the Great Red Spot, which is about as large in surface area as Earth. Astronomers think that the Great Red Spot, which is about eight km high, is a storm or other disturbance in Jupiter's atmosphere.

Photographs of Jupiter from spacecraft have shown that its gaseous surface consists of alternating light- and dark-colored bands. These bands run parallel to Jupiter's equator. Astronomers think that the dark bands are regions of sinking gas, whereas the lighter-colored bands are regions of rising gas.

*Saturn* is named after the Roman god of agriculture and is the most distant planet that can be seen without a telescope. Like Jupiter, Saturn's surface has light and dark bands that are paralleled to its equator. Viewed through a telescope, Saturn is the most beautiful of all the planets because of three concentric (one inside the other) rings that encircle the planet's equator. Actually, these rings consist of about 1000 tiny ringlets of dust and frozen gas particles.

*Uranus* is named after the Greek god of the heavens. Until 1781, astronomers thought that Uranus was a star. In that year, the British astronomer *Sir William Herschel* (1738–1822) noticed that the position of Uranus changed among the stars. Thus, Herschel concluded that Uranus was a planet.

Uranus is about 19 times farther from the sun than Earth is. As a result, the surface temperature of Uranus is about −200°C. However, recent exploration of the planet's atmosphere by spacecraft has revealed a surprising discovery. The side of Uranus facing away from the sun is not cooler than the side facing the sun. Consequently, astronomers think that there must be strong atmospheric winds circling Uranus, which distribute its heat.

*Neptune* was named after the Roman god of the sea because of its blue color. The blue color is caused by methane in Neptune's atmosphere. Neptune was discovered as a result of careful measurements made of the orbit of Uranus. Measurements showed that Uranus did not follow the orbit it should according to the laws of planetary motion discovered by Newton and Kepler. Therefore, astronomers concluded that there must be a planet beyond Uranus that was disturbing its orbit. So, based upon mathematical calculations, Neptune was discovered in 1846.

*Pluto* is named after the Roman god of the underworld and is the smallest planet in the solar system. Because it is so small and far from Earth, Pluto was not discovered until 1930. Although it is usually the farthest planet from the sun, part of Pluto's orbit crosses inside the orbit of Neptune. The tiny, frozen planet is thought to have a small, rocky core surrounded by a thick mantle of ice water. Recently, astronomers have realized that Pluto is one of many small, planet-like bodies in the outer solar system. This region, called the Kuiper Belt, may be the source of some comets. In May 2002, an object about half the size of Pluto was discovered farther out in the Kuiper Belt.

## COMETS

*Comets* have been called "dirty snowballs" and "dirty snowdrifts." That is because a **comet** is an accumulation of dust particles and frozen gases. As a comet approaches the sun, the sun's heat causes the comet's icy surface to evaporate. As the comet evaporates, escaping dust particles and gases glow brightly. The glowing dust and gases form a "tail" that streams away from the comet.

The solar wind exerts a force that pushes the comet's tail away from its head. Consequently, a comet's tail always points away from the sun. Like Earth and the other planets, a comet travels around the sun in an elliptical orbit.

The best-known comet is Halley's Comet, which has a well-known orbit (see Fig. 20-11). It makes one revolution of the sun every 76 years. Halley's comet was first observed in 1758, and has returned in 1835, 1910, and 1986. During its 1986 appearance, Halley's comet did not appear as bright as in previous observations. Perhaps this is because most of its icy surface has evaporated into space.

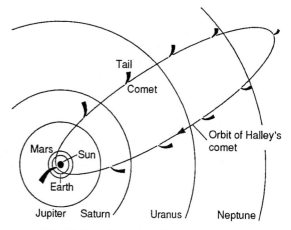

**Fig. 20-11. The orbit of Halley's comet.**

**Fig. 20-12. Meteor Crater in Arizona.**

## METEOROIDS, METEORS, AND METEORITES

The solar system also contains countless particles of rock and metal called **meteoroids.** Most meteoroids range in size from a grain of sand to a large boulder; they are always smaller than asteroids, that is, less than one km in diameter. Meteoroids may actually be pieces of asteroids that collided with each other.

When a meteoroid enters Earth's atmosphere, the meteoroid passes mainly through molecules of gaseous nitrogen and oxygen. Friction between the meteoroid and these atmospheric gases causes the meteoroid to heat up rapidly and begin to burn.

The streak of light produced by a burning meteoroid is called a **meteor** (or *shooting star*). A meteor's brightness is related to the size of the particular meteoroid entering Earth's atmosphere. For example, a large meteoroid produces a bright meteor upon entering the atmosphere. Meteoroids usually burn up completely, high in the atmosphere. Occasionally, a meteoroid is so large that it strikes Earth's surface before being completely vaporized.

A meteoroid that passes through the atmosphere and strikes Earth's surface is called a **meteorite.** A meteoroid must weigh more than 1000 kilograms to survive passage through Earth's atmosphere and become a meteorite. It has been estimated that meteorites add more than 900 metric tons to Earth's weight daily.

An example of a huge impact crater caused by a meteorite is *Meteor Crater* (also called Barringer Crater) in Arizona. Meteor Crater,

formed about 50,000 years ago, is a circular depression more than one km wide and 175 m deep (see Fig. 20-12). According to scientists, the meteorite, less than 100 m in diameter, exploded with great force on impact. Small, iron-rich meteorites were scattered beyond the crater for as far as 500 km.

Most scientists think that even larger meteorites have struck Earth's surface in the past. Two other craters possibly caused by meteorites were discovered in Canada, on the eastern shore of Hudson Bay. One of the craters has a diameter of about 30 km and the other of about 20 km. Some scientists theorize that an enormous meteorite struck Earth near the Yucatán about 65 million years ago, which may have led to the extinction of the dinosaurs. The largest impact in recent years occurred in Russia in 1908 when a meteorite exploded in Siberia, leaving no crater but devastating several hundred square kilometers of forest.

## ORIGIN OF THE SOLAR SYSTEM

Using telescopes, spectroscopes, and other devices, astronomers have collected a great deal of information about the solar system. However, the origin of the solar system is still a mystery among astronomers. The two most widely held hypotheses on the origin of the solar system are called the *tidal hypothesis* and the *condensation hypothesis*.

According to the tidal hypothesis, the solar system was produced by a near collision between the sun and a much larger star (see Fig. 20-13). As the sun and the larger star passed

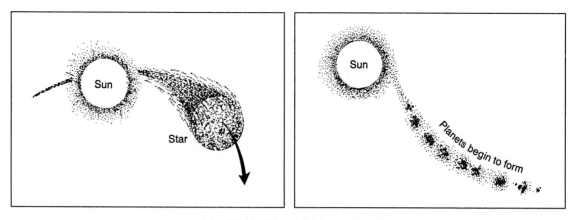

Fig. 20-13. The tidal hypothesis.

each other, the gravitational attraction of the passing star caused an enormous mass of hot gases to be drawn away from the sun. This mass of hot gases began to orbit the sun, gradually cooling and condensing into nine planets. Today, few astronomers support this hypothesis.

According to the more widely accepted condensation hypothesis, the solar system formed from a rotating cloud of dust and gas (see Fig. 20-14). The dust and gases in the cloud's center condensed to form the sun, while the smaller parts of the cloud condensed to form the planets and other objects in the solar system. Astronomers estimate that about 99 percent of the original cloud became the sun. The remaining one percent condensed into planets, moons, asteroids, comets, and meteoroids.

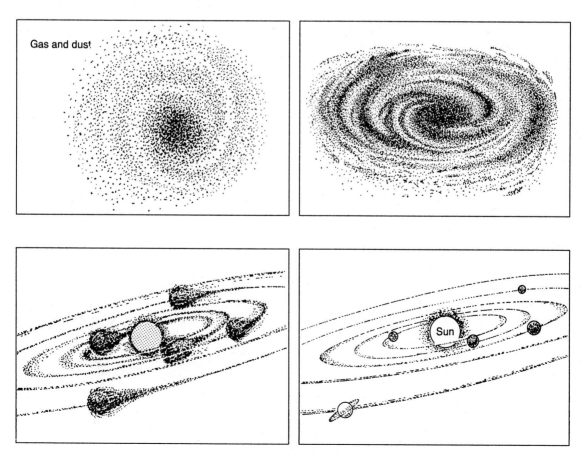

Fig. 20-14. The condensation hypothesis.

# The Universe

Astronomers refer to all of the matter and energy in space as the **universe.** The universe includes the solar system, the stars, and all of the other objects in space. Since Galileo first explored the sky with a simple refracting telescope more than 350 years ago, astronomers have learned a great deal about the size and composition of the universe.

## SIZE OF THE UNIVERSE

The universe is so vast that units of distance, such as kilometers, are inadequate to describe its extent. As a result, astronomers use an extremely large unit of distance called a *light-year* to help them describe distances between Earth and the rest of the universe. A **light-year** is the distance that light travels in one year, or about 10 trillion km.

For example, the star *Alpha Centauri* is 43 trillion km from Earth, or 4.3 light-years away. The sun is 150 million km from Earth, slightly more than eight light-minutes away. Thus, it takes sunlight about eight minutes to travel from the sun to Earth.

There are stars in the universe that are many billions of light-years away from Earth. The light you see when looking at one of these distant stars has been traveling through space for billions of years!

## COMPOSITION OF THE UNIVERSE

In addition to individual stars, early astronomers observed hazy-looking regions in the night sky. These hazy regions were called *nebulae*, from a Latin word meaning clouds. In time, astronomers using telescopes discovered that many of the hazy objects were actually enormous groups of stars.

Today, astronomers refer to an enormous collection of stars as a **galaxy.** (The term **nebula** now means a cloud of dust and gas in space that sometimes glows by reflected starlight.) The two galaxies closest to Earth are the "Magellanic Clouds" (100,000 light-years away) and the Andromeda Galaxy (2,000,000 light-years away).

## EARTH'S GALAXY: THE MILKY WAY

The sun is a star within the **Milky Way** galaxy, which is about 100,000 light-years in diameter and about 10,000 light-years thick (see Fig. 20-15a). Our galaxy is shaped like a flattened spiral and consists of billions of stars. From a distant galaxy, the Milky Way would look like a huge disk with long, thin arms spiraling outward (see Fig. 20-15b).

There are billions of stars in the Milky Way galaxy. Yet, there is an enormous amount of space between individual stars. For example, a spacecraft launched from Earth would have to travel at the speed of light for about 300 years just to reach the North Star.

The space between stars and galaxies is not entirely empty. There are dust particles and molecules of light gases, especially hydrogen, that escaped from inner planets after they formed. However, the dust particles and gas molecules are so far apart that astronomers consider space almost a total vacuum.

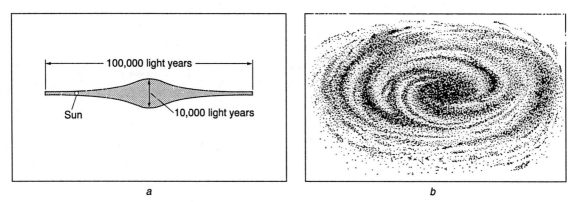

*a*          *b*

**Fig. 20-15. The Milky Way galaxy.**

## CONSTELLATIONS

When ancient people looked at the night sky, they observed that groups of stars formed patterns. Some of these patterns were named for animals and humans, such as the swan, the bear, and the hunter. People still use the ancient names when referring to these groups of stars, which astronomers call **constellations.**

In the Northern Hemisphere, two familiar constellations are *Ursa Major* (big bear) and *Ursa Minor* (little bear). The *Big Dipper* is a smaller constellation within Ursa Major, and the *Little Dipper* is a smaller constellation inside Ursa Minor. Both are referred to as *circumpolar constellations* because their positions make them visible throughout the year, and they appear to revolve around *Polaris* (the North Star). Polaris is located almost directly above the North Pole, forming the first star in the handle of the Little Dipper (see Fig. 20-16).

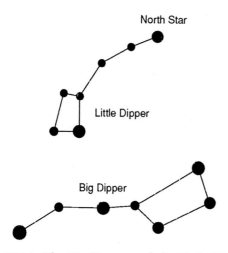

Fig. 20-16. **The Big Dipper and the Little Dipper are circumpolar constellations.**

## STARS

To many people, all stars appear similar. Stars, however, differ in size, brightness, color, composition, and age.

**Sizes of Stars.** Stars range in diameter from less than 100 km to several billion km. The smallest stars are called *dwarfs*, and the largest stars are called *supergiants*.

The main difference between a dwarf and a supergiant is density. Dwarfs are very dense. The matter in a dwarf is so compressed that a cubic centimeter of a dwarf weighs more than one ton. In contrast, a supergiant is made of matter that is spread so far apart, it is thinner than our atmosphere. Consequently, the density of a supergiant is much less than the density of a dwarf.

**Magnitudes of Stars.** Ancient Greek astronomers classified stars numerically according to their brightness. The brightest stars that could be seen from Earth were called *first-magnitude stars. Sixth-magnitude stars* were the faintest stars that could be seen on Earth.

Astronomers still use this ancient Greek system when comparing the brightness of stars by *apparent magnitude,* or how bright a star appears from Earth. To an observer on Earth, a first-magnitude star is 100 times brighter than a sixth-magnitude star. With the latest reflecting telescopes, astronomers can now see stars that are one million times dimmer than a sixth-magnitude star.

The apparent magnitudes of stars brighter than first magnitude are expressed as values less than 1.0. Some stars have apparent magnitudes of less than zero. For example, the brightest star in the night sky, Sirius, has an apparent magnitude of nearly $-1.5$. Apparent magnitudes are also used to describe the brightness of planets and moons. The apparent magnitudes of the sun, full moon, Venus, and various stars are given in Table 20-2.

**Temperature of Stars.** If you look at stars on a clear night, you will observe that they appear either red, yellow, or blue. The color of a star depends on its surface temperature. Red stars have a surface temperature of about 3000°C, whereas blue stars may have a surface temperature of more than 30,000°C. Yellow stars, like the sun, have a surface temperature of about 5500°C.

**Composition of Stars.** Astronomers using spectroscopes have discovered that many of the elements found on Earth are also present in stars. Stars consist mainly of hydrogen and helium, with traces of other elements. For ex-

### TABLE 20-2. APPARENT MAGNITUDES OF SOME CELESTIAL BODIES

| Celestial Body | Apparent Magnitude |
| --- | --- |
| Sun | $-26.5$ |
| Full moon | $-12.5$ |
| Venus (at brightest) | $-4.0$ |
| Sirius | $-1.5$ |
| Canopus | $-0.7$ |
| Alpha Centauri | $-0.3$ |
| Altair | $1.0$ |
| Polaris (North Star) | $2.0$ |

ample, the sun is made of 82% hydrogen and 17% helium. The other one percent of the sun's mass is composed of many other elements.

## DISTANCES TO STARS

A procedure astronomers often use to determine a star's distance from Earth is called the *parallax method*. To demonstrate the parallax method, hold a pencil at arm's length, and then observe the pencil by alternately opening and closing each eye. Repeat the same procedure with the pencil held at half an arm's length. You will notice that the farther away you hold the pencil, the smaller is the distance the pencil shifts against the background. The closer the pencil, the greater the shift (see Fig. 20-17).

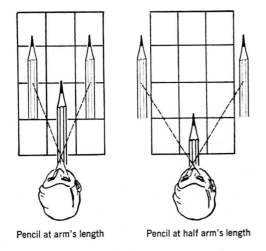

Pencil at arm's length    Pencil at half arm's length

**Fig. 20-17. Demonstration of parallax.**

Astronomers use the parallax method to measure the change in position of nearby stars. For example, an astronomer first takes a photograph of a nearby star against the background of distant stars. Another photograph of the nearby star is taken six months later, when Earth has moved to the opposite end of its orbit (see Fig. 20-18). By comparing the star's position in the two photographs and knowing how much Earth's position in space has changed, an astronomer can calculate the star's distance to Earth.

The distance that Earth moves between the two photographs is 300,000,000 kilometers. When nearby stars are observed by the parallax method, their change of position is large enough to be measured. The farther a star is from Earth, the smaller is the change in position. Stars more than 300 light-years from Earth do not show a measurable change in po-

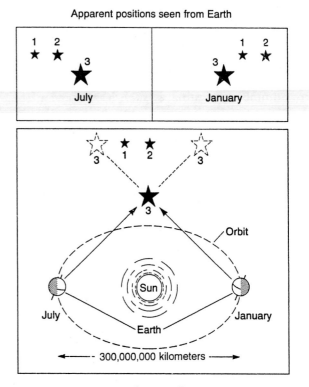

Apparent positions seen from Earth

**Fig. 20-18. The parallax method.**

sition. Consequently, as Earth orbits the sun, the nearby stars appear to change position against the background of the more distant stars. The closer a star is to Earth, the greater the apparent change of position.

## LIFE CYCLE OF A STAR

Astronomers think stars form from clouds of dust and gas in outer space. The dust and gas particles that make up such a cloud exert a gravitational force on each other. As the gas and dust particles are pulled together, the cloud begins to contract and grow warmer. Over millions of years, the cloud of dust and gas eventually becomes hot enough to glow.

A star passes through several stages after it begins to glow. A newly formed star appears dull red because it is still relatively cool. As a young star contracts, it becomes hotter. The star's color changes from dull red to bright red, to orange, to yellow, and finally to a bright blue-white.

When the star's interior temperature reaches several million degrees Celsius, the hydrogen atoms inside the star are changed into helium atoms by thermonuclear fusion. As you read earlier in this chapter, the sun is a star currently in this stage. In this stage, the star stops contracting and glows steadily.

Finally, a point is reached when there is not enough hydrogen left for thermonuclear fusion to continue. At this point, the core of the star collapses, generating tremendous heat. The heat causes the star to expand by millions of kilometers, becoming a much cooler red giant or supergiant star.

In a relatively short time, the red giant collapses completely, with a bright explosion in which its outer layer is removed. The bright explosion of a red giant is called a *nova*. The star then shrinks to become a white dwarf. In time, the star contracts even further into a small, cold body of heavier elements such as iron and carbon.

Red supergiants may explode violently and collapse into objects so dense that their gravitational forces do not allow light rays to escape. These invisible objects are called **black holes.** Astronomers think that black holes may exist at the centers of many galaxies.

According to estimates made by astronomers, the life cycle of a star is probably about 10 billion years. Our sun is about halfway through its life cycle. Thus, the sun is expected to continue generating heat and light at its current rate for at least another 5 billion years.

## ORIGIN OF THE UNIVERSE

Astronomers think that there are billions of galaxies in the universe. All of them appear to be moving away from one another at tremendous speeds. There are two main hypotheses that attempt to explain the origin of the universe based on the outward movement of galaxies: (1) the *big bang theory* and (2) the *steady-state theory.*

**Big Bang Theory.** Astronomers have determined that the stars and galaxies are moving away from Earth. Some scientists think that this outward movement indicates that all galaxies originated from the violent explosion of a densely packed mass of hydrogen. According to the **big bang theory,** a violent explosion spread hydrogen in all directions with a "big bang" 15 to 20 billion years ago. The expanding cloud of hydrogen gas, mixed with dust particles and other gases, gradually condensed into the galaxies. Recent information gathered by astronomers concerning the temperature of gases in the universe appears to confirm this theory.

The *oscillating universe hypothesis* is related to the big bang theory. This hypothesis states that the universe periodically *oscillates*, or expands and contracts. According to the oscillating universe hypothesis, galaxies now moving away from one another will eventually slow down and stop. Then, the galaxies will begin to move toward one another. In time, all of the galaxies will form a huge mass of densely packed matter, which will again undergo a violent explosion, or "big bang."

**Steady-State Theory.** A less widely accepted theory about the origin of the universe, called the **steady-state theory,** states that the universe is continually being renewed. According to this theory, as the galaxies move outward, new galaxies take their place. However, there is no evidence that new matter is being created between the galaxies.

# CHAPTER REVIEW

## Science Terms

*The following list contains all of the boldfaced words found in this chapter and the page on which each appears.*

asteroids (p. 306)
astronomers (p. 299)
big bang theory (p. 313)
black holes (p. 313)
chromosphere (p. 303)
comet (p. 307)
constellations (p. 311)
corona (p. 303)
galaxy (p. 310)
light-year (p. 310)

meteor (p. 308)
meteorite (p. 308)
meteoroids (p. 308)
Milky Way (p. 310)
nebula (p. 310)
photosphere (p. 302)
planet (p. 304)
radio telescope (p. 300)
reflecting telescope (p. 300)
refracting telescope (p. 299)

solar flare (p. 302)
solar system (p. 301)
solar wind (p. 302)
spectroscope (p. 301)
spectrum (p. 301)
steady-state theory (p. 313)
sunspots (p. 302)
thermonuclear fusion (p. 303)
universe (p. 310)

# Matching Questions

*On the blank line, write the letter of the item in column B that is most closely related to the item in column A.*

| Column A | Column B |
|---|---|
| ____ 1. refracting telescope | *a.* systems composed of millions or billions of stars |
| ____ 2. spectroscope | *b.* telescope that gathers light by using lenses |
| ____ 3. radio telescope | *c.* telescope that uses mirrors in addition to a lens |
| ____ 4. solar system | *d.* method used to measure a star's distance from Earth |
| ____ 5. parallax | *e.* instrument used to determine the composition of stars |
| ____ 6. steady-state theory | *f.* small groups of stars, named for their patterns |
| ____ 7. big bang theory | *g.* dense objects that do not let light rays escape |
| ____ 8. constellation | *h.* universe began with a huge explosion |
| ____ 9. galaxy | *i.* universe continually renews its matter |
| ____ 10. black holes | *j.* telescope that collects invisible waves from distant objects |
|  | *k.* the sun and all of its surrounding heavenly bodies |

# Multiple-Choice Questions

*On the blank line, write the letter preceding the word or expression that best completes the statement.*

**1.** The branch of science that studies the universe is called
   *a.* astrology   *b.* astronomy   *c.* geology   *d.* aerospace science          1 ____

**2.** A light-year is equal to a distance of about
   *a.*   1 billion km                    *c.*   1 trillion km
   *b.* 10 billion km                    *d.* 10 trillion km                        2 ____

**3.** The first person to see the craters of the moon through a telescope was
   *a.* Newton   *b.* Ptolemy   *c.* Galileo   *d.* Copernicus                      3 ____

**4.** The world's largest refracting telescope is only about 100 cm in diameter because a larger lens would
   *a.* be too heavy to move          *c.* sag and produce distorted images
   *b.* be too difficult to keep steady   *d.* be too expensive to build            4 ____

**5.** The instrument that makes use of a glass prism to study the stars is a
   *a.* range finder                   *c.* beam splitter
   *b.* infrared telescope             *d.* spectroscope                            5 ____

**6.** The stars in a galaxy can number well into the
   *a.* hundreds   *b.* thousands   *c.* millions   *d.* billions                   6 ____

7. The name of the galaxy that holds our solar system is the
   *a.* Milky Way   *b.* Andromeda   *c.* Great Spiral   *d.* Orion

   7 ____

8. Giant stars are
   *a.* the hottest stars
   *b.* blue-white stars
   *c.* made of densely packed matter
   *d.* made of loosely packed matter

   8 ____

9. Radio telescopes *cannot* be used to
   *a.* study stars in the daytime
   *b.* study stars in cloudy weather
   *c.* take photographs of stars
   *d.* detect stars beyond optical telescope range

   9 ____

10. The color of the hottest type of star is
    *a.* dull red   *b.* yellow   *c.* bright red   *d.* blue-white

    10 ____

11. The two most abundant elements in stars are
    *a.* hydrogen and helium
    *b.* hydrogen and iron
    *c.* hydrogen and calcium
    *d.* helium and iron

    11 ____

12. The parallax method of determining the distance of a star depends on the
    *a.* brightness of the star when viewed from different positions
    *b.* shift in position of the star when viewed from different positions
    *c.* apparent size of the star when using telescopes that are parallel
    *d.* temperature of the star at different times of the year

    12 ____

13. The sun is believed to be about
    *a.* 5 million years old
    *b.* 500 million years old
    *c.* 5 billion years old
    *d.* 10 billion years old

    13 ____

14. A heavenly body that is *not* part of our solar system is
    *a.* the sun   *b.* a comet   *c.* a meteoroid   *d.* a galaxy

    14 ____

15. Of all the mass in the solar system, the sun contains about
    *a.* 99%   *b.* 80%   *c.* 75%   *d.* 50%

    15 ____

16. The force that keeps planets from leaving their orbits around the sun is
    *a.* magnetism   *b.* gravity   *c.* inertia   *d.* momentum

    16 ____

17. The diameter of the sun is about
    *a.* 13,000 km
    *b.* 140,000 km
    *c.* 1,390,000 km
    *d.* 300,000 km

    17 ____

18. Shortly after the sun has strong sunspot activity, Earth experiences
    *a.* interruptions in radio communications
    *b.* more tornadoes and hurricanes
    *c.* changes in weather patterns
    *d.* more meteors in the sky

    18 ____

19. The outer part of the sun's atmosphere, consisting of thin gases, is called the
    *a.* neon-light area   *b.* aurora   *c.* corona   *d.* afterglow

    19 ____

20. The lowest part of the sun's atmosphere is called the
    *a.* chromosphere   *b.* troposphere   *c.* solar sphere   *d.* photosphere

    20 ____

21. In thermonuclear fusion within the sun, hydrogen atoms
    *a.* split apart to form helium atoms
    *b.* join to form helium atoms
    *c.* gain mass by heating up
    *d.* burn to produce light

    21 ____

22. The mass lost by the sun during thermonuclear fusion is converted into
    *a.* energy   *b.* helium   *c.* oxygen   *d.* magnetism

    22 ____

**23.** The laws that describe planetary motion were first stated by
   *a.* Copernicus   *b.* Galileo   *c.* Ptolemy   *d.* Kepler          23 ____

**24.** A comet's tail always points away from the sun due to the force of
   *a.* solar wind   *b.* gravity   *c.* fusion   *d.* radio waves       24 ____

**25.** According to the laws of planetary motion, the fastest orbiting planets are the
   *a.* largest planets              *c.* planets closest to the sun
   *b.* smallest planets             *d.* planets farthest from the sun   25 ____

# Modified True-False Questions

*In some of the following statements, the italicized term makes the statement incorrect. For each incorrect statement, write the term that must be substituted for the italicized term to make the statement correct. For each correct statement, write the word "true."*

**1.** The two main types of optical telescopes are the reflecting telescope and the *radio* telescope.                1 _____

**2.** A *spectroscope* is an instrument used by astronomers to determine the composition of stars.                 2 _____

**3.** The hazy patches of light that may be seen at night are clouds of dust and gas called *meteors*.                3 _____

**4.** Supergiant stars can have diameters as large as several *billion* kilometers.                                4 _____

**5.** The fusion reaction that takes place in stars involves the conversion of hydrogen atoms into *helium* atoms.      5 _____

**6.** The diameter of the sun is almost *ten* times greater than the diameter of Earth.                            6 _____

**7.** *Hydrogen* is the most abundant element in the sun.        7 _____

**8.** Sunspots are more than 1000°C *hotter* than the surrounding surface of the sun.                              8 _____

**9.** The red, glowing layer of gases thousands of kilometers above the sun's surface is called the *corona*.          9 _____

**10.** The sun changes about 4,000,000 tons of matter into energy every *day*.                                   10 _____

# Testing Your Knowledge

**1.** Is it possible that a star you see in the sky no longer exists? Explain.

_____

_____

**2.** Name the three main types of telescopes, and briefly describe each one.

*a.* _____

_____

*b.* _____

_____

*c.* _____

_____

**3.** Identify two reasons why an astronomer would attach a camera to a telescope.

*a.* _____

*b.* _____

**4.** What is the difference between a galaxy and a constellation? _____

_____

_____

**5.** Describe two ways in which it is thought that the universe may have begun.

*a.* _____

*b.* _____

**6.** How does sunspot activity affect Earth's atmosphere? _____

_____

_____

**7.** Where does the sun's energy come from? _____

_____

_____

**8.** The sun continues to shine despite the loss of trillions of tons of mass each year. Explain.

_____

_____

**9.** Describe how the planets move, according to the laws of Kepler. _____

_____

**10.** Describe two differences between meteors and comets.

*a.* _____

_____

*b.* _____

_____

# Attack of the Killer Asteroids: Should We Worry?

Invisible to the human eye, more than one million chunks of rock and metal spin and tumble between Jupiter and the sun. These are the solar system's debris; bits and pieces that were destined to become a tenth planet, but for one reason or another never got together to do so.

Called *asteroids*, some are huge: Ceres, the giant among asteroids, measures 933 km across its middle, roughly equal to the distance between New York City and Toledo, Ohio. A few other asteroids join Ceres in the 100-km-plus class, but most are between 1 km to 2 km wide.

Asteroids have been whizzing around the solar system almost since its birth some 4.5 billion years ago. One very large family has taken up residence between the orbits of Mars and Jupiter. Other families have found homes nearer to the sun. Some, called the Apollo and Atens families, have made their homes in our "neighborhood," near Earth. And some of *these* have a particularly troubling habit of crossing our planet's orbit.

It doesn't take a mathematical genius to figure out that, over hundreds of millions of years, some of these crossings have caused collisions. As a matter of fact, scientists estimate that a collision between Earth and an asteroid 500 meters or more in diameter occurs once every 10,000 years. And the collisions with smaller asteroids occur much more often.

What are the effects of such collisions? Even a small asteroid one-half kilometer across would explode on the Earth with a force equal to more than one million tons of TNT! That's equivalent to 50 times the force of the atom bomb that completely wiped out the Japanese city of Hiroshima at the end of World War II.

Has such an explosion ever rocked Earth? Yes, about 65 million years ago, say many scientists. If the scientists are correct, and evidence of various kinds supports their view, such a collision and explosion may have caused the extinction of thousands of species of living things, including the dinosaurs.

This collision, the scientists point out, would have caused tremendous clouds of dust and smoke to rise into the atmosphere, blocking sunlight for months or years. Temperatures on Earth would have plummeted. Many plants would have died off, and along with them both the animals that fed on plants and the animals that ate the plant-eaters.

Sixty-five million years ago is far in the past. Hardly worth worrying about? A smaller asteroid hitting Earth might not be as devastating as the one that may have wiped out the dinosaurs, but it could still destroy an entire city, a whole state, or a good part of a country.

Such an object plowed into Siberia near the Tunguska River on June 30, 1908. Over three hundred square kilometers of forest burst into flames; the trees were flattened like toothpicks and destroyed along with the animals that lived there. Cows over 100 kilometers away were even knocked down by the force of the blast! Fortunately, the area held few people, so not many human lives were lost. Impacts like this, and still smaller ones, happen every 100 years or so. What's more, near misses occur even more frequently. In January of 1991 an asteroid zipped past Earth only about 160,000 km away. What's even more frightening, the asteroid was discovered only a few hours before its brush with our planet.

For all of these reasons, many astronomers are urging that an asteroid watch system be

set up. This would consist of five special telescopes positioned around the world. The telescopes would be designed to spot moving objects like asteroids. The orbits of these objects would be quickly determined to find out if they were headed our way.

Such a system of asteroid observatories might give us a few years' warning of an asteroid whose path led directly to Earth. Armed with this information we might be able to head off the collision. How?

A number of ideas have been proposed. Among these would be sending a space vehicle to the asteroid. The vehicle could deposit a rocket on the asteroid which, when fired, would alter the asteroid's orbit so it missed Earth. Another idea would be to explode a nuclear bomb on one side of the asteroid, again pushing it out of its normal path toward Earth.

Setting up a tracking system to spot "doomsday asteroids" would take millions of dollars to do—your family's tax dollars. Should we spend this money to try to avoid a possibly catastrophic event that might not even occur in your lifetime, or even in the lifetime of your children or grandchildren? What do you think?

1. *The diameter of most asteroids is*

   a. more than 500 km.
   b. between 200 km to 500 km.
   c. between 10 km to 100 km.
   d. between 1 km to 2 km.

2. *Scientists are concerned about Apollo and Atens asteroids because*

   a. they are very large.
   b. they orbit between Mars and Jupiter.
   c. their orbits cross that of Earth's.
   d. they are made of metal or stone.

3. *Scientists estimate that a collision between a 500-meter-wide asteroid and Earth occurs on average once every*

   a. 10 years.      c. 1000 years.
   b. 100 years.     d. 10,000 years.

4. *What do many scientists believe happened*

   *about 65 million years ago?* _____

   _____

   _____

5. *On a separate sheet of paper, express your views on whether an expensive asteroid-watching system should be installed around the world. Back up your views with appropriate data.*

319

# Exploring Space

## LABORATORY INVESTIGATION

### ACTION AND REACTION

**Note:** This experiment should be performed as a teacher demonstration.
**A.** Study the apparatus shown in Fig. 21-1. In particular, note the position of the rubber band, the weight to be propelled, and the position of the backstop. *The backstop* (a wastebasket) *is a safety device that will catch the weight when it is propelled backward.* **CAUTION: All observers should wear safety goggles.**

**B.** Prepare the apparatus for firing by following these steps:
(1) Tie one end of a piece of string (about 35–40 centimeters [cm] long) to the rubber band, using a double knot.
(2) Pull the rubber band back with one hand, and pull the string back with your other hand. Have a student hold the platform in place. Holding the rubber band at the 3-cm mark, pull the string tight, and wind the loose end of the string about a dozen times around the nail.

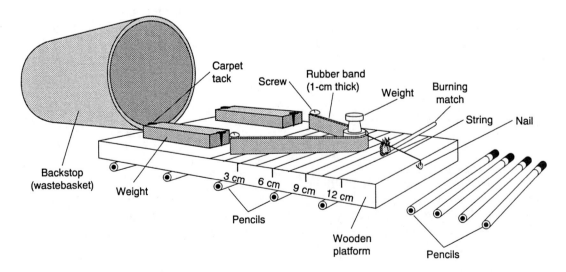

**Fig. 21-1.**

(3) Place the platform on 4 or 5 round pencils as shown in the diagram. Position several more pencils in front of the platform.

(4) Place the 50-gram weight inside the rubber band.

(5) Standing clear, light a wood splint. Hold it beneath the taut string, and observe what happens. (Do not forget to blow out the burning splint.)

(6) Measure the distance the platform traveled. Record the results in the table below.

(7) Cut the burned string, and remove it from the platform.

(8) Repeat the procedure, but this time pull the rubber band to the 6-cm mark. Record your observations.

(9) Repeat the procedure using the 100-gram weight. Fire the weight twice—once from the 3-cm mark and once from the 6-cm mark. Record the results in the table.

| Weight | Stretched Rubber Band | Distance Platform Travels |
|--------|----------------------|---------------------------|
| 50-gram | 3-cm mark | |
| 50-gram | 6-cm mark | |
| 100-gram | 3-cm mark | |
| 100-gram | 6-cm mark | |

**1.** Describe the effect on the platform when

   *a.* the rubber band is pulled back farther. _____

_____

   *b.* a heavier weight is propelled. _____

_____

**2.** Identify two factors that affect how far the platform travels.

   *a.* _____

   *b.* _____

**3.** If the weights on the platform and the propelled weight remain the same, how can you make the platform travel farther? _____

_____

# FLIGHT

In mid-December 1903, in North Carolina, the Wright brothers (Wilbur and Orville) made history by becoming the first people to fly an airplane successfully. Just over 50 years later, in October 1957, the Soviet Union (now Russia) launched the first artificial satellite, *Sputnik I*, in orbit around Earth. Four months later, the United States placed an artificial satellite, *Explorer I*, in orbit around Earth. These first two flights into space were followed by many more, some involving astronauts.

On July 20, 1969, two U.S. astronauts, Neil Armstrong and Edward Aldrin, became the first human beings to walk on the moon. This voyage by these astronauts and their spacecraft, *Apollo 11*, to the moon and back was a dramatic step in the continuing human exploration of the solar system. The knowledge and experience gained from almost a century of air flight have paved the way for these and other explorations of space.

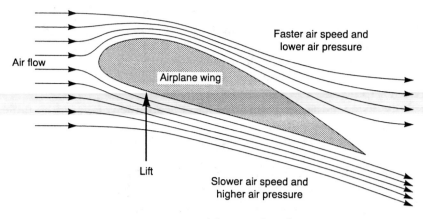

**Fig. 21-2. How lift is produced.**

**Types of Air Flight.** There are two types of air flight: *lighter-than-air flight* and *heavier-than-air flight*. Balloons and dirigibles ("blimps") are examples of lighter-than-air aircraft. A balloon or dirigible rises because the gas inside is lighter than the air in Earth's atmosphere.

Gliders and airplanes are examples of heavier-than-air aircraft. The wings of gliders and airplanes are designed to produce a force called *lift*. A glider or airplane rises and remains airborne because the lift produced by its wings is greater than its weight (see Fig. 21-2). To take off, an airplane must travel fast enough for the lift of its wings to overcome its weight. Airplanes have either propellers or jet engines that provide enough speed for the plane to rise off the ground and move through the air. Gliders which do not have engines, usually are towed into the air by airplanes.

**Space Flight.** Like a glider and an airplane, a spacecraft is a heavier-than-air vehicle that travels through space. In some ways, flying through air is similar to space flight. Spacecraft that carry a crew into space, however, face additional problems. For example, a spacecraft must carry sufficient oxygen for its crew, protect the crew from harmful solar radiations, carry sufficient food and water into space, and provide for the recycling of body wastes.

## NAVIGATION

The act of steering a vehicle to its destination is called **navigation**. Navigating in your city or even a strange city by automobile is relatively easy because streets are identified by signs. However, there are no signs over most of Earth's surface and none at all in space.

**Navigation in the Air and on Earth's Surface.** Maps and the system of lines of latitude and longitude (described in Chapter 11) help navigators of land vehicles, ships, and airplanes. For example, a navigator needs only to calculate latitude, longitude, and the four main directions—north, south, east, and west—to determine his or her exact location.

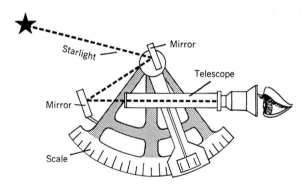

**Fig. 21-3. A sextant.**

**Determining Latitude.** To determine latitude, a navigator uses an instrument called a **sextant**. The sextant consists of a telescope, a system of mirrors, and a scale (see Fig. 21-3). For example, an airplane navigator in the Northern Hemisphere would use a sextant to find the altitude of the North Star in degrees above the horizon. This number is equal to the latitude of the navigator's plane. The closer a navigator is to the North Pole, the greater the altitude of the North Star (see Fig. 21-4).

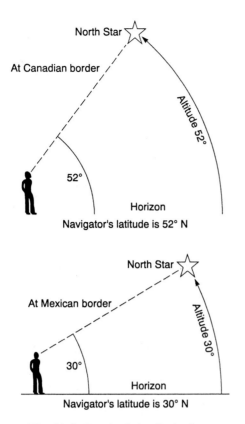

Fig. 21-4. Determining latitude.

**Determining Longitude.** Longitude is measured in degrees east or west of the prime meridian, which passes through Greenwich, England. Longitude can be determined by comparing the time at the prime meridian (Greenwich time), with 12 noon at the navigator's location.

A special clock, called a **chronometer,** is carried by a navigator. A chronometer always shows Greenwich time. When it is 12 noon at the navigator's location, the navigator reads the chronometer to find Greenwich time. If Greenwich time is later than 12 noon, the navigator is west of the prime meridian.

As you learned in Chapter 11, each hour of difference between Greenwich time and local time is equal to 15 degrees (15°) of longitude. As shown in Fig. 21-5a, if Greenwich time is 5:00 P.M. when it is 12 noon at the navigator's location, the navigator is 75° west of the prime meridian (15° × 5 = 75°).

If Greenwich time is earlier than 12 noon, the navigator is located east of the prime meridian. As shown in Fig. 21-5b, if Greenwich time is 10:00 A.M. when it is 12 noon at the navigator's location, the navigator is 30° east of the prime meridian (15° × 2 = 30°).

**Determining North.** In the United States, there are several ways of determining north. If you face the sun at 12 noon, north is directly behind you. At night, north can be determined by finding the North Star, which is located by following the two pointer stars of the Big Dipper (refer back to Fig. 20-16).

You can also determine north by using a magnetic compass. In the United States, the needle of a compass will not usually point to the true North Pole, or true north. Rather, the compass needle will point to the magnetic north pole in the Hudson Bay region of Canada. True north is determined by finding magnetic north and then referring to special charts that contain the correction factor to calculate true north.

Navigators aboard ships and airplanes use a special instrument called a **gyrocompass** to determine true north. This instrument is a combination of a *gyroscope* and a *compass*. A spinning gyroscope always keeps its axis pointing in the same direction. Thus, a gyrocompass continues pointing to magnetic north regardless of the course followed by a ship or

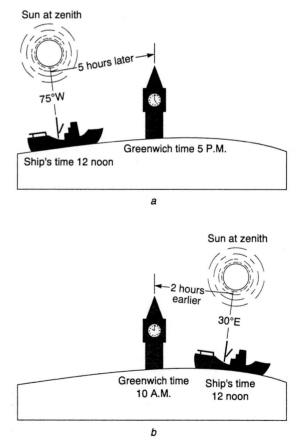

Fig. 21-5. Determining longitude.

plane. Once true north is determined, the other compass directions are easily determined. If you face north, south is behind you, east is on your right, and west is on your left.

**Navigation in Space.** To navigate in space, a spacecraft uses an *inertial guidance system.* This system consists of computers, gyroscopes, sextants, and other devices that guide the spacecraft to its destination. The inertial guidance system determines the spacecraft's position relative to Earth, the moon, and the stars. Once the spacecraft escapes Earth's gravity, the inertial guidance system automatically keeps the spacecraft from veering off the planned course.

In space flight, a destination is a "moving target." For example, astronauts on the *Apollo* missions to the moon had to consider that Earth was revolving in its orbit around the sun, and that the moon was revolving in its orbit around Earth. Thus, the *Apollo* astronauts had to navigate their spacecraft to the location in space where the moon would be located several days after they left Earth (see Fig. 21-6). When they left the moon, the astronauts guided the spacecraft to the point in space where they calculated Earth would be in several days.

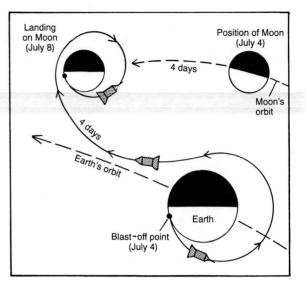

**Fig. 21-6. Traveling to the moon.**

Navigators must be extremely careful when setting a course. A small error made by a navigator on a ship or airplane may result in the vehicle being steered several kilometers off course. A similar error made by a navigator on a spacecraft could place the vehicle several thousand kilometers off course.

# Rockets and Space Exploration

The history of rockets spans almost 1000 years. The Chinese built rockets propelled by gunpowder as early as the year 1040. They used these rockets as weapons and for fireworks displays. In 1926, the American physicist *Robert H. Goddard* (1882–1945) launched the first rocket propelled by liquid fuel. By 1930, Goddard had built liquid-propelled rockets that could travel about 800 kilometers (km) per hour and reach an altitude greater than 600 meters (m).

During World War II, German scientists built rocket weapons that travelled 5000 km per hour and carried one-ton bombs a distance of almost 250 km. By the late 1950s, the United States and the Soviet Union had built rockets that could travel into space.

**How a Rocket Works.** A rocket moves forward by propelling gases backward. You can demonstrate this type of movement with an ordinary balloon. First, inflate the balloon, and tie its mouth. The balloon stays inflated because the air inside pushes against every part of the balloon's interior with equal force (see Fig. 21-7*a*). Because the air pushes equally in all directions, the forces inside the balloon are *balanced.*

However, when you untie the balloon's mouth, the escaping air upsets this balance.

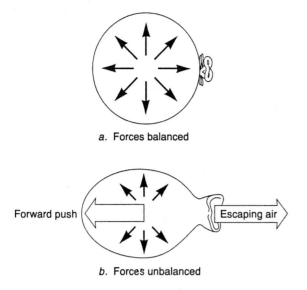

a. Forces balanced

Forward push | Escaping air

b. Forces unbalanced

**Fig. 21-7. How a rocket moves.**

As air rushes out of the balloon, the force exerted against its mouth decreases rapidly. The part of the balloon opposite the mouth still has force pushing against it. As a result of these *unbalanced forces*, the balloon travels forward (see Fig. 21-7*b*).

This demonstration is an example of Newton's third law, which states that for every *action* there is an equal and opposite *reaction*. The action is the air escaping through the balloon's mouth. The reaction is the balloon's movement in the opposite direction. You observed another example of Newton's third law in your laboratory investigation in which a wooden platform moved forward by propelling a weight in the opposite direction.

The reaction (movement) caused by air escaping from the mouth of a balloon is similar to the way in which a rocket engine produces a reaction force. In a rocket engine, fuel is burned inside a combustion chamber (see Fig. 21-8). The burning fuel releases hot, expanding gases that push against the walls of the combustion chamber. Gases escape at high speeds from the rear of the engine, which re-

duces the pressure at the open end of the combustion chamber. This causes the forces inside the chamber to become unbalanced. The expanding gases push against the front, or closed end, of the combustion chamber, producing forward movement. This forward movement is called **thrust.**

The amount of thrust produced by a rocket depends on the pressure inside the combustion chamber and the speed at which the gases are escaping from the nozzle (see Fig. 21-8). The greater the velocity of escaping gases, the greater the thrust.

As a rocket rises above Earth, its thrust becomes more effective for the following reasons:

1. A rocket burns tons of fuel during each second that it rises. Thus, as a rocket rises, it becomes lighter, and the engines have less mass to push forward.
2. The farther an object is from Earth's center, the smaller is the pull of Earth's gravity on the object. For this reason, you weigh slightly less on a moutaintop than at sea level. Likewise, a rocket engine weighs less as it moves away from Earth.

    The pull of Earth's gravity at sea level is referred to as 1 g (g = acceleration due to Earth's gravity). About 2500 km above Earth, the pull of Earth's gravity is about 0.5 g. At this altitude, a rocket weighs half its weight at sea level. At a distance of 25,000 km, the pull of Earth's gravity is only .04 g. Here, a rocket would weigh only about 1/25 its weight at sea level.
3. As a rocket rises through the thin air of Earth's upper atmosphere, there is less air to produce resistance (friction) to the rocket's forward movement. That is because thin air at high altitudes offers less resistance to gases escaping from the nozzles of a rocket engine than does the denser air at lower altitudes. Consequently, thrust increases as the altitude of a rocket increases.

**Jet Engines and Rocket Engines.** The more powerful a jet engine, the faster the jet can fly. However, a supersonic jet flying at 3200 km per hour does not have enough force to overcome the pull of Earth's gravity and fly off into outer space. Even if a jet were able to reach outer space, at about 32 km above Earth, the atmosphere contains so little oxygen that the jet engine's fuel could not burn. Thus, the jet engine would stop functioning.

In fact, the main difference between a jet engine and a rocket engine is the source of ox-

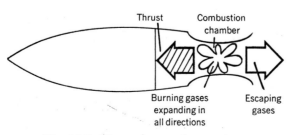

Thrust   Combustion chamber

Burning gases expanding in all directions

Escaping gases

**Fig. 21-8. Thrust in a rocket engine.**

ygen. A **jet engine** takes in air from Earth's atmosphere and uses the oxygen in the air to burn a fuel such as kerosene (see Fig. 21-9). A **rocket engine** carries its own liquid oxygen (LOX), which is combined with a liquid fuel in the combusion chamber. Thus, a rocket can travel into space where oxygen is not present.

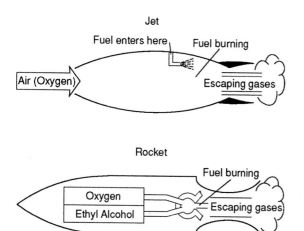

**Fig. 21-9. Comparison of jet and rocket engines.**

## TYPES OF ROCKETS

The most common type of rocket is a **chemical rocket,** which uses either liquid or solid fuel. Chemical rockets produce enough thrust to lift a spacecraft into orbit around Earth or travel to distant planets.

**Solid-propellant Rockets.** A **solid-propellant rocket** is similar to the rockets used in fireworks displays, which are made of several cardboard cylinders stuffed with *gunpowder* and closed at one end. Gunpowder consists of a fuel and an oxidizing agent, or *oxidizer*, that permits burning of a fuel. As the fuel burns, hot gases escape rapidly from the open end of the cardboard cylinder. Solid-propellant rockets used in exploring space are made of metal cylinders filled with a special mixture of fuel and oxidizer.

The advantages of solid-propellant rockets are that (1) they can be stored indefinitely and (2) their engines have few moving parts, which reduces the chance of a malfunction. The disadvantages of solid-propellant rockets are that (1) they cannot be turned off once they are ignited, (2) they provide less thrust than liquid-propellant engines, and (3) an explosion may occur because the fuel and oxidizer are always in contact with each other.

**Liquid-propellant Rockets.** In a **liquid-propellant rocket,** the fuel and the oxidizer are stored in separate tanks. A system of pipes, valves, and pumps brings the fuel and oxidizer together in the combustion chamber, where the mixture is ignited. Ethyl alcohol, kerosene, and liquid hydrogen are fuels commonly used as liquid-propellants. The oxidizer may be either liquid oxygen, liquid fluorine, or nitric acid.

The advantages of liquid-propellant rockets are that (1) they produce much more thrust than solid-propellant rockets, (2) their engines can be turned "off" and "on," and (3) the rate at which their fuel is burned can be controlled. The disadvantages of liquid-propellant rockets are that (1) their complex system of pipes, valves, and pumps increases the chance of engine malfunction and (2) they cannot be stored for long periods of time because liquid propellants must be kept refrigerated at extremely low temperatures.

**Nuclear-powered Rockets.** Solid-propellant and liquid-propellant rockets are satisfactory for sending spacecraft to the moon or the inner planets. However, they are not practical for long space voyages because they use up their fuels rapidly and their propellants make up too much of the total weight of a spacecraft. For these reasons, **nuclear-powered rockets** are now being considered for long-range voyages.

The main problem involved with using nuclear-powered rockets for launching spacecraft is their weight. That is because nuclear-powered rockets would require heavy lead shielding to protect astronauts from harmful nuclear radiations. However, small nuclear-powered rockets may be used in combination with the liquid-propellant or solid-propellant rockets to power spacecraft on missions to distant planets or stars.

## LAUNCHING SPACECRAFT INTO ORBIT

At about 160 km above Earth, a spacecraft must travel about 8 km per second to avoid being pulled back to Earth by gravity. At this speed, the spacecraft begins to orbit Earth. The spacecraft is then referred to as an *artificial satellite* (see Fig. 21-10). A spacecraft orbiting Earth is continuously being pulled downward by the pull of gravity. However, the

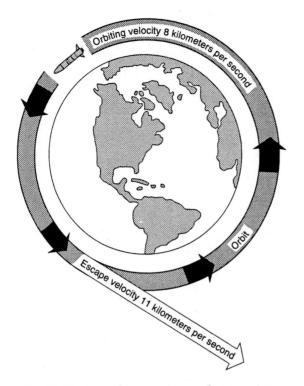

**Fig. 21-10. Launching a spacecraft into orbit.**

escape Earth's gravity (see Fig. 21-10). This velocity is called **escape velocity.**

The U.S. *Apollo* missions to the moon during the late 1960s and early 1970s were powered by the huge *Saturn V* rocket. The *Saturn V* rocket was about 90 m tall and weighed about 2,700,000 kilograms (kg). The extremely high thrust of this rocket enabled the *Apollo* spacecraft to escape Earth's gravity and reach the moon.

The rocket engines of the *Saturn V* and other launch vehicles of the 1960s and 1970s required several stages to reach escape velocity and orbit Earth. The *Saturn V* rocket engine was fired in three stages. After the first stage had burned all of its fuel, its engine was *jettisoned*, or released, from the rocket. The removal of the first stage reduced the overall weight of the rocket and attached spacecraft. Its speed increased further as the second-stage and third-stage engines were fired and jettisoned (see Fig. 21-11).

force propelling the spacecraft forward at a certain velocity balances the downward pull of Earth's gravity. This velocity, called **orbiting velocity,** keeps the spacecraft in a stable orbit.

Spacecraft in stable orbits around Earth follow a curved path. However, an increase or decrease in a spacecraft's velocity will cause it to leave orbit. A spacecraft's velocity can be decreased by firing *retro-rockets*. Retro-rockets apply reverse thrust, or force in a direction opposite to the direction in which the spacecraft is moving. This causes the spacecraft to slow down and lose altitude.

In time, the spacecraft reenters the atmosphere and is slowed further by friction with air. Friction with the atmosphere would cause an unprotected spacecraft to burn up like a meteoroid. For this reason, spacecraft carrying astronauts have special *heat shields* that protect the vehicle and its crew upon reentry into Earth's atmosphere.

A spacecraft supplied with sufficient thrust can break away completely from the pull of Earth's gravity. When the velocity of the spacecraft reaches 11 km per second, it can

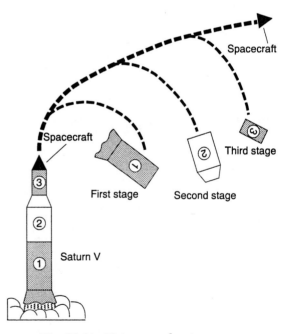

**Fig. 21-11. Firing a rocket in stages.**

Today, the *Space Shuttle* uses two reusable solid-fuel boosters and a disposable, liquid-propelled rocket engine to achieve orbit. The only part of a Space Shuttle not recycled is the empty liquid fuel tank, which drops into the ocean after launch (see Fig. 21-12). Unlike other spacecraft, the *delta-wing section* of the

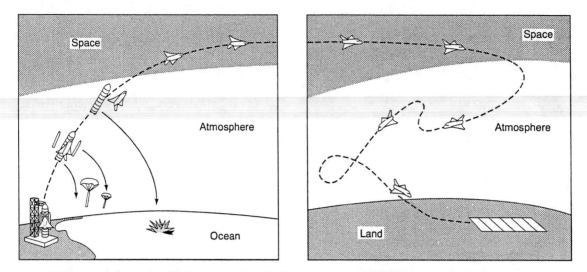

**Fig. 21-12. After takeoff, the space shuttle drops its reusable solid-fuel boosters and empty liquid fuel tank. The orbiter continues to fly and later returns by landing on Earth.**

Space Shuttle (holding the instruments, crew, and laboratories) can land like a glider after returning from space. This reusable section, which is about 40 m long, is called the *orbiter*. The orbiter is protected from burning when it reenters Earth's atmosphere by a heat shield made of about 30,000 silica tiles, which cover the fuselage and wings.

## TYPES OF SPACECRAFT

In less than 40 years, the United States, the Soviet Union, and other countries have launched thousands of spacecraft. All spacecraft can be classified into two general categories, based on whether or not they carry humans. Spacecraft that carry humans into space are usually referred to as "manned." For example, the Space Shuttle is a manned spacecraft. Space vehicles carrying only instruments, such as space probes, are "unmanned" spacecraft.

**Artificial Satellites.** The moons of all planets in the solar system are natural satellites. A spacecraft built by humans and placed into orbit to perform a specific task is called an **artificial satellite.** Most artificial satellites are unmanned. However, the United States, the Soviet Union, and the European Space Agency have successfully orbited manned satellites around Earth and the moon.

Unmanned satellites orbiting Earth are used for communications, weather forecasting, navigation, and even spying. For example, *communication satellites* are unmanned relay stations in space that receive and transmit telephone, radio, and TV signals around the world. *Weather satellites* photograph global weather patterns and transmit them and other data for use by weather forecasters. *Navigation satellites* transmit precise location data that allow navigators of oceangoing ships and planes to pinpoint their exact position on a map. Other artificial satellites are used to map remote parts of Earth's surface, to observe environmental conditions, and to detect underground mineral resources.

**Space Probes.** An unmanned spacecraft that leaves Earth's orbit and explores other parts of the solar system is called a **space probe.** American and Soviet space probes have orbited and landed on Venus and Mars. Other space probes have flown past Mercury, Jupiter, Saturn, Uranus, and Neptune (see Table 21-1).

**Space Laboratories.** In 1973, the United States launched *Skylab*, the first **space laboratory** to orbit Earth. Skylab, which was equipped with living quarters for a crew of three astronauts and a laboratory for performing experiments, remained in orbit until 1979. Since then, research in space has been conducted aboard the United States' fleet of Space Shuttles, the Russian Space Station *Mir*, which orbited Earth from 1986 to 2001, and, more recently, the International Space Station.

## TABLE 21-1. SOME SPACE PROBES AND THEIR ACCOMPLISHMENTS

| Name | Country | Accomplishment and Date |
|---|---|---|
| Mariner IV | United States | Photographed surface of Mars, 7/65. |
| Luna 9 and Luna 10 | Soviet Union | First spacecrafts to orbit and land on the moon, 1966. |
| Venera 4 | Soviet Union | First spacecraft to collect data on the atmosphere of Venus, 6/67 |
| Surveyor 5 | United States | Landed on the moon; transmitted data on soil back to Earth for analysis, 9/67. |
| Luna 16 | Soviet Union | First unmanned spacecraft to return to Earth with samples of lunar crust, 9/70. |
| Venera 7 | Soviet Union | First spacecraft to transmit data to Earth from the surface of Venus, 12/70. |
| Mariner IX | United States | First space probe to orbit Mars, 11/71. |
| Mars 3 | Soviet Union | Carried capsule that made the first soft landing on Mars, 12/71. |
| Pioneer X | United States | First spacecraft to cross the asteroid belt; flew past Jupiter and transmitted data back to Earth, 12/73; first craft to travel beyond the orbit of Pluto, 1983. |
| Mariner X | United States | Photographed and collected data on Venus, 2/74, and Mercury, 3/74. |
| Pioneer-Saturn | United States | Photographed and collected data on Jupiter, 12/74, and Saturn, 9/79. |
| Vikings I and II | United States | Landed on Mars; transmitted photographs and data back to Earth, 7/76 and 9/76. |
| Pioneer Venus I | United States | Orbited Venus and transmitted radar images of its surface, 12/78. |
| Voyager 1 | United States | Flew past Jupiter, 3/79, and Saturn, 11/80; passed beyond Pluto's orbit, 5/88. |
| Voyager 2 | United States | Flew past and photographed Jupiter, 7/79, Saturn, 8/81, Uranus, 1/86, and Neptune, 8/89; passed beyond Pluto's orbit 8/90. |
| Venera 13 | Soviet Union | Landed on Venus; transmitted color photographs and data on soil samples back to Earth, 3/82. |
| Giotto | European Space Agency | Flew within 540 km of Halley's comet and transmitted data back to Earth, 3/86. |
| Magellan | United States | Orbited Venus from 1990 to 1994 and used high-resolution radar to map 98% of its surface. |
| Galileo | United States | First spacecraft to orbit Jupiter, 12/95; took photographs and collected data on Jupiter, its rings and moons; released probe into Jupiter's atmosphere. |
| Mars Pathfinder | United States | Landed on Mars, 7/97; collected and analyzed rock and soil samples, collected data on winds and weather, took photographs of surface. |

The laboratory aboard the Space Shuttle allows scientists to make observations and perform many experiments that cannot be carried out on Earth. For example, astronomers can use telescopes to study distant planets and stars without interference from Earth's atmosphere. Biologists can study the effects of weightlessness on plants and animals, and geologists and industrial scientists can study the growth of mineral crystals in space. By the beginning of the twenty-first century, a fleet of Space Shuttles is scheduled to erect a large space laboratory as part of a permanently orbiting space station.

## HUMANS IN SPACE

Manned spacecraft must provide astronauts with the basic requirements for life. Astronauts require oxygen, water, food, a suitable air pressure and temperature, as well as protection from harmful radiations.

**Oxygen, Water, and Food.** Each day, a human being needs about 0.5 kg of concentrated food, 1 kg of oxygen, and 2 kg of water. On missions of several weeks or months, sufficient amounts of oxygen, water, and food can be carried aboard spacecraft. However, supplying a crew of astronauts with oxygen, water, and food on long space voyages presents problems. For example, a trip of a year or more would require many tons of oxygen, food, and water. Scientists are now trying to develop a system in which oxygen, water, and food would be totally recycled from the waste products given off by the astronauts.

One recycling system being considered for use aboard spacecraft on very long voyages would be similar to a balanced aquarium (see Fig. 21-13), in which plants and animals continuously provide each other with the basic requirements for life. Aboard these spacecraft, for example, algae would be grown in a tank.

By using artificial light to carry out photosynthesis, algae would release oxygen for use by the astronauts. They also would absorb carbon dioxide given off by the astronauts, grow, and reproduce. In time, some of the algae could be harvested, treated, and prepared to supply food for the astronauts. In this system, algae provide the astronauts with oxygen and food, while the astronauts provide algae with carbon dioxide.

Water in exhaled air, perspiration, and urine could be recovered and reused to provide sufficient water for both the algae and the astronauts. Water vapor from exhaled air and evaporated perspiration could be condensed into a liquid. This water, combined with urine, would then be distilled (purified) into a pure, germ-free liquid suitable for drinking. During the water purification process, minerals would be removed from the water. These minerals would then be supplied to the algae for proper growth.

**Radiations in Space.** Beyond Earth's atmosphere, astronauts must protect themselves against harmful radiations such as cosmic rays and ultraviolet light. The spacecraft and space suits worn by the astronauts block these radiations. If the moon or other planets are ever colonized, a protective enclosure will have to be constructed to shield humans from such radiations.

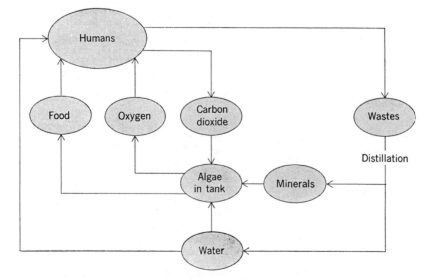

**Fig. 21-13. Scheme of a spacecraft "aquarium" recycling system.**

**Meteoroids in Space.**  As you learned in Chapter 20, most meteoroids burn up before reaching Earth's surface. In space, however, a spacecraft and its crew are in constant danger of being struck by meteoroids. A meteoroid the size of a golf ball could penetrate the shell of a spacecraft. This would leave a gaping hole through which air inside the spacecraft would escape with explosive force.

Like some automobile tires, spacecraft today have a self-sealing substance in their outer shell. This reduces the danger of explosions and air loss in case the spacecraft's outer shell is punctured by a meteoroid. When astronauts leave the spacecraft, they wear *kevlar* (a material also used in bullet-proof vests) to protect themselves from tiny meteoroids, or *micrometeoroids.*

**Maintaining Air Pressure.**  When leaving the spacecraft, an astronaut must use a special chamber called an *air lock.* An air lock is a separate, closed chamber in which proper air pressure is maintained. If the astronaut enters the air lock from outside the spacecraft, air pressure in the air lock would be zero. The air lock is then sealed, and the astronaut increases the air pressure until it is equal to the air pressure maintained inside the living quarters of the spacecraft.

**Maintaining Gravity.**  On its launching pad, a spacecraft is affected by the pull of Earth's gravity—a force equal to 1 g. When the spacecraft takes off, the astronauts inside are subjected to an increase in g force. You probably have noticed a similar effect when sitting in a car or carnival ride that accelerates suddenly. As the car moved forward, you were pushed back forcefully against your seat.

The maximum force a seated pilot can tolerate and still be able to control aircraft or spacecraft is about 8 g. If lying down, however, a pilot can tolerate about twice this force. For this reason, spacecraft have special couches so that astronauts are lying down during takeoff.

Weightlessness presents special problems aboard a spacecraft. For example, unfastened solid objects will float or remain suspended in the spacecraft. And liquids must always be capped tightly, or they will float out of their containers. When drinking, an astronaut must force the water directly into his or her mouth.

The reduced weight of astronauts exploring the moon also may cause problems. On the moon, a person weighs only one sixth of his or her Earth weight because the force of gravity on the moon's surface is only one sixth the force of gravity on Earth. Astronauts may find it awkward to work in this low-gravity environment because their sense of balance would not be working normally. They would have to depend on their sight, rather than balance, to remain upright.

Although a great deal has been learned about the effects of increased gravity on humans, little is known about the effects of prolonged weightlessness on the human body. One problem that may be related to prolonged weightlessness is calcium loss. When an astronaut returns to Earth, the calcium level in the astronaut's body is noticeably lower than normal. However, it gradually returns to normal.

Other effects that may be related to prolonged weightlessness include weakening of muscles, loss of balance, and changes in the blood and blood vessels. In time, scientists may be able to overcome the problems caused by weightlessness by producing artificial gravity within spacecraft. One way of accomplishing artificial gravity is to build a spacecraft that can spin like a top while traveling through space.

**The Future of Space Exploration.**  The Space Shuttle will continue to be used for conducting scientific research on space flight, for placing artificial satellites into orbit, and for retrieving and repairing malfunctioning satellites already in orbit. By the year 2004, construction of the International Space Station, which is already being used to conduct ground-breaking research, is expected to be complete.

The International Space Station will help astronomers observe and learn more about the effects of weightlessness on humans and other living things. The space station may also be used as a refueling stop for spacecraft on interplanetary voyages. Other space missions being planned for the near future include (1) exploration of the moon and near-Earth asteroids for valuable minerals, (2) collection and return to Earth of rock and soil samples from Mars and Venus, and (3) an unmanned probe to visit Pluto and its moon, Charon.

As humans travel farther into space, they will encounter new problems. For example, scientists wonder how prolonged space travel to distant planets will affect an astronaut's physical and mental health. How long can several people live together in isolation from the rest of the world? What new diseases might astronauts bring back to Earth? These and other questions will be answered as space exploration continues through the twenty-first century.

# CHAPTER REVIEW

## Science Terms

*The following list contains all of the boldfaced words found in this chapter and the page on which each appears.*

artificial satellite (p. 328)
chemical rocket (p. 326)
chronometer (p. 323)
escape velocity (p. 327)
gyrocompass (p. 323)
jet engine (p. 326)
liquid-propellant rocket (p. 326)
navigation (p. 322)

nuclear-powered rockets (p. 326)
orbiting velocity (p. 327)
rocket engine (p. 326)
sextant (p. 322)
solid-propellant rocket (p. 326)
space laboratory (p. 328)
space probe (p. 328)
thrust (p. 325)

## Matching Questions

*On the blank line, write the letter of the item in column B that is most closely related to the item in column A.*

### Column A

_____ 1. forward movement produced in a combustion chamber

_____ 2. cylinders are filled with a special fuel and oxidizer mixture

_____ 3. special clock used by navigator to determine longitude

_____ 4. speed that keeps spacecraft in a stable orbit

_____ 5. speed that enables spacecraft to break away from Earth's gravity

_____ 6. instrument used by navigator to determine true north

_____ 7. fuel and oxidizer are stored in separate tanks

_____ 8. instrument used by navigator to determine latitude

_____ 9. spacecraft placed into orbit to perform a specific task

_____ 10. spacecraft that leaves Earth's orbit to explore the solar system

### Column B

a. liquid-propellant rockets
b. sextant
c. nuclear-powered rockets
d. space probe
e. artificial satellite
f. solid-propellant rockets
g. gyrocompass
h. chronometer
i. thrust
j. escape velocity
k. orbiting velocity

# *Multiple-Choice Questions*

*On the blank line, write the letter preceding the word or expression that best completes the statement.*

**1.** The main problem in navigating to the moon is that the moon
   *a.* is much smaller than Earth     *c.* revolves around Earth
   *b.* rotates on its axis     *d.* has no atmosphere     1 ____

**2.** Compared to the force of gravity on Earth, the force of gravity on the moon is
   *a.* the same   *b.* 6 times less   *c.* 6 times more   *d.* 10 times more     2 ____

**3.** Rocket engines can be used in outer space because they are
   *a.* lighter than jet engines
   *b.* able to carry their own supply of oxygen
   *c.* more powerful than jet engines
   *d.* more streamlined than jet engines     3 ____

**4.** Before a spacecraft can orbit Earth, it must reach a speed of about
   *a.* 2 km per second     *c.* 8 km per second
   *b.* 4 km per second     *d.* 11 km per second     4 ____

**5.** The only type of rocket engine that is capable of lifting a spacecraft into orbit at the present time is the
   *a.* nuclear engine     *c.* turbo-jet engine
   *b.* ion engine     *d.* chemical engine     5 ____

**6.** The latitude of an observer who measures the altitude of the North Star as 30 degrees in the Northern Hemisphere is
   *a.* 0 degrees N   *b.* 30 degrees N   *c.* 45 degrees N   *d.* 60 degrees N     6 ____

**7.** A clock used by navigators to determine longitude is called a
   *a.* sextant   *b.* coronagraph   *c.* chronometer   *d.* odometer     7 ____

**8.** If a navigator determines that Greenwich time is 12:00 noon and that the local time is 8:00 A.M., the longitude of the ship is
   *a.* 15 degrees E   *b.* 60 degrees E   *c.* 15 degrees W   *d.* 60 degrees W     8 ____

**9.** The amount of oxygen that a person needs every day is about
   *a.* 1 kg   *b.* 2 kg   *c.* 3 kg   *d.* 4 kg     9 ____

**10.** In a spacecraft recycling system in which the water is reprocessed, oxygen inhaled by the crew could be supplied by
   *a.* oxygen tanks built into the walls of the spacecraft
   *b.* chemicals containing oxygen
   *c.* air pumps
   *d.* algae     10 ____

**11.** An air lock is used on a spacecraft to
   *a.* allow an astronaut to enter and leave safely
   *b.* lighten the spacecraft
   *c.* provide an emergency living area
   *d.* store an emergency supply of oxygen     11 ____

**12.** One known effect of prolonged weightlessness on the human body is
   *a.* anemia
   *b.* a great loss of weight
   *c.* an increased number of dizzy spells
   *d.* a temporary loss in calcium     12 ____

# Modified True-False Questions

*In some of the following statements, the italicized term makes the statement incorrect. For each incorrect statement, write the term that must be substituted for the italicized term to make the statement correct. For each correct statement, write the word "true."*

1. The Chinese invented rockets about *100* years ago.

   1 _____

2. The first liquid-fueled rockets were built by *1900*.

   2 _____

3. Newton's *third law* states that for every action there is an equal and opposite reaction.

   3 _____

4. If you weigh 60 kg on Earth, you would weigh about *20* kg on the moon.

   4 _____

5. A proposed rocket engine that would be valuable for very long voyages through space is the *liquid-fueled* rocket.

   5 _____

6. To an observer at the *equator*, the North Star appears to be on the horizon.

   6 _____

7. If you are located west of Greenwich, England, your local time is *earlier* than Greenwich time.

   7 _____

8. When you face north, *east* is on your right.

   8 _____

# Testing Your Knowledge

1. Describe four problems encountered by astronauts that are not encountered by airplane pilots.

   a. _____

   b. _____

   c. _____

   d. _____

2. How do gases escaping from a rocket engine produce forward thrust? _____

   _____

   _____

3. Jet aircraft cannot go into orbit around Earth. Explain. _____

   _____

   _____

**4.** Briefly describe three reasons why a rocket engine works more efficiently as it rises through the atmosphere.

*a.* _____

_____

*b.* _____

_____

*c.* _____

_____

**5.** Describe two ways in which solid-propellant rockets are different from liquid-propellant rockets.

*a.* _____

_____

*b.* _____

_____

**6.** How could you determine your latitude at night? _____

_____

**7.** How does use of a chronometer help a navigator determine longitude? _____

_____

**8.** Describe two different ways to locate true north.

*a.* _____

_____

*b.* _____

_____

# *Glossary*

**abrasion** the grinding action of rock fragments scraping against a streambed and one another

**absolute dating** the use of radioactivity to make accurate determinations of Earth's age

**acid rain** rain or snow that contains droplets of sulfuric acid and nitric acid

**air current** air moving toward or away from Earth's surface

**air mass** a large section of the troposphere in which the temperature and humidity are similar

**air pressure belts** belts of high or low air pressure that encircle Earth

**alluvial fan** deposit formed by a stream that flows from steep mountains out onto a broad, flat plain

**altimeter** a barometer that indicates altitude

**altitude** the height (elevation) above sea level

**amber** fossil tree resin that sometimes contains preserved specimens of insects

**anemometer** instrument used to measure wind speed

**aneroid barometer** a type of barometer that does not contain mercury, water, or any other liquid

**anthracite** rock formed when bituminous coal is pressed and folded by moving blocks of the crust

**anticyclone** a high-pressure system in which the wind blows clockwise

**aquifer** the water-bearing layer of permeable rock in an artesian system

**argon** chemically inert (inactive) element found in the atmosphere; used in light bulbs

**artesian system** layers of impermeable rock that enclose a layer of permeable rock, forming a natural pipe for groundwater

**artificial satellite** a spacecraft built by humans and placed into orbit to perform a specific task

**asteroids** thousands of rocky fragments that revolve around the sun in circular orbits

**astronomers** scientists who study the sun, moon, planets, and stars

**atmosphere** the mixture of gases and solids that surrounds Earth's lithosphere and hydrosphere

**atmospheric pressure** the force exerted by the outside air pressure

**barograph** an instrument that keeps a continuous record of changes in atmospheric pressure

**barometer** an instrument that measures atmospheric pressure

**barycenter** the common center of gravity around which Earth and the moon rotate

**basalt** a dark-colored, smooth igneous rock; very abundant near Earth's surface

**batholith** a huge mass of coarse-grained igneous rock that forms deep in the cores of mountain ranges

**bayou** swampy area that is formed when sediments fill in an oxbow lake

**bench mark** a round metal plate that indicates the ground's elevation above sea level; indicated on a topographic map by BM

**big bang theory** idea that all galaxies originated from the violent explosion of a densely packed mass of hydrogen; most widely accepted theory about origin of the universe

**bimetallic bar** the part of a thermograph that reacts to changes in temperature; made of two different metals welded together into one strip

**black holes** invisible objects in space, so dense that their gravitational forces do not let light escape;

form when red supergiant stars explode and collapse

*carbon dating* method of dating a substance, based upon the rate of decay of radioactive carbon-14 into nitrogen

*carbon dioxide* element produced by combustion of fuels, it does not support combustion; plants take it in to carry out photosynthesis; thought to cause warming of climate

*carbonic acid* weak acid formed when carbon dioxide and water combine chemically; forms caverns by dissolving calcite from rocks

*caverns* large underground chambers and passages, formed by groundwater that dissolves limestone

*Cenozoic Era* most recent era, including the present; called age of mammals (the dominant life forms)

*chemical compound* a substance that consists of two or more elements combined in fixed proportions by weight

*chemical element* simplest kind of substance; cannot be broken down by ordinary means

*chemical rocket* the most common rocket type, it uses either liquid or solid fuel

*chemical weathering* the breaking down of rocks due to changes in their chemical composition

*chromosphere* a region of glowing hydrogen gas that extends thousands of kilometers outward from the sun's photosphere

*chronometer* special clock used by navigators to determine longitude (by comparing with the time at the prime meridian)

*cinder cone* steep, cone-shaped hill formed by the explosive eruption of lava that hardens and falls as cinders and ash

*cirque lake* forms when the glacier that carved a cirque melts, filling it up with water

*cirques* bowl-shaped depressions carved into a mountainside by glaciers

*cleavage* physical property that causes a mineral to split along planes that meet at characteristic angles

*climate* the average of the general weather conditions in a region over many years

*clouds* huge masses of water droplets or ice crystals that form when water vapor in the atmosphere condenses

*coal* rock formed from the remains of dead plants and animals; made up of carbon

*cold front* weather boundary having cold, dry air behind it, which forces its way under warm, moist air that is usually ahead of it

*comet* an accumulation of dust particles and frozen gases that travels around the sun in an elliptical orbit

*composite volcano* a mountain formed by layers of both lava flows and cinder and ash falls

*condensation* process by which, when air cools, water vapor changes from a gas into a liquid or solid

*condensation nuclei* tiny particles in the atmosphere on which most water vapor condenses, such as soil, salt, smoke, and dust

*conduction* the process by which heat energy is transferred from molecule to molecule in a substance

*conglomerate* sedimentary rock made up of rounded pebbles cemented together by sand, silt, or clay

*conic projection* a mapmaking process in which a model of Earth is projected onto a paper cone; more accurate than Mercator projection

*conservation* the practice of preserving natural resources and using them wisely

*constellations* groups of stars that form recognizable patterns

*continental drift* the idea that crustal movements take place on a global scale; proposed by Wegener

*continental glacier* covers a vast area of land with an ice sheet

*continental shelves* the shallow, submerged edges of a continent

*continental slope* the steeply sloping region at the outer edge of a continental shelf; several kilometers deep

*contour interval* the difference in elevation between two contour lines on a map

*contour lines* on a topographic map, these lines represent the shape and elevation of Earth's surface

*contour maps* see *topographic map*

*contour plowing* method of farming in which deep furrows are dug across the slope of a hill to reduce water runoff and soil erosion

*control group* in an experiment, the subject group that is not exposed to the variable

*controlled experiment* the process by which a hypothesis is tested; includes a control group and an experimental group

*convection* the transfer of heat through a substance (gas or liquid) by currents moving inside the substance itself

*convection cell* the circular movement of air in convection currents

*convection currents* movements of a gas or liquid in a continuous cycle as the heated part rises and the cooler part sinks

*coral islands* form when coral animals deposit their hard outer parts on top of submerged volcanoes that later rise above sea level

*Coriolis effect* the influence of Earth's rotation on the direction of winds

*corona* a layer of very thin gas that extends millions of kilometers outward from the sun's chromosphere

*crater* the bowl-shaped depression at the top of a volcano

**crop rotation** method of growing different kinds of crops in the same field each year to preserve minerals in the soil

**crust** Earth's outermost layer, made up of solid rock

**crystal** a mineral solid that has flat surfaces arranged in a definite geometric shape

**current** a mass of seawater that flows continuously through a large region of the ocean

**cyclone** a low-pressure system in which the wind blows counterclockwise

**dam** a structure made of boulders, sand, branches, or boards that is placed across a gully to reduce water flow and soil erosion

**data** scientific observations

**day** the amount of time it takes Earth to rotate once on its axis

**degree of longitude** the distance equal to 1/360 of Earth's circumference

**delta** fan-shaped deposit of sand and clay sediments at the mouth of a river that empties into a lake or bay

**deposit** an accumulation of sediments left by a stream or river

**depression contour** a contour line that has hachures on it

**desalination** the process of removing salt from seawater

**dew** water vapor that condenses into droplets on or near the ground

**dew point** the air temperature at which water vapor begins to condense

**diabase** a dark-colored igneous rock, coarser than basalt

**dike** a layer of igneous rock that forms when magma forces its way into a fracture that cuts across layers of sedimentary rock

**doldrums** the region directly surrounding the equator; an area of low air pressure

**dome mountain** forms when a large amount of magma pushes up, but doesn't break through, the crust, forcing the overlying rock layers to arch upward

**drift** a mass of ocean water that is wider, shallower, and slower than an ocean current

**drumlins** small, rounded, oval-shaped hills that consist of glacial deposits

**dust particles** sands, clays, soils, sea salts, and ashes that are in the atmosphere

**earthquake** a sudden movement of the crust, with noticeable shaking of the ground

**easterly trades** the northeast trades and the southeast trades together; see also *trade winds*

**English system** the system of measurement that uses the foot for length, pound for weight, and quart for volume

**epicenter** the point on Earth's surface that is directly above the focus (source) of an earthquake

**equinox** meaning "night of equal length" occurs when the most direct rays of sun fall on the equator; marks the beginning of spring and of fall

**erosion** the process by which weathered rock material is removed and carried away by running water, glaciers, and wind

**erratics** large boulders that have been carried long distances and then deposited by glaciers

**escape velocity** velocity of 11 km per second that is needed to supply a spacecraft with enough thrust to break away from the pull of Earth's gravity

**estuaries** submerged river valleys located along a coast

**evaporation** the process by which a liquid changes into a gas

**exfoliation** the peeling away of large outer layers of a rock, due to frost wedging and release of pressure

**experimental group** in an experiment, the subject group that is exposed to the variable

**fault** a fracture in the crust, along which sections of the crust slide past each other

**fault-block mountains** formed by blocks of crust that are pushed or tilted upward

**fiord** a drowned glacial valley, found along the coasts of Alaska, Norway, and Greenland

**floodplain** a flat area of fine-grained sediments, deposited on a valley floor by an overflowing river

**fluorescence** a property of minerals that glow with colors when exposed to X rays and/or ultraviolet radiation

**focus** the underground point that is the source of an earthquake's seismic waves; where rocks first fracture in an earthquake

**fog** an accumulation of tiny water droplets light enough to float in the air

**fold mountains** produced by the folding together of (sedimentary) rock layers

**fossil** the remains or trace of an ancient organism that has been preserved naturally in Earth's crust

**fossil cast** formed by sediments or minerals that fill the fossil mold of a shell or bone and later harden into that shape

**fossil fuels** nonrenewable natural resources that formed over millions of years from the remains of plants and animals; coal and natural gas

**fossil mold** hollow space in a rock that retains the shape of a bone or shell that once occupied it

**fossil tracks** form when mud containing an animal's footprints or trail hardens into stone

**fracture** the break in a mineral that does not exhibit cleavage

**frost** form of condensation that occurs when water vapor changes directly into a solid, forming ice crystals, near Earth's surface

**frost wedging** process by which water that freezes

into ice inside cracks in rocks forces the rocks apart

*gabbro* a dark-colored igneous rock, coarser than basalt or diabase

*galaxy* an enormous collection of stars

*geologic time scale* the division of Earth's history into eons, eras, periods, and epochs, based upon major geological and biological events

*geologists* scientists who study Earth's structure

*geyser* a type of hot spring that periodically shoots a fountain of hot water and steam into the air

*glaciers* large masses of slow-moving ice

*gneiss* rock formed from the continued metamorphism of granite or schist; often used as a building stone

*granite* a light-colored igneous rock having large crystals; makes up most of Earth's crust

*greenhouse effect* the trapping of heat energy by water vapor, carbon dioxide, and other gases in Earth's atmosphere

*Gregorian calendar* adopted in 1582, this calendar currently in use drops 10 extra days per year from the Julian calendar and loses three leap-year days every 400 years

*ground moraine* an irregular deposit of rock debris left behind by the melting of a glacier's main body

*groundwater* all of the water located below Earth's surface

*gullies* deep channels that are carved into Earth's surface by running water

*gypsum* sedimentary mineral deposits formed by evaporated bodies of water; forms plaster of paris and alabaster

*gyrocompass* an instrument used by navigators to determine true north

*hachures* on a contour map, tiny comblike markings on a contour line that indicate a depression in the ground

*hail* frozen precipitation that forms when a raindrop repeatedly freezes and melts; made up of concentric layers of ice

*half-life* the time required for one half of a given amount of a radioactive element to decay into another element

*halite* see *rock salt*

*hardness* a physical property of a mineral, determined by how easily it scratches, or is scratched by, other minerals

*hard water* groundwater that contains a large amount of dissolved calcium and magnesium

*heat energy* describes how fast the atoms and molecules in a substance are moving

*helium* second lightest chemical element; inert and does not burn or support combustion; used in balloons and blimps

*highlands* the light-colored mountainous, crater-filled areas on the moon

*horns* sharp pyramid-shaped mountains formed due to erosion by valley glaciers

*horse latitudes* a region of high air pressure at 30° latitude, characterized by calm weather and light winds

*humidity* the amount of water vapor in the atmosphere

*humus* the decayed remains of plants and animals, found in the topsoil

*hurricane* a large, rotating, low-pressure system with heavy rains and strong winds; also called tropical cyclone or typhoon

*hydrogen* lightest chemical element; very active, it can burn in the air; part of water, fuels, oils, and acids

*hydrolysis* chemical weathering that occurs when water reacts with certain minerals in rocks

*hydrosphere* all bodies of water on Earth, that is, the lakes, oceans, and rivers

*hygrometers* instruments that measure the relative humidity of air

*hypothesis* a possible solution to a scientific problem

*ice age* period of time in which glaciers advance over a large part of a continent

*igneous rocks* form as molten material cools

*index fossils* fossils characteristic of certain rock strata that are used to date their relative ages

*inner core* deepest part of Earth, probably solid

*International Date Line* an imaginary north-south line in the middle of the Pacific Ocean at which the date changes from one day to the next day; the 180th meridian

*island arc* a curving chain of volcanic islands, formed when one oceanic plate plunges beneath another one

*isobars* lines on a weather map that connect places in the atmosphere having the same air pressure

*jet engine* takes in air from the atmosphere and uses the oxygen to burn its fuel

*jet stream* a band of swiftly moving westerly winds in the middle latitudes; found near the tropopause

*Julian calendar* adopted in 46 B.C., based on a year of 365.25 days, this calendar introduced every fourth year as a leap year; no longer in use

*kettle lakes* form when kettle holes fill with water

*kettles* depressions in the ground that form when blocks of glacial ice melt away

*lagoon* the body of water between an offshore bar (sandbar) and the mainland

*land breeze* a breeze that blows from the land to the ocean or a large lake

**landslide** sudden movement of a mass of rock and soil that breaks loose from the side of a cliff or mountain

**lateral moraines** form along the sides of a glacier; consist of rock fragments from valley walls

**lava** molten rock that has poured out onto Earth's surface

**law** a theory that has been proven true by repeated successful testing

**levees** long, low walls of sediment deposited along the banks of a river that repeatedly overflows

**lightning bolt** a discharge of electricity (in a cumulonimbus cloud) from cloud to cloud or from cloud to ground

**light-year** the distance that light travels in one year; approximately 10 trillion kilometers

**limestone** sedimentary rock made up of calcium carbonate (calcite)

**liquid-propellant rocket** the fuel and the oxidizer are stored in separate tanks; they are mixed together and ignited in the combustion chamber

**liquid thermometer** a sealed glass tube with a liquid-containing bulb at one end; measures the temperature

**lithosphere** the solid part of Earth (rocks)

**loess** unlayered deposits of windblown silt; from glacial rock flour, it weathers to produce good topsoil

**lunar eclipse** occurs when the moon moves into Earth's umbra (shadow)

**luster** physical property that describes how a mineral shines or reflects light

**magma** molten rock inside Earth

**mainstream** a very large stream formed by the coming together of many smaller streams

**mantle** Earth's thickest layer, located just below the crust

**map scale** shows how the size of a topographic map compares to the actual distance on Earth

**marble** rock formed by the metamorphism of limestone; used for sculptures

**maria** the large, darker-colored areas on the moon, composed of great basins and plains

**meanders** S-shaped curves in the course of a mature river

**medial moraine** forms when the inner lateral moraines of two glaciers meet as they merge into one larger glacier

**Mercator projection** a mapmaking process in which a model of Earth is projected onto a cylinder; useful to navigators

**mercury barometer** the weight of a column of mercury is balanced against the atmospheric pressure to determine any changes in it; invented by Torricelli

**meridians of longitude** 360 equal divisions of Earth's circumference around the equator; on a map, these lines converge at the poles

**mesosphere** layer of the atmosphere above the stratosphere, up to about 80 km high; has extremely strong winds

**Mesozoic Era** era of Earth's history known as age of reptiles; dinosaurs were dominant life forms

**metamorphic rocks** changed from their original form by heat, pressure, or chemical action

**meteor** the streak of light produced by a burning meteoroid; also called a "shooting star"

**meteorite** a meteoroid that passes through the atmosphere and strikes Earth's surface

**meteoroids** particles of rock and metal in space, less than one kilometer in diameter

**metric system** the system of measurement that uses the meter for length, gram for mass, and liter for volume; based on multiples of ten

**mid-ocean ridges** long, underwater mountain ranges that wind around the world

**millibar** the unit used by meteorologists to measure atmospheric pressure

**Milky Way** our spiral-shaped galaxy, which is about 100,000 light-years in diameter and consists of billions of stars

**mineral** a substance that has a definite chemical composition of one or more elements

**mineral ores** nonrenewable natural resources (metals) such as iron, copper, and lead, that are used by industrialized nations

**monsoon** a huge land and sea breeze that reverses direction in different seasons; brings either cold, dry weather or very wet weather

**months** units of time, each lasting about 30 days, into which the year is divided; based on reappearance of the full moon

**moraine-dammed lakes** form when a moraine is deposited across a valley, thus blocking the flow of streams or meltwater through it

**moraines** large ridges and mounds of rock debris carried by a moving glacier or left behind by a melting glacier

**mountain** a part of Earth's crust that has been raised high above the surrounding landscape

**mountain breeze** cooler, denser air that moves into a valley from a mountain slope during the nighttime

**navigation** the act of steering a vehicle to its destination

**nebula** a cloud of dust and gas in space that glows by reflected starlight

**neon** chemically inert element in the atmosphere; used in electric signs

**nitrogen** the most abundant element in the atmosphere, it does not burn or support combustion

**nodules** round lumps of metallic ores found on the ocean floor

**nonrenewable resources** resources that, once removed, cannot be replenished for millions of years; for example, coal, oil, and mineral ores

**nuclear-powered rockets** engines being considered for use for long-range space voyages

**obsidian** glassy-textured igneous rock

**occluded front** the boundary between two cold air masses and a warm, moist air mass trapped above them

**ocean basins** the depressions in Earth's crust that contain the oceans

**orbit** the path Earth follows in its movement around the sun

**orbiting velocity** velocity of about 8 km per second that is needed by a spacecraft to counteract the pull of Earth's gravity and stay in a stable orbit

**organic limestone** rock made up of remains of shell-forming animals cemented together; coquina, chalk, and coral limestone

**outer core** molten layer of Earth, located below the mantle

**outwash plains** flat areas formed by large quantities of fine sediments, carried beyond the terminal moraine by glacial meltwater

**oxbow lake** forms when sediments cut off a meander from an old river

**oxidation** chemical reaction in which a substance, such as iron, combines with oxygen; causes the chemical weathering of rocks

**oxygen** abundant element in Earth's crust, ocean, and atmosphere; it does not burn, but it does support combustion

**ozone** a form of oxygen that is found in abundance in a thin layer of the atmosphere; protects living things from ultraviolet radiation

**Paleozoic Era** era of Earth's history divided into three ages: invertebrates, fishes, and amphibians

**parallels of latitude** on a map, the parallel lines that divide the distance from the equator to both the north and south poles into 90 equal parts each

**pegmatite** light-colored igneous rock having very large crystals, some of commercial value

**penumbra** the lighter, longer, outer shadow that covers a widening space, cast by Earth and the moon

**petrified wood** forms when groundwater containing minerals seeps into wood buried in sediments, replacing the wood with minerals

**phases** daily changes in the moon's appearance (from full moon to new moon to full moon again)

**photosphere** the innermost part of the sun's atmosphere, a layer of hydrogen gas about 500 km thick

**physical weathering** process by which rocks are broken down without undergoing a change in chemical composition

**plain** a broad, flat region, composed of horizontal layers of rock or sediment lower than the land around it

**planet** a large, solid body that shines by reflected light and revolves around the sun in stable orbit

**plateau** a broad, flat-topped region that rises abruptly above the rest of the landscape; formed by crustal uplifting of horizontal layers of rock

**plate tectonics** theory that combines ideas of continental drift and seafloor spreading; states that Earth's crust is broken into a number of large plates that move over a layer of hot rock in the upper mantle

**polar easterlies** the dense, cold air moving away from both poles

**Precambrian** first four billion years of Earth's history, divided into three eons; early simple life-forms appeared

**precipitate** a curdlike deposit that forms when soap reacts with the minerals in hard water

**precipitation** water that falls to the ground in the form of rain, sleet, hail, or snow

**prevailing westerlies** the winds that move northward and southward from the horse latitudes

**prime meridian** the town of Greenwich, England, which has been assigned the value of zero degrees longitude

**projection** process by which mapmakers transfer the features of Earth's surface onto a sheet of paper

**pumice** lightweight, spongy-looking rock that forms when lava cools

**quartzite** rock formed by the metamorphism of quartz sandstone

**radiant energy** the sun emits heat in this form of energy

**radiation** how the sun's radiant energy travels through outer space as rays

**radioactive** a property of minerals that give off invisible radiation; detected by a geiger counter

**radioactive decay** the continuous breakdown of the atoms of radioactive elements; radiations are emitted, and a new element is formed as the old one decays

**radioactive elements** those whose atoms give off radiations (rays and particles)

**radio telescope** collects invisible radio waves (generated by the sun and other stars) and converts them into electrical signals

**rain** large water droplets that fall from the atmosphere to the ground

**rain gauge** special container that is used to measure the amount of rain that falls

**reflecting telescope** collects light by use of one or more concave mirrors and magnifies the image through an eyepiece lens

**refracting telescope** collects light through one convex lens and magnifies the image through a second convex (eyepiece) lens

**relative humidity** percentage representing the amount of water vapor in the air as compared with the maximum amount it can hold at a given temperature

**relief** on a map, the differences in height (elevation) of the land above sea level

**renewable resources** resources, such as minerals in the soil, that can be replaced (with proper care) by natural processes

**reservoir** an artificial lake formed by a dam built across a river

**revolution** the movement of Earth around the sun

**river system** formed by a mainstream and all of its tributaries

**rock cycle** the continual and gradual process of creation, change, and destruction of the three rock types on Earth

**rocket engine** carries its own liquid oxygen, which is then combined with a liquid fuel in the combustion chamber

**rock salt** sedimentary mineral deposits formed by evaporated bodies of water; usually found with gypsum deposits

**rotation** the spinning motion of Earth on its axis

**salinity** the measure of the weight of salts in a given volume of water

**sandbar** deposit of sediments inside a curve in a riverbank, where the water flow slows down; strips of land that accumulate offshore from ocean-carried sediments

**sand dune** a large hill of wind-deposited sand grains; usually forms on a boulder

**sandstone** gritty sedimentary rock, consisting mainly of quartz sand grains cemented together

**saturated** condition of a parcel of air when it is holding as much water vapor as it can, that is, at its capacity

**schist** rock formed from the continued metamorphism of slate or basalt; makes up bedrock in some regions

**scientific method** an organized approach used for problem-solving

**sea breeze** the movement of cool ocean air toward the land

**sea cliffs** highlands that are formed when ocean waves pound and erode rocky coastlines

**seafloor spreading** process by which the ocean floor is splitting apart at the Mid-Atlantic Ridge, where molten rock rises into the rift and forms new ocean crust along each side

**seamounts** underwater mountains formed by volcanoes on the ocean floor

**sedimentary rocks** form when sediments are compressed or cemented together

**sediments** particles of sand, silt, and clay that are transported and deposited by moving water

**seismograph** an instrument that can detect and record seismic waves from earthquakes

**sextant** an instrument used by navigators to determine latitude

**shale** sedimentary rock consisting of clay and silt particles packed closely together

**shield volcano** broad, massive mountain with gentle slopes; formed by repeated lava flows from quiet eruptions

**sidereal day** exact length of a day based upon the successive passage of a star across a meridian of longitude; about 23 hours, 56 minutes, and 4 seconds long

**sill** a sheet of igneous rock that forms when magma flows (parallel) between layers of sedimentary rock

**sinkholes** round depressions in the ground, found in regions with limestone bedrock

**slate** metamorphic rock formed when shale is compressed and heated; used for roofs and chalkboards

**sleet** pellets of ice that form when falling rain freezes

**smelting** a chemical method of separating a metal mineral from its ore; requires high temperatures

**smog** a combination of smoke and fog; frequently occurs over cities

**snow** frozen precipitation that forms when water vapor in clouds changes directly into a solid; usually hexagonal-shaped ice crystals

**soil** consists of particles formed by the weathering of rock, such as sand, clay, and minerals

**soil erosion** problem that occurs when soil is carried away by wind and water faster than it can be formed again

**solar day** the time it takes the sun to return to the meridian (highest point in its path across the sky) at noon; about 24 hours

**solar eclipse** occurs when the moon's umbra (shadow) falls on Earth, blocking the sun's light

**solar flare** hot gases that erupt from a sunspot, causing ejection of ions and radiation

**solar system** the sun and all the natural objects in space that orbit it

**solar wind** electrically charged particles (ions) and radiation ejected from sun during a solar flare

**solid-propellant rocket** made of metal cylinders filled with a special mixture of fuel and an oxidizer

**solstice** meaning "sunstop" refers to the time when the sun seems to stop moving higher in the sky each day; marks the beginning of summer and of winter

**solution** the dissolving of minerals from rocks by streams

**space laboratory** an orbiting spacecraft equipped with living quarters for a crew and a laboratory for performing experiments; now part of the Space Shuttles

**space probe** an unmanned spacecraft that leaves Earth's orbit to explore other parts of the solar system

**specific gravity** a measure that tells how much

heavier or lighter a substance is compared to an equal volume of water

*specific heat* describes the heat storage ability of substances

*specific humidity* the actual amount of water vapor in Earth's atmosphere

*spectroscope* an attachment on telescopes that separates white light into different colors (by the wavelengths)

*spectrum* the series of colors of light produced by a spectroscope

*spring* occurs wherever the water table meets the surface of the ground

*stalactites* icicle-shaped deposits of calcite that hang from the ceiling in caves

*stalagmite* blunt, cone-shaped deposits of calcite that accumulate from groundwater dripping on the floor of caves

*stationary front* occurs when the front separating a cold air mass and a warm air mass does not change position

*steady-state theory* idea that the universe is being continually renewed by new galaxies; not a widely accepted theory

*strata* layers of rock

*stratosphere* layer of the atmosphere above the troposphere; contains the ozone layer in its upper regions

*streak* a physical property, it is the color of a mineral when ground into a powder

*strip-cropping* farming method in which rows of clover or soybeans are planted between rows of corn or cotton to reduce soil erosion

*subduction* process in which a plate of dense ocean crust collides with a lighter plate of continental crust and slides under it, plunging down into the mantle

*subduction zone* site at which subduction of a plate of oceanic crust occurs; forms a deep depression on the seafloor

*sublimation* a type of condensation in which water vapor changes directly from a gas into a solid

*subsoil* second layer of soil, found under the topsoil; has more clay than topsoil

*sunspots* dark areas on the sun's photosphere that are cooler than the surrounding area

*talus* the mound of rock fragments that accumulates at the base of a cliff or slope

*temperature inversion* a condition of upside-down air temperature in which the temperature rises for several hundred meters of increasing altitude before dropping again at higher altitudes

*terminal moraine* the long ridge of rock debris that marks the farthest advance of a glacier

*terracing* farming method in which a series of flat platforms are dug across a slope to reduce soil erosion

*theory* a scientific explanation of natural events

*thermograph* a device that writes a continuous record of temperature changes over time

*thermonuclear fusion* the process whereby, under great heat and pressure, hydrogen atoms in the sun's core are changed into helium atoms

*thermosphere* uppermost layer of the atmosphere; made up of the ionosphere (where auroras are produced) and the exosphere

*thrust* forward movement produced in the combustion chamber of a rocket engine

*thunder* the explosion caused by the expansion of air resulting from a lightning bolt

*thunderstorm* a brief, local storm produced by the rapid upward movement of warm, moist air inside a cumulonimbus cloud

*tides* the daily rise and fall of the oceans, seen along coastlines

*time zones* based on a standard time, Earth is divided into 24 time zones, each of which extends through 15° of longitude

*topographic map* shows the physical features and elevation of a small section of Earth's surface; also called contour maps

*topsoil* the top layer of soil, in which most plants grow; contains humus

*tornado* a severe storm with rapidly swirling winds

*trade winds* the winds moving toward the equator from the north and south (horse latitudes region); see also *easterly trades*

*transpiration* the process by which water evaporates from plants through pores in their leaves

*trenches* deep, elongated depressions on the ocean floor, formed by subduction of oceanic plates

*tributaries* all of the streams that supply a mainstream

*troposphere* layer of the atmosphere closest to Earth; region where the weather is produced, it contains most of the water vapor and the dust particles

*tsunamis* fast-moving ocean waves produced by earthquakes on the seafloor

*umbra* the darker, shorter, inner shadow that covers a cone-shaped space, cast by Earth and the moon

*underwater canyons* elongated depressions in the ocean floor that extend outward from the coasts into deeper water

*universe* all of the matter and energy in space, that is, all stars and other objects

*uranium dating* method of dating very old rocks, based upon the rate of decay (half-life) of uranium-238 into lead-206

*valley breeze* a wind that blows up a mountainside during the daytime

*valley glaciers* found in stream-eroded valleys in high mountain ranges

**variable** the factor that is changed in a controlled experiment

**vent** an opening within the crater of a volcano, through which lava pours

**volcanic island** forms when the top of an underwater volcano rises above sea level

**volcano** an opening in the crust through which molten materials and rocks reach the surface, pour out, and pile up

**vulcanism** the movement of magma through the crust, or its emergence as lava onto Earth's surface

**warm front** weather boundary in which moist, cool air forces moist, warm air upward along a gentle slope

**water cycle** continuous process in which water evaporates, condenses into clouds, and falls to Earth as precipitation

**waterfall** occurs in a young river where there is a sudden drop in the riverbed's elevation

**watershed** the area of land drained by a river system

**water table** the upper surface of the ground's zone of saturation

**water vapor** water as a gas in the atmosphere

**waves** form due to friction generated by the wind dragging the top layers of ocean water, causing ripples to form on the surface

**weather** the hourly and daily changes in the atmosphere over a region

**weather front** the boundary between air masses that have different characteristics

**weathering** breaking down of rocks at or near Earth's surface into smaller pieces

**wells** holes dug from the surface into the zone of saturation to obtain groundwater

**wind** air moving parallel to Earth's surface

**wind abrasion** erosion of boulders and large rock outcrops caused by flying sand grains

**windbreaks** rows of trees or shrubs planted next to a farm field to reduce wind erosion of soil; also called shelter belts

**winter storms** severe storms that have precipitation in the form of snow and ice

**year** the amount of time it takes Earth to orbit the sun; 365.25 days long

# Index